AF581747

Sustainable Food Production and Consumption

Sustainable Food Production and Consumption

Editors

Ada Margarida Correia Nunes Da Rocha
Belmira Almeida Ferreira Neto

MDPI • Basel • Beijing • Wuhan • Barcelona • Belgrade • Manchester • Tokyo • Cluj • Tianjin

Editors

Ada Margarida Correia Nunes Da Rocha
University of Porto
Portugal

Belmira Almeida Ferreira Neto
University of Porto
Portugal

Editorial Office
MDPI
St. Alban-Anlage 66
4052 Basel, Switzerland

This is a reprint of articles from the Special Issue published online in the open access journal *Sustainability* (ISSN 2071-1050) (available at: https://www.mdpi.com/journal/sustainability/special_issues/food_production).

For citation purposes, cite each article independently as indicated on the article page online and as indicated below:

LastName, A.A.; LastName, B.B.; LastName, C.C. Article Title. *Journal Name* **Year**, *Volume Number*, Page Range.

ISBN 978-3-0365-4853-1 (Hbk)
ISBN 978-3-0365-4854-8 (PDF)

Contents

About the Editors . . . . **vii**

Ada Rocha and Belmira Neto
Special Issue "Sustainable Food Production and Consumption"
Reprinted from: *Sustainability* **2022**, *14*, 8508, doi:10.3390/su14148508 . . . . **1**

Iva Pires, Jerusa Machado, Ada Rocha and Margarida Liz Martins
Food Waste Perception of Workplace Canteen Users—A Case Study
Reprinted from: *Sustainability* **2022**, *14*, 1324, doi:10.3390/su14031324 . . . . **5**

Ronja Teschner, Jessica Ruppen, Basil Bornemann, Rony Emmenegger and Lucía Aguirre Sánchez
Mapping Sustainable Diets: A Comparison of Sustainability References in Dietary Guidelines of Swiss Food Governance Actors
Reprinted from: *Sustainability* **2021**, *13*, 12076, doi:10.3390/su132112076 . . . . **17**

Kieran Magee, Joe Halstead, Richard Small and Iain Young
Valorisation of Organic Waste By-Products Using Black Soldier Fly (*Hermetia illucens*) as a Bio-Convertor
Reprinted from: *Sustainability* **2021**, *13*, 8345, doi:10.3390/su13158345 . . . . **39**

Margarida Liz Martins, Ana Sofia Henriques and Ada Rocha
Evaluation of Food Waste at a Portuguese Geriatric Institution
Reprinted from: *Sustainability* **2021**, *13*, 2452, doi:10.3390/su13052452 . . . . **57**

Taíse Portugal, Susana Freitas, Luís Miguel Cunha and Ada Margarida Correia Nunes Rocha
Evaluation of Determinants of Food Waste in Family Households in the Greater Porto Area Based on Self-Reported Consumption Practices
Reprinted from: *Sustainability* **2020**, *12*, 8781, doi:10.3390/su12218781 . . . . **65**

Philipp Schepelmann, An Vercalsteren, José Acosta-Fernandez, Mathieu Saurat, Katrien Boonen, Maarten Christis, Giovanni Marin, Roberto Zoboli and Cathy Maguire
Driving Forces of Changing Environmental Pressures from Consumption in the European Food System
Reprinted from: *Sustainability* **2020**, *12*, 8265, doi:10.3390/su12198265 . . . . **77**

Xuezhen Guo, Jan Broeze, Jim J. Groot, Heike Axmann and Martijntje Vollebregt
A Worldwide Hotspot Analysis on Food Loss and Waste, Associated Greenhouse Gas Emissions, and Protein Losses.
Reprinted from: *Sustainability* **2020**, *12*, 7488, doi:10.3390/su12187488 . . . . **107**

Se-Hak Chun and Ariunzaya Nyam-Ochir
The Effects of Fast Food Restaurant Attributes on Customer Satisfaction, Revisit Intention, and Recommendation Using DINESERV Scale
Reprinted from: *Sustainability* **2020**, *12*, 7435, doi:10.3390/su12187435 . . . . **127**

Nouf Sahal Alharbi, Malak Yahia Qattan and Jawaher Haji Alhaji
Towards Sustainable Food Services in Hospitals: Expanding the Concept of 'Plate Waste' to 'Tray Waste'
Reprinted from: *Sustainability* **2020**, *12*, 6872, doi:10.3390/su12176872 . . . . **147**

Kai Victor Hansen and Lukasz Andrzej Derdowski
Sustainable Food Consumption in Nursing Homes: Less Food Waste with the Right Plate Color?
Reprinted from: *Sustainability* **2020**, *12*, 6525, doi:10.3390/su12166525 . . . . **155**

Manuel Navarro Gausa, Silvia Pericu, Nicola Canessa and Giorgia Tucci
Creative Food Cycles: A Cultural Approach to the Food Life-Cycles in Cities
Reprinted from: *Sustainability* **2020**, *12*, 6487, doi:10.3390/su12166487 . . . . **165**

Brent Stoffle, Richard Stoffle and Kathleen Van Vlack
Sustainable Use of the Littoral by Traditional People of Barbados and Bahamas
Reprinted from: *Sustainability* **2020**, *12*, 4764, doi:10.3390/su12114764 . . . . **181**

Miguel Vigil, Maria Pedrosa Laza, Henar Moran-Palacios and JV Alvarez Cabal
Optimizing the Environmental Profile of Fresh-Cut Produce: Life Cycle Assessment of Novel Decontamination and Sanitation Techniques
Reprinted from: *Sustainability* **2020**, *12*, 3674, doi:10.3390/su12093674 . . . . **207**

Gerald C. Shurson
"What a Waste"—Can We Improve Sustainability of Food Animal Production Systems by Recycling Food Waste Streams into Animal Feed in an Era of Health, Climate, and Economic Crises?
Reprinted from: *Sustainability* **2020**, *12*, 7071, doi:10.3390/su12177071 . . . . **225**

About the Editors

Ada Margarida Correia Nunes Da Rocha

PhD in Biotechnology at High School of Biotechnology, Portuguese Catholic University. Graduated in Nutrition and Food Sciences, Faculty of Nutrition and Food Sciences, University of Porto (FCNAUP); Portugal Associate professor at FCNAUP, Portugal (1998–2022) Chairman of the Council of Representatives (2018–2022) Member of the FCNAUP Scientific Council Lecturer in Food Technology, Management and Food Services, Food Preservation, Safety of Technological Processes, Food Production Systems. Director of Master in Food Service since 2007. Coordinator of the specialty of Food Service at the Portuguese Association of Nutrition and of the Portuguese Nutritionist Council (2010–2019). European Specialist Dietetic Network (ESDN) for Food Service/Catering (2020-. . .) Specialist in Food Service by Portuguese Nutrition Council (since 2021). Areas of interest: optimization of food production systems; quality control and development of new alternatives nutritional balanced; safe and sustainable with a good relationship cost/benefit; role of food service unit's health of consumers in different scenarios; supporting tools for the management of food service units; optimization of food production systems namely evaluation and control food waste.

Belmira Almeida Ferreira Neto

Assistant professor at the Faculty of Engineering at the University of Porto and Senior Researcher at the ALiCE/LEPABE—Laboratory for Process Engineering, Environment, Biotechnology and Energy. She has a PhD degree in Environmental Sciences from the University of Wagenigen (the Netherlands). She is a member of the Network for Sustainable Campus and leads the Working Group Sustainable Food Production and Consumption. She is responsible for the Sustainable Food Production and Consumption course at the Faculty of Engineering at the University of OPorto (Portugal). She supervises several Master and PhD students in the topic of Life Cycle Assessment and Sustainability. Her research topic is on Life Cycle Assessment & Sustainability. Her main research interests are in the field of environmental systems analysis, essentially on developing and application of management tools based on life cycle thinking, such as life cycle assessment, life cycle sustainability assessment, and carbon and water footprints.

Editorial

Special Issue "Sustainable Food Production and Consumption"

Ada Rocha [1,*] and Belmira Neto [2,3,*]

[1] GreenUPorto—Sustainable Agrifood Production Research Centre, Faculty of Nutrition and Food Sciences, University of Porto, Rua do Campo Alegre, 823, 4150-180 Porto, Portugal
[2] LEPABE—Laboratory for Process Engineering, Environment, Biotechnology and Energy, Faculty of Engineering, University of Porto, Rua Dr. Roberto Frias, 4200-465 Porto, Portugal
[3] ALiCE—Associate Laboratory in Chemical Engineering, Faculty of Engineering, University of Porto, Rua Dr. Roberto Frias, 4200-465 Porto, Portugal
* Correspondence: adarocha@fcna.up.pt (A.R.); belmira.neto@fe.up.pt (B.N.)

1. Editorial on Special Issue

This Special Issue (SI) "Sustainable Food Production and Consumption" intends to be the union of multidisciplinary areas of knowledge, under the sustainability pillar, based on knowledge about one of the most relevant agents for overall environmental impacts: food production and consumption.

The SI aims to highlight sustainability assessment within agri-food production, food consumption, and food waste reduction to meet the needs of updating knowledge and developing new skills required by multiple social and economic agents. The purpose is to shine a light on the significance of research and practical initiatives engaged in the United Nations Agenda 2030 for Sustainable Development, specifically in protecting the planet by promoting sustainability in food production and consumption aiming at informing and influencing policy and practice globally.

The research needs to move on towards combined efforts to sustainable food systems that are translated by the articles presented below.

Reducing food loss and waste is prioritized in the UN sustainable development goals (SDG) target 12.3 to contribute to "ensure sustainable consumption and production patterns". It is expected to significantly improve global food security and mitigate greenhouse gas (GHG) emissions.

Food waste occurs in all stages of the food supply chain, namely at home and in the food service sector and has a huge impact on loss of sustainability. This special issue includes 8 out of 14 papers dedicated to different approaches of food waste in different settings.

The paper from Iva Pires et al. (contribution number 1) stated that Portuguese canteen users showed an accurate perception of their plate waste and excessive portions were identified by consumers as the main reason for plate waste.

Ronja Teschner et al. (contribution number 2) analyzed how and to what extent different state and non-state actors in Switzerland incorporate sustainability aspects in their dietary guidelines. It examines how these DGs account for different dimensions at the basis of sustainability thinking, including the classic environmental, economic, and social dimensions as well as issues of health and governance.

Margarida Liz Martins et al. (contribution number 3) reported that food waste at care institutions is a matter of great concern, that requires regular monitoring, representing 36.1% of meals served, composed of 24.1% leftovers and 12.0% plate waste. The wasted meals would be enough to feed 1486 older adults and would correspond to annual losses of approximately €107,112. High values of leftovers are related to the food service system and staff, pointing to the need for improvements during the planning and processing of meals. On the other hand, high plate waste values are associated with consumers, indicating the low adequacy of the menu in reagards to older adults' habits and preferences.

Citation: Rocha, A.; Neto, B. Special Issue "Sustainable Food Production and Consumption". *Sustainability* **2022**, *14*, 8508. https://doi.org/10.3390/su14148508

Received: 2 July 2022
Accepted: 5 July 2022
Published: 12 July 2022

In Norway, Kai Victor Hansen et al. (contribution number 4) referred that approximately 992.6 tons of food per year could potentially be saved with only a single change, ultimately ameliorating the unsustainable food consumption problem among residents of nursing homes.

The paper from Nouf Sahal Alharbi et al. (contribution number 5) reported that tray waste arising at the ward level of hospitals across Saudi Arabia, equated to 4831 tons of food, 3535 tons of plastic, 1414 tons of paper, and 235 tons of metal each year. As all of this waste ends up in landfills, without any form of recycling, the paper published proposes the need for a more comprehensive, political approach that unites all food system stakeholders around a shared vision of responsible consumption and sustainable development.

Another paper by Se-Hak Chun and Ariunzaya Nyam-Ochir from Mongolia (contribution number 6) showed that food quality, service quality, price, and the atmosphere of a restaurant positively influences customer satisfaction and their intention to revisit global fast-food restaurants.

Also, Taíse Portugal et al. (contribution number 7) evaluated household food waste by Portuguese families and reported a positive attitude concerning buying, consumption, and wastage, revealing a particular awareness of food waste and its social and environmental impact.

Xuezhen Guo et al. (contribution number 8) stated that findings with policy implications is that priorities for Food Loss and Waste reduction vary, dependent on prioritized sustainability criteria (e.g., GHG emissions versus protein losses).

The paper from Philipp Schepelmann et al. (contribution number 9) provided an integrated assessment of environmental and socio-economic effects arising from the final consumption of food products by European households by applying environmentally extended input–output analysis (EE-IOA). Results shows that European food consumption generates relatively less environmental pressures outside Europe (due to imports) than average European consumption. The results highlight the importance of directing specific research and policy efforts towards food consumption to support the transition to a more sustainable food system in line with the objectives of the EU Farm to Fork Strategy.

The work from Manuel Navarro Gausa et al. (contribution number 10) describes tools to engage different stakeholders, such as architects, product designers, and citizens, from a cultural point of view. The research focus and educational campaign and an open platform where prototypes, new materials, and products are developed as inspiration for change. The Creative Food Cycles (CFC) project is described to address the topic of food as a cross-cutting factor and powerful accelerator toward the co-design of sustainability in cities. Food waste and food losses are shown to be a powerful tool for raising awareness of sustainable development at the community level.

The paper from Brent Stoffle et al. (contribution number 11) describes the sustainable adaptations to the littoral, which included both marine and terrestrial components, by the people of Barbados and The Bahamas, in the Caribbean that have lived in a sustainable way for five generations. The analysis is based on interviews conducted towards the practices of sustainable food use and environmental preservation. The findings document the need for local gardens and exchanges to prepare for perturbations caused by climate change, economic withdrawal, and development intrusion.

The research from Miguel Vigil et al. (contribution number 12) presented techniques to address two major issues regarding fresh-cut vegetables washing operations: the current low water recirculation rate and the use of chlorinated compounds as sanitizing agents. The authors perform a life cycle assessment (LCA) to assess the environmental effects of these new solutions and to compare those impacts with the burden derived from the current strategy. The novel technologies show to decrease the environmental burden, mainly due to the enhanced water recirculation and the subsequent decrease in energy consumption for pumping and cooling the water stream.

At the end, Gerald C. Shurson et al. (contribution number 13) looked at food waste as a major barrier to achieving global food security and environmental sustainability. The author identifies the potentials from repurposing food waste streams into animal feed, from pre-harvest to post-consumer stages of supply chains. The need for risk assessments is explicitly mentioned together with the development of extensive biosecurity protocols, especially for pathogenic viruses, to minimize the risk of pathogen and prion transmission. Overall, it is mentioned the need for a wide range of society agents (as governments, citizens and entrepreneurs) to build food waste collection and processing infrastructures economically and environmentally sustainable.

2. List of Contributions

1. Pires, I.; Machado, J.; Rocha, A.; Liz Martins, M. Food Waste Perception of Workplace Canteen Users—A Case Study. *Sustainability* **2022**, *14*, 1324.
2. Teschner, R.; Ruppen, J.; Bornemann, B.; Emmenegger, R.; Sánchez, L.A.; Mapping Sustainable Diets: A Comparison of Sustainability References in Dietary Guidelines of Swiss Food Governance Actors. *Sustainability* **2021**, *13*, 12076.
3. Liz Martins, M.; Henriques, A.S.; Rocha, A. Evaluation of Food Waste at a Portuguese Geriatric Institution. *Sustainability* **2021**, *13*, 2452.
4. Hansen, K.V.; Derdowski, L.A. Sustainable Food Consumption in Nursing Homes: Less Food Waste with the Right Plate Color? *Sustainability* **2020**, *12*, 6525.
5. Sahal Alharbi, N.; Yahia Qattan, M.; Haji Alhaji, J. Towards Sustainable Food Services in Hospitals: Expanding the Concept of 'Plate Waste' to 'Tray Waste'. *Sustainability* **2020**, *12*, 6872.
6. Chun, S.H.; Nyam-Ochir, A. The Effects of Fast Food Restaurant Attributes on Customer Satisfaction, Revisit Intention, and Recommendation Using DINESERV Scale. *Sustainability* **2020**, *12*, 7435.
7. Portugal, T.; Freitas, S.; Cunha, L.M.; Rocha, A.M.C.N. Evaluation of Determinants of Food Waste in Family Households in the Greater Porto Area Based on Self-Reported Consumption Practices. *Sustainability* **2020**, *12*, 8781.
8. Guo, X.Z.; Broeze, J.; Groot, J.J.; Axmann, H.; Vollebregt, M. A Worldwide Hotspot Analysis on Food Loss and Waste, Associated Greenhouse Gas Emissions, and Protein Losses. *Sustainability* **2020**, *12*, 7488.
9. Schepelmann, P.; Vercalsteren, A.; Acosta-Fernandez, J.; Saurat, M.; Boonen, K.; Christis, M.; Marin, G.; Zoboli, R.; Maguire, C. Driving Forces of Changing Environmental Pressures from Consumption in the European Food System. *Sustainability* **2020**, *12*, 8265.
10. Navarro Gausa, M.; Pericu, S.; Canessa, N.; Tucci, G. Creative Food Cycles: A Cultural Approach to the Food Life-Cycles in Cities. *Sustainability* **2020**, *12*, 6487.
11. Stoffle, B.; Stoffle, R.; Van Vlack, K. Sustainable Use of the Littoral by Traditional People of Barbados and Bahamas. *Sustainability* **2020**, *12*, 4764.
12. Vigil, M.; Pedrosa Laza, M.; Moran-Palacios, H.; Alvarez Cabal, J.V. Optimizing the Environmental Profile of Fresh-Cut Produce: Life Cycle Assessment of Novel Decontamination and Sanitation Techniques. *Sustainability* **2020**, *12*, 3674.
13. Shurson, G.C. "What a Waste"—Can We Improve Sustainability of Food Animal Production Systems by Recycling Food Waste Streams into Animal Feed in an Era of Health, Climate, and Economic Crises? *Sustainability* **2020**, *12*, 7071.

Funding: This work was financially supported by LA/P/0045/2020 (ALiCE), UIDB/00511/2020 and UIDP/00511/2020 (LEPABE), funded by national funds through FCT/MCTES (PIDDAC) and within the scope of UIDB/05748/2020 e UIDP/05748/2020.

Acknowledgments: The authors acknowledge the financial supported by LA/P/0045/2020 (ALiCE), UIDB/00511/2020 and UIDP/00511/2020 (LEPABE), funded by national funds through FCT/MCTES (PIDDAC) and within the scope of UIDB/05748/2020 e UIDP/05748/2020.

Conflicts of Interest: The authors declare no conflict of interest.

 sustainability

MDPI

Article

Food Waste Perception of Workplace Canteen Users—A Case Study

Iva Pires [1], Jerusa Machado [2], Ada Rocha [3,*] and Margarida Liz Martins [3,4,5]

1 Interdisciplinary Centre of Social Sciences—CICS.NOVA, Universidade Nova de Lisboa, 1069-061 Lisbon, Portugal; pimm@fcsh.unl.pt
2 Faculty of Food Science, University of Porto, 4099-002 Porto, Portugal; jerusa.sm.nutri@gmail.com
3 GreenUPorto-Sustainable Agrifood Production Research Centre, Faculty of Nutrition and Food Sciences, University of Porto, Rua do Campo Alegre, 823, 4150-180 Porto, Portugal; mliz@utad.pt
4 Centre for the Research and Technology of Agro-Environmental and Biological Sciences, University of Trás-os-Montes and Alto Douro (CITAB-UTAD), Quinta de Prados, 5000-801 Vila Real, Portugal
5 CBQF—Centro de Biotecnologia e Química Fina—Laboratório Associado, Universidade Católica Portuguesa, 4169-005 Porto, Portugal
* Correspondence: adarocha@fcna.up.pt; Tel.: +351-122-5074-320

Abstract: Background: Food waste occurs in all stages of the food supply chain, namely in the food service sector. Understanding how much and why food is wasted and whether consumers are aware of it is essential to design effective interventions in this setting. This case study aims to compare the food waste perception by consumers and measure plate waste in a Portuguese workplace canteen in order to recognize if trained consumers can estimate his/her food waste. Methods: Data were collected from 160 users randomly selected attending a workplace canteen during one month. Plate waste was evaluated by the weighing method. Visual estimation was performed by each participant to evaluate food waste perception at the end of the meal. Consumers were also asked about reasons for wasting food. Results: Plate waste was 8.4% for soup, 9.0% for the main course, and 4.0% for dessert. These values follow the same trend of waste perceived by consumers for soup ($R = 0.722$; $p < 0.001$), main course ($R = 0.674$; $p < 0.001$), and dessert ($R = 0.639$; $p < 0.001$), showing a high relation between self-assessment and measured plate waste. Excessive portions (46.1%), dislike of meal flavor (18.6%), cooking method (8.8%), and texture (3.9%) were identified as the main causes for plate waste. Conclusions: Canteen users showed an accurate perception of their plate waste for all meal components. Excessive portions were identified by consumers as the main reason for plate waste.

Keywords: consumer perception; food service; food waste; plate waste; workplace canteen

Citation: Pires, I.; Machado, J.; Rocha, A.; Liz Martins, M. Food Waste Perception of Workplace Canteen Users—A Case Study. *Sustainability* **2022**, *14*, 1324. https://doi.org/10.3390/su14031324

Academic Editor: Piergiuseppe Morone

Received: 13 December 2021
Accepted: 21 January 2022
Published: 25 January 2022

Publisher's Note: MDPI stays neutral with regard to jurisdictional claims in published maps and institutional affiliations.

1. Introduction

Each year tonnes of food are lost or wasted, leading to economic, social, and environmental impact [1]. This is unacceptable when 750 million people in the world suffer from severe levels of food insecurity. Additionally, the COVID-19 pandemic may have added between 83 and 132 million people to the total number of undernourished in the world in 2020 [2].

In the EU, around 88 million tonnes of food waste are generated annually, of which 53% at the consumer level, with associated costs estimated at EUR 143 billion [1,3]. However, a recent study concluded that it might be underestimated by a factor greater than 2 (214 Kcal/day/capita versus 527 Kcal/day/capita), presenting strong evidence of a link between food waste (FW) and consumer affluence (affluence elasticity of waste). This means that consumers in developed countries may waste more food than previous estimations [4].

Although food waste occurs along the food supply chain, in developed countries, it is higher in the final steps, closer to the consumer [1]. At the final stages, when the food has already been produced, processed, transported, distributed, and cooked, the negative impacts for the environment, the economy, and society are even higher [5].

The food service sector is responsible for producing and supplying meals to satisfy consumers' nutritional needs, considering safety, sensorial and sociocultural constraints.

The number of out-of-home meals has risen in the past decades as a leisure activity or linked to work and academic life, and it is expected to increase. Drivers that explain the trend towards out-of-home food consumption include urbanization, growing incomes, conviviality, lack of time to prepare meals, and working commitments [6–10].

An increasing number of meals served in the food service sector implies that plate waste represents a relevant issue for this setting. Food service accounts for 12% of all food waste in the European Union, producing around 11 million tons of waste, which corresponds to 21 kg per person per year [5,11–15].

Food waste control in food service is a difficult task since it involves the consumers and their relationship with meals, both factors that may vary on a daily basis [10,16–18]. Both service-related aspects, service quality and personal factors, are relevant to explain food waste [19]. Users' reasons for wasting vary in relation to their relationship with food, food preferences, consumers' emotional state, and appetite at the mealtime [10,17,18], which can introduce some variability and unpredictability in the amount of food wasted. Some consumers may experience contradictory feelings regarding food waste.

On one side, consumers feel pressured by social desirability to avoid throwing food away. On the other hand, they do not want to eat food they dislike nor to eat excessive portions that are usually served [13]. Bell et al. used the theory of planned behaviour (TPB) to analyze the influence of canteen workers' behavioral intentions toward food waste reduction. Loss of appetite, poor taste, low appeal, or lack of time to have meals presented a positive impact on food waste behavior [20]. The same TPB was used to analyze food waste behaviors in mass catering services and concluded that a greater intention not to leave edible food and a higher degree of perceived control over this behavior reduced the quantity of food waste [21].

Food waste by consumers results from an interaction between individual and social factors, requiring motivation and means to avoid waste, which implies the need to raise consumer awareness of the impact of behavioral changes [16–18]. A study conducted within 253 urban households found that a higher perception of food waste was associated with a smaller amount of food waste [9]. Carvalho et al. in 2015 demonstrated that the consumers' perception of food waste is inaccurate either in quantity or in type of food wasted [22]. Nevertheless, consumers were worried about the environmental, social, and economic impacts of food waste [23].

The European Commission recommends stepping up the fight against food waste by proposing legally binding targets across the EU member states by 2023. This is part of the ambitious EU Green Deal and its Farm to Fork strategy that includes an objective of reducing food loss and waste as key to a sustainable food system [24–26].

Understanding how much and why food is wasted at the food service is vital to implementing food waste minimization actions. For that, it is essential to understand the reasons and how much users waste and whether consumers are aware of their food waste.

Most food service waste studies have been developed on school canteens [27–30], social centers [31], and hospitals [32,33], and only a few studies have performed empirical research on food waste at workplace canteens [20,21,34,35]. In Portugal, the business structure of the accommodation, restaurant, and similar sectors (Canal HoReCa) corresponds to 7.8% of the total registered companies. In 2017, the sector reached a turnover of more than EUR 10.1 billion, representing 2.8% of the total economic activity and employing 293.478 individuals, which corresponded to 8.0% in relation to the active population [36].

In a recent systematic literature review of food waste in hospitality and food services (HaFS), only 2 out of 63 selected studies focused on workplace canteens [10,21].

This case study intends to contribute to filling this gap, aiming to compare food waste perception by consumers and measured plate waste in a Portuguese workplace canteen in order to recognize if a trained consumer can estimate his/her food waste.

2. Materials and Methods

2.1. Ethics and Sampling Characteristics

The present case study was developed at a workplace canteen in the North of Portugal accessible to about 3000 public employees of a Municipality [37].

The canteen provides daily lunch on working days to an average of 120 regular users, without previous meal booking. The catering service is outsourced to a private company.

The sample included all the canteen users that agreed to participate after knowing the objectives and procedures involved. The study complied with all of the principles of the Declaration of Helsinki [38]. Written consent from canteen users, the private company, and Institution was previously obtained.

This study involved 160 participants. The sociodemographic characteristics of the respondents are presented in Table 1. The respondents were equally distributed between genders. The majority of them were aged between 41 and 60 years old and graduated. Our sample is biased towards more graduated respondents.

Table 1. Participants sociodemographic characteristics (n = 160).

	Female (%)	Male (%)
Participants	49.4	50.6
Age group (years)		
17–30	5.6	5.0
31–40	8.8	13.8
41–50	16.9	13.8
51–60	9.4	11.3
61+	0.6	1.3
No response	5.0	8.5
Education level		
Basic education (4 years)	0.0	1.3
2nd/3rd cycles (6–9 years)	1.2	2.6
Secondary school (12 years)	15.6	21.3
Graduation	27.5	20.0
Master/Ph.D.	5.0	5.6

Approximately 35.6% of participants had meals at the canteen 4–5 times a week, while another 29.4% had lunch at this place 1–2 times a week.

On data collection days, 40.6% of the participants chose meat main course, 25.6% selected fish main course, and 17.5% diet main course, with vegetarian the least selected option by only 15% of the participants.

2.2. Meal Characteristics

Monthly menus were available on the institutional website.

The menus included: (1) a vegetable soup, presenting two options: with and without potatoes; (2) the main course, to be selected from four options: fish, meat, vegetarian, and "diet" (low-fat meal) combined with a carbohydrate source (rice, pasta, potato, or pulses) and a vegetable component. Composed main courses are those presenting the main protein source in fractions mixed with other components; non-composed main courses have the main protein source separated from the carbohydrate and vegetable components; (3) dessert or fruit (three different options of each one available daily).

2.3. Methods

Food waste evaluation by individual weighing was chosen due to its accuracy, allowing comparisons between meal components, despite its high logistical burden being a relevant disadvantage and the possibility to disrupt normal food service operations [39,40].

Visual estimation was used to evaluate food waste perception by consumers. It is well recognized that this method could overestimate plate waste values compared to the weighing method; it is more difficult to interpret, namely for aggregated food groups. This occurs due to the use of a non-continuous scale in visual estimation [39,40]. However, it is a non-invasive method that is not too time consuming, which makes it a valuable method in food waste studies, highly correlated with actual weighed food [39,40].

2.4. Data Collection

Field work was performed during one month and included five weekly menus. At least six consumers were randomly selected each working day in order to ensure participant follow-up and plate waste determination procedures.

The study flow was organized in four steps, starting with the recruitment of participants who were informed on the objectives of the study and ending with the weighing of plate waste (Figure 1).

A printed form was developed to collect individual plate waste during mealtime, and a questionnaire was designed to allow the collection of information about canteen frequency attendance, meal acceptance, portion served, plate waste (visual estimation and causes), and consumers' perception of the sensory characteristics of meals (flavor, taste, texture, and appearance), as well as social demographic information. Finally, the plate with remaining food was weighed, and the amount of food wasted was determined by subtracting the empty plate weight allowing for the comparison between the self-assessment of plate waste and the measured one.

At least 6 consumers randomly selected per day. A. It was ensured that the same person was not evaluated twice.

one month, five weekly menus
160 participants

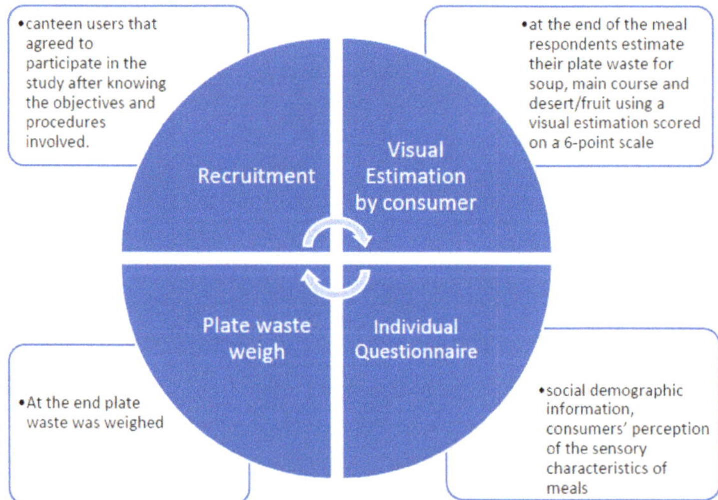

Figure 1. Study flow.

2.4.1. Visual Estimation by Consumer (Plate Waste Perception)

Visual estimation was performed by each participant at the end of the meal in order to estimate individual plate waste for soup, main course, and dessert/fruit. Keeping in mind the appearance of full servings and their variability, the amount of food wasted on each plate was scored on a 6-point scale, adapted from the one by Comstock et al. [39]. If the plate was left untouched, a score of 6 was given; if at least one bite was eaten, a score of 5 was attributed, corresponding to 75% of plate waste; if half the food remained, a score of 4 was scored; if one-quarter remained corresponding to 25% of plate waste, the score for the food item was 3; and if less than one quarter was wasted, a score of 2 was given; a score of 1 was assigned when no plate waste occurred [39] (Figure 2).

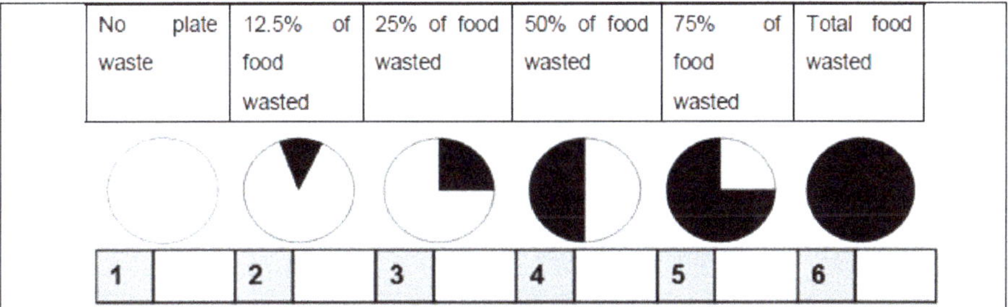

Figure 2. Plate waste estimation. Adapted from Comstock [39].

2.4.2. Plate Waste Determination

For plate waste determination, initially, the plate was weighed empty. Individual meal components were weighed separately after plating (soup, main course, and dessert/fruit). The serving size was determined by subtracting the empty plate weight. At the end of the meal, the plate was collected, non-edible items were removed, the plate with remaining food was weighed, and the amount of food wasted was determined by subtracting the empty plate weight. Aggregated waste across food items was collected when composed main courses were served, while individual plate waste was collected for non-composed main courses, according to the methodology described by Liz Martins et al. [40]. The empty plate was weighed, and the percentage of plate waste was calculated by the ratio of edible food (food available for consumption after removing bones, peels, and stones) discarded per edible food served. Plate waste is referred to in this research as food waste. The weighing was performed on a digital scale accurate to the nearest gram (SECA, model 851, Hamburger, Germany).

2.5. Data Analysis

Statistical software package IBM SPSS Statistics, version 21.0, and Excel Microsoft Office Program Professional Plus 2010 were used for data analyses. Mean, standard deviations (SD), maximum and minimum values were used to provide descriptive analysis. Mann–Whitney test, Kruskal–Wallis tests, and Spearman correlation were used to correlate data collected. The confidence level was set at 95%.

3. Results

3.1. Plate Waste by Meal Components

The percent of plate waste varied from 4.0% for dessert to 9.0% in the main course (Table 2). The proportion of food wasted ranged from 7.1% for vegetables to 11.6% for carbohydrate components (Table 3).

Table 2. Portion served and plate waste according to the meal component.

	Soup (*n* = 102)			**Main Course (*n* = 108)**			**Dessert (*n* = 103)**		
	Mean ± SD	**Max**	**Min**	**Mean ± SD**	**Max**	**Min**	**Mean ± SD**	**Max**	**Min**
Portion served (g)	286.3 ± 64.6	420.0	72.0	382.6 ± 89.4	680.0	158.0	162.4 ± 50.9	268	22
Plate waste (g)	22.4	146.0	0.0	39.9 ± 55.4	248.0	0.0	6.9 ± 17.0	40.0	0.0
Plate waste (%)	8.4			9.0			4.0		

SD—standard deviation; Max—maximum value; Min—minimum value.

Table 3. Portion served and plate waste according to main course component.

Mean ± SD	**Protein Component**	**Carbohydrate Component**	**Vegetable Component**
Portion served (g)	161.9 ± 57.8	169.8 ± 66.2	43.6 ± 29.5
Plate waste (g)	14.1 ± 30.8	18.8 ± 29.9	3.6 ± 10.4
Plate waste (%)	7.9 ± 15.7	11.6 ± 17.9	7.1 ± 14.2

SD—standard deviation.

Portions served and food wasted according to main course type are presented in Table 4.

Table 4. Portion served and plate waste according to main course type.

	Non-Composed Main Course (*n* = 83)			**Composed Main Course (*n* = 25)**		
	Mean ± SD	**Max**	**Min**	**Mean ± SD**	**Max**	**Min**
Portion served (g)	375.3 ± 86.0	680.0	230.0	412.6 ± 85.1	650	236
Plate waste (g)	39.7 ± 52.4	236.0	0.0	43.7 ± 64.4	248.0	0.0

SD—standard deviation; Max—maximum value; Min—minimum value.

The mean weight of food waste was 39.7 g for the non-composed main courses (SD = 52.4) and 43.7 g for the composed main courses (SD = 64.4).

A synthesis of weight plate waste according to meal type, main course component, and main course type is provided in Figure 3.

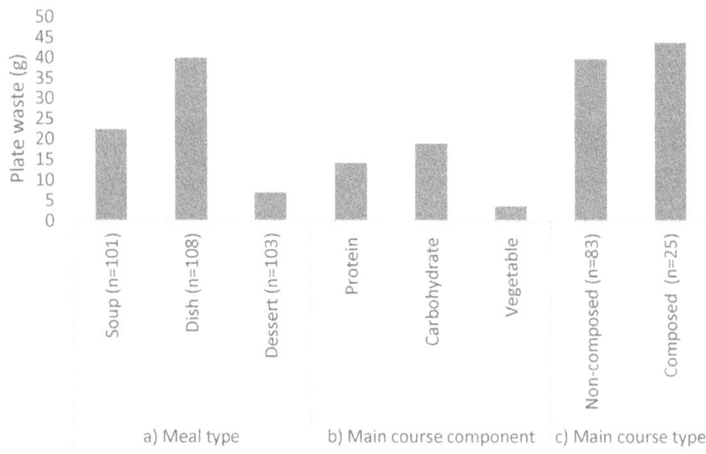

Figure 3. Plate waste (grams) according to meal type, main course component, and main course type.

3.2. Food Waste Perception by Canteen Users

Visual estimation of food waste perceived by consumers is presented in Table 5. According to the visual estimation scale, approximately about half of the participants (46.3%) reported no waste for soup, 54.4% no waste for the main course, and 58.1% no waste for dessert. Considering meal components, 26.9% of participants reported waste of carbohydrate component, 11.2% of protein component, and 1.9% of vegetable component. There are some missings in this question.

About 87% of participants reported overall satisfaction with meal sensory characteristics, namely flavor, taste, texture, and appearance. Participants identified the main causes for food waste excessive portions (46.1%), dislike of meal flavor (18.6%), cooking method (8.8%), and texture (3.9%).

Table 5. Visual estimation of plate waste perceived by consumers for soup, main course, and dessert (n = 160).

	No Food Waste	12.5% of Food Wasted	25% of Food Wasted	50% of Food Wasted	75% of Food Wasted	Total Food Wasted
Soup	46.3%	0.6%	3.1%	2.5%	1.3%	0.0%
Main course	54.4%	27.5%	12.2%	1.3%	2.5%	0.6%
Dessert	58.1%	5.0%	0.0%	0.6%	0.0%	0.6%

The majority of participants indicated wasting food once or twice a week (30.6%), while 44.4% indicated that they never wasted food.

3.3. Measured Food Waste versus Food Waste Perception by Consumers

Reported food waste by participants was similar to measured food waste for all meal components (Table 6).

Table 6. Measured food waste and consumers perceived food waste according to meal components.

Reported Food Waste by Consumers		Measured Food Waste (%) Mean ± SD	p *
Did you waste soup?	Yes (n = 25)	28.3 ± 16.2	<0.001
	No (n = 75)	1.9 ± 7.5	
Did you waste main course?	Yes (n = 70)	17.6 ± 12.4	<0.001
	No (n = 88)	2.1 ± 5.6	
Did you waste dessert?	Yes (n = 11)	23.4 ± 17.4	<0.001
	No (n = 92)	1.6 ± 5.2	

* p value according to the Mann–Whitney test at a confidence level of 95%; SD—standard deviation.

Measured food waste follows the same trend of waste perceived by consumers for soup ($R = 0.722$; $p < 0.001$), main course ($R = 0.674$; $p < 0.001$), and dessert ($R = 0.639$; $p < 0.001$).

3.4. Food Waste According to Sociodemographic Characteristics and Meal Cost

Age and education level had no influence on plate waste values for different meal components. Women wasted more of the main course than men (11.3% ± 12.7% versus 6.7% ± 11.0%; $p = 0.003$). For other meal components, no differences were observed between genders.

The frequency of canteen attendance did not influence plate waste value for soup ($R = 0.057$; $p = 0.572$), main course ($R = 0.002$; $p = 0.978$), and dessert ($R = 0.134$; $p = 0.176$).

Consumers that wasted more main course also wasted more soup ($R = 0.211$; $p = 0.035$) and dessert ($R = 0.196$; $p = 0.005$). Menu option (meat, fish, vegetarian, and diet) had no influence on food waste ($p = 0.343$).

4. Discussion

Our findings showed that participants of this case study have a good perception of plate waste for all meal components. The amount of food waste reported by canteen users, through the visual estimation method, is in line with the food waste assessed by the weighing method. Those reporting waste of soup, main course, and dessert also presented high food waste values for all meal components. In the case of the main course, the plate waste represents 9% of the portion served; the overall waste value of the main course is lower than the acceptable limit of 10% for plate waste, in line with other studies [41–47].

One of the reasons contributing to this low plate waste may be that 87% of participants reported overall satisfaction with meal sensory characteristics, which was already identified as a factor that influences food waste in food services [43].

According to visual estimation, 46.3%, 54.4%, and 58.1% of the respondents mentioned "no food waste" for soup, main course, and dessert, which is in line with another study where even higher values of 72% of consumers indicated that they had no food waste [10]. High standard deviations observed on plate waste may result from consumers' preferences, meal characteristics, and appetite.

A preference for meat main courses was observed, in line with results found by others in different settings [22,28,44]. Nevertheless, in this study, no significant differences were found between plate waste for different menus. The type of the main course chosen was not a trigger for waste in the form of plate waste. This contradicts the results found in a study carried out in a Brazilian university canteen, where menus influenced plate waste, pointing to the low variety of menus served as a contributor to food waste [34]. A study on the attitudes of employees toward food offered in staff canteens indicated a desire for variety, including vegetarian dishes, as well as health-promoting dishes [16]. The possibility of choosing between fish and meat and other options, as is the case of the workplace canteen where the study was conducted, allows higher satisfaction of consumer expectations and consequently lower waste values.

Bell's study (2020) reported a relationship between food waste and level of education [20]; on the opposite, no significant relationship was found between food waste and education level and frequency of attendance to the canteen in the present study, probably due to the homogeneity of participants' graduation level.

An influence of gender on main course plate waste was observed, with higher waste values for women, which can be explained by standardization of portions served, frequently excessive for women [21,48].

Results from these studies are relevant to support awareness campaigns and other strategies to reduce food waste as there is a solid business case for reducing food loss and waste at the food service sector with a triple win: for the economy, for food security, and for the environment [49,50].

A study conducted in Brazil in different restaurant configurations revealed that variable price buffet service had an average plate waste of 23.9 g/plate, while the fixed price buffet/canteen service had 45.8 g/plate. Interviews conducted with users show they are sensitive to monetary incentives, like paying according to quantity served, and tend to be more careful and accurate when selecting the quantity of the food to put on their plates. The same study concluded that when dessert is not included in the meal and consumers have to pay for it separately, no waste of desserts was observed, while when dessert was offered for free, it was the most wasted food product during that day [51]. Considering that saving money was reported to be a primary motivator for food waste reduction, along with moral values [23], the variable price based on the amount of food served could be a solution in workplace canteens, as well as using smaller plate size [52]. This is similar to household research that showed money was a key motivator to reduce food waste and that using it as a key part of an intervention yielded significant food waste reduction results [53]. Another solution could be to offer different plate sizes and let the consumer choose which plate would suit him or her best [21,54].

Similar to the results of a study on household food waste that concluded that the higher household consumer perception, the smaller the amount of food waste per meal [55]; the high accuracy of the perceived food waste of the majority of this workplace canteen canteens users can also contribute to explain the lower volumes of plate waste.

In the present study, the main reason reported for plate waste was the excessive amount of food served, mentioned by 46.1% of the participants, pointing to the need of evaluating and monitoring platting regularly in order to reduce unnecessary overproduction and plate waste [49]. The second most cited reason for wasting food was dislike of the taste, mentioned by 18.6% of the participants, that is related to the acceptability of the menu and meal sensory characteristics. This is in line with other studies that identified taste as having the greatest direct impact on food waste behavior in a catering company [10] and that recommended focus on interventions designed to improve food quality and make portion sizes more flexible [21]. Reasons identified in other studies, like lack of appetite, not enough time for having the meal, or appearance of plate were rarely mentioned by the participants of this study.

The information collected could be an indicator that increasing consumers' perception of their waste will promote waste reduction. This could be used in awareness campaigns, helping to change consumers' behavior.

There is also a solid business case for reducing food loss and waste in the food service sector with a triple win: for the economy, for food security, and for the environment [50]. Taking into account that thousands of people eat in workplace canteens every day, these results can help to design more personalized awareness campaigns that, in addition to contributing to combat food waste, can also play a role to encourage the adoption of healthier eating habits [16,56,57]. In addition, to be considered a good place to introduce healthy eating habits, canteens could also be used to promote consumer's behavior changes related to food waste. In this sense, awareness campaigns towards increasing consumers' perception of their plate waste and of its environmental and economic impact could contribute to that aim.

5. Conclusions

This study uses primary data on measured and perceived food waste at a workplace canteen. A key finding of this case study was the alignment of self-reported versus measured food waste for all meal components, as the amount of food waste assessed by the weighing method confirmed the food waste reported through the visual estimation method. The higher perception of plate waste was associated with higher measured plate waste. Study participants reporting waste of soup, main course, and dessert also presented high food waste values for all meal components.

Excessive portions were identified by consumers as the main reason for plate waste. The information gathered in this study has practical implications for more efficient management of canteen service; namely, they reinforce the need to implement campaigns focusing on the user's awareness about the importance of controlling food waste and choose adequate portions according to their needs and appetite. Considering that consumers seem to have a good perception of their food waste but, nevertheless, still waste food, even if it was below the limit of acceptability for food waste in this setting, we hypothesized the need, on the one hand, to reinforce food waste campaigns and also to develop strategies to counteract the identified barriers and, on the other hand, the need to regularly monitor the quality and the portions served in order to reduce food waste.

We hypothesized awareness campaigns developed by the Municipality supported by research on self-reported versus measured food waste are effective in the reduction of food waste.

Data collection was performed only by one researcher, impairing the possibility to follow more participants on the same day. Additional data collected at other workplace canteens are required to further add more insights to these results.

Author Contributions: Conceptualization, M.L.M., I.P. and A.R.; methodology, J.M. and M.L.M.; formal analysis, J.M. and M.L.M.; investigation, J.M.; data curation, J.M. and M.L.M.; writing—original draft preparation, M.L.M., I.P. and A.R.; writing—review and editing, I.P., M.L.M. and A.R. All authors have read and agreed to the published version of the manuscript.

Funding: This work is partially financed by national funds through FCT—Foundation for Science and Technology, I.P., within the scope of the project «UIDB/50016/2020» (CBQF—Centro de Biotecnologia e Química Fina), «UIDB/05748/2020» and «UIDP/05748/2020» (GreenUPorto), «UIDB/04033/2020» (CITAB—Centre for the Research and Technology of Agro-Environmental and Biological Sciences), and UIDB/04647/2020 «UIDB/04647/2020» (CICS.NOVA—Interdisciplinary Centre of Social Sciences of Universidade Nova de Lisboa). AgriFood XXI I&D&I project (NORTE01-0145-FEDER-000041) co-financed by European Regional Development Fund (ERDF).

Institutional Review Board Statement: Not applicable.

Informed Consent Statement: Not applicable.

Data Availability Statement: Not applicable.

Acknowledgments: The authors thank Phil Lonsdale for the English grammar and structure revision of the manuscript.

Conflicts of Interest: The authors declare no conflict of interest.

References

1. Food and Agriculture Organization of the United Nations. *The State of Food and Agriculture—Moving Forward on Food Loss and Waste Reduction*; Food and Agriculture Organization of the United Nations: Rome, Italy, 2019.
2. FAO; IFAD; UNICEF; WFP; WHO. The State of Food Security and Nutrition in the World 2020. In *Transforming Food Systems for Affordable Healthy Diets*; FAO: Rome, Italy, 2020.
3. Gustavsson, J.; Cederberg, J.; Sonesson, C.; Otterdijk, R.; Meybeck, A. *Global Food Losses and Food Waste*; FAO: Rome, Italy, 2011.
4. Verma, M.; de Vreede, L.; Achterbosch, T.; Rutten, M.M. Consumers discard a lot more food than widely believed: Estimates of global food waste using an energy gap approach and affluence elasticity of food waste. *PLoS ONE* **2020**, *15*, e0228369. [CrossRef] [PubMed]
5. Stenmarck, Â.; Jensen, C.; Quested, T.; Moates, G.; Buksti, M.; Cseh, B.; Juul, S.; Parry, A.; Politano, A.; Redlingshofer, B.; et al. *Estimates of European Food Waste Levels*; IVL Swedish Environmental Research Institute: Stockholm, Sweden, 2016.
6. d'Angelo, C.; Gloinson, E.; Draper, A.; Guthrie, S. *Food Consumption in the UK. Trends, Attitudes and Drivers*; RAND Corporation: Santa Monica, CA, USA; Cambridge, UK, 2020.
7. Lund, T.B.; Kjaernes, U.; Holm, L. Eating out in four Nordic countries: National patterns and social stratification. *Appetite* **2017**, *119*, 23–33. [CrossRef] [PubMed]
8. Paddock, J.; Warde, A.; Whillans, J. The changing meaning of eating out in three English cities 1995–2015. *Appetite* **2017**, *119*, 5–13. [CrossRef] [PubMed]
9. Zhang, X.Y.; Jeong, E.; Olson, E.D.; Evans, G. Investigating the effect of message framing on event attendees' engagement with advertisement promoting food waste reduction practices. *Int. J. Hosp. Manag.* **2020**, *89*, 102589. [CrossRef]
10. Lorenz, B.; Hartmann, M.; Hirsch, S.; Kanz, O.; Langen, N. Determinants of Plate Leftovers in One German Catering Company. *Sustainability* **2017**, *9*, 807. [CrossRef]
11. Corrado, S.; Sala, S. Food waste accounting along global and European food supply chains: State of the art and outlook. *Waste Manag.* **2018**, *79*, 120–131. [CrossRef] [PubMed]
12. Dhir, A.; Talwar, S.; Kaur, P.; Malibari, A. Food waste in hospitality and food services: A systematic literature review and framework development approach. *J. Clean. Prod.* **2020**, *270*, 122861. [CrossRef]
13. Silvennoinen, K.; Nisonen, S.; Pietilainen, O. Food waste case study and monitoring developing in Finnish food services. *Waste Manag.* **2019**, *97*, 97–104. [CrossRef]
14. Beretta, C.; Stoessel, F.; Baier, U.; Hellweg, S. Quantifying food losses and the potential for reduction in Switzerland. *Waste Manag.* **2013**, *33*, 764–773. [CrossRef]
15. Eriksson, M.; Persson Osowski, C.; Malefors, C.; Bjorkman, J.; Eriksson, E. Quantification of food waste in public catering services—A case study from a Swedish municipality. *Waste Manag.* **2017**, *61*, 415–422. [CrossRef]
16. Czarniecka-Skubina, E.; Gorska-Warsewicz, H.; Trafialek, J. Attitudes and Consumer Behavior toward Foods Offered in Staff Canteens. *Int. J. Environ. Res. Public Health* **2020**, *17*, 6239. [CrossRef]

17. Lorenz-Walther, B.A.; Langen, N.; Göbel, C.; Engelmann, T.; Bienge, K.; Speck, M.; Teitscheid, P. What makes people leave LESS food? Testing effects of smaller portions and information in a behavioral model. *Appetite* **2019**, *139*, 127–144. [CrossRef] [PubMed]
18. Rohm, H.; Oostindjer, M.; Aschemann-Witzel, J.; Symmank, C.; Lalmli, V.; De Hooge, I.E.; Normann, A.; Karantininis, K. Consumers in a Sustainable Food Supply Chain (COSUS): Understanding Consumer Behavior to Encourage Food Waste Reduction. *Foods* **2017**, *6*, 104. [CrossRef] [PubMed]
19. Liz Martins, M.; Veiga, M.J.; Rocha, A. Desperdício alimentar numa população sem abrigo: Um estudo de caso. *Acta Port. Nutr.* **2021**, *24*, 50–54. (In Portuguese)
20. Bell, A.E.; Ulhas, K.R. Working to Reduce Food Waste: Investigating Determinants of Food Waste amongst Taiwanese Workers in Factory Cafeteria Settings. *Sustainability* **2020**, *12*, 9669. [CrossRef]
21. Sebbane, M. Food Leftovers in Workplace Cafeterias: An Investigation of Beliefs and Psychosocial Factors. Available online: https://hal.archives-ouvertes.fr/hal-01616614/ (accessed on 15 May 2021).
22. Carvalho, J.G.; Lima, J.P.M.; Rocha, A. Desperdício alimentar e satisfação do consumidor com o serviço de alimentação da Escola de Hotelaria e Turismo de Coimbra. *DEMETRA Aliment. Nutr. Saúde* **2015**, *10*, 405–418. [CrossRef]
23. Neff, R.A.; Spiker, M.L.; Truant, P.L. Wasted Food: U.S. Consumers' Reported Awareness, Attitudes, and Behaviors. *PLoS ONE* **2015**, *10*, e0127881. [CrossRef]
24. United Nations. *Resolution Adopted by the General Assembly 70/1. Transforming Our World: The 2030 Agenda for Sustainable Development*; United Nations: New York, NY, USA, 2015.
25. Haines, A.; Scheelbeek, P. European Green Deal: A major opportunity for health improvement. *Lancet* **2020**, *395*, 1327–1329. [CrossRef]
26. European Commission. *Communication from the Commission to the European Parliament, the Council, the European Economic and Social Committee and the Committee of the Regions. A Farm to Fork Strategy for a Fair, Healthy and Environmentally-Friendly Food System*; European Commission: Brussels, Belgium, 2020.
27. Liz Martins, M.; Rodrigues, S.S.; Cunha, L.M.; Rocha, A. Strategies to reduce plate waste in primary schools—Experimental evaluation. *Public Health Nutr.* **2016**, *19*, 1517–1525. [CrossRef]
28. Liz Martins, M.; Rodrigues, S.S.P.; Cunha, L.M.; Rocha, A. Factors influencing food waste during lunch of fourth-grade school children. *Waste Manag.* **2020**, *113*, 439–446. [CrossRef]
29. Garcia-Herrero, L.; De Menna, F.; Vittuari, M. Food waste at school. The environmental and cost impact of a canteen meal. *Waste Manag.* **2019**, *100*, 249–258. [CrossRef] [PubMed]
30. Kowalewska Maria, T. Food, nutrient, and energy waste among school students. *Br. Food J.* **2018**, *120*, 1807–1831. [CrossRef]
31. Liz Martins, M.; Henriques, A.S.; Rocha, A. Evaluation of Food Waste at a Portuguese Geriatric Institution. *Sustainability* **2021**, *13*, 2452. [CrossRef]
32. Dias-Ferreira, C.; Santos, T.; Oliveira, V. Hospital food waste and environmental and economic indicators—A Portuguese case study. *Waste Manag.* **2015**, *46*, 146–154. [CrossRef] [PubMed]
33. Oliveira, D.; Liz Martins, M.; Rocha, A. Food waste index as an indicator of menu adequacy and acceptability in a Portuguese mental health hospital. *Acta Port. De Nutr.* **2020**, *20*, 14–18. (In Portuguese)
34. Ricarte, M.P.R.; Fé, M.A.B.M.; da Silva Santos, I.H.V.; Lopes, A.K.M. Avaliação do desperdício de alimentos em uma unidade de alimentação e nutrição institucional em Fortaleza-CE. *Saber Cient.* **2008**, *1*, 159–175.
35. Augustini, V.; Kishimoto, P.; Tescaro, T.; Almeida, F. Avaliação do Índice de Resto-Ingesta e Sobras em Unidade de Alimentação e Nutrição (UAN) de uma empresa metalúrgica na cidade de Piracicaba/SP. *Simbio-Logias* **2008**, *1*, 99–110.
36. Europe Food Service. European Industry Overview. 2017. Available online: http://www.foodserviceeurope.org/en/european-industry-overview (accessed on 20 May 2021).
37. Instituto Nacional de Estatística. *PORDATA Bases de Dados Portugal Contemporâneo*; Instituto Nacional de Estatística: Lisboa, Portugal, 2018.
38. World Medical Association (WMA). Helsinki Declaration for Medical Research in Human Subject. Available online: https://www.wma.net/policies-post/wma-declaration-of-helsinki-ethical-principles-for-medical-research-involving-human-subjects (accessed on 20 May 2021).
39. Comstock, E.M.; St Pierre, R.G.; Mackiernan, Y.D. Measuring individual plate waste in school lunches. Visual estimation and children's ratings vs. actual weighing of plate waste. *J. Am. Diet. Assoc.* **1981**, *79*, 290–296. [CrossRef]
40. Liz Martins, M.; Cunha, L.M.; Rodrigues, S.S.; Rocha, A. Determination of plate waste in primary school lunches by weighing and visual estimation methods: A validation study. *Waste Manag.* **2014**, *34*, 1362–1368. [CrossRef]
41. Engstrom, R.; Carlsson-Kanyama, A. Food losses in food service institutions. Examples from Sweden. *Food Policy* **2004**, *29*, 203–213. [CrossRef]
42. Steen, H.; Malefors, C.; Roos, E.; Eriksson, M. Identification and modelling of risk factors for food waste generation in school and pre-school catering units. *Waste Manag.* **2018**, *77*, 172–184. [CrossRef] [PubMed]
43. van Geffen, E.J.; van Herpen, H.W.I.; van Trijp, J.C.M. *Causes & Determinants of Consumers Food Waste: Project Report, EU Horizon 2020 REFRESH*; Wageningen University and Research: Wageningen, The Netherlands, 2016.
44. Ferreira, M.; Liz Martins, M.; Rocha, A. Food waste as an index of foodservice quality. *Br. Food J.* **2013**, *115*, 1628–1637. [CrossRef]
45. Conselho Federal de Nutricionistas. *Resolução CFN N° 380/2005*; Conselho Federal de Nutricionistas: Brasília, Brasil, 2005.

46. de Miranda Luciano Rabelo, N.; Alves, T.C.U. Avaliação do percentual de resto-ingestão e sobra alimentar em uma unidade de alimentação e nutrição institucional. *Rev. Bras. Tecnol. Agroindustrial.* **2016**, *10*, 2039–2052.
47. Betz, A.; Buchli, J.; Gobel, C.; Muller, C. Food waste in the Swiss food service industry—Magnitude and potential for reduction. *Waste Manag.* **2015**, *35*, 218–226. [CrossRef] [PubMed]
48. Wardle, J.; Haase, A.M.; Steptoe, A.; Nillapun, M.; Jonwutiwes, K.; Bellisle, F. Gender differences in food choice: The contribution of health beliefs and dieting. *Ann. Behav. Med.* **2004**, *27*, 107–116. [CrossRef]
49. Vizzoto, F.; Tessitore, S.; Iraldo, F.; Testa, F. Passively concerned: Horeca managers' recognition of the importance of food waste hardly leads to the adoption of more strategies to reduce it. *Waste Manag.* **2020**, *107*, 266–275. [CrossRef]
50. Clowes, A.; Hanson, C.; Mitchell, P. The Business Case for Reducing Food Loss and Waste: Catering. A Report on Behalf of Champions 12.32018. Available online: https://champions123.org/sites/default/files/2020-07/business-case-reducing-food-loss-waste-catering.pdf (accessed on 15 May 2021).
51. Matzembacher, D.; Brancoli, P.; Moltene Maia, L.; Eriksson, M. Consumer's food waste in different restaurants configuration: A comparison between different levels of incentive and interaction. *Waste Manag.* **2020**, *114*, 263–273. [CrossRef]
52. Cunningham, E. What impact does plate size have on portion control? *J. Am. Diet. Assoc.* **2011**, *111*, 1438. [CrossRef]
53. van der Werf, P.; Seabrook, J.A.; Gilliland, J.A. "Reduce Food Waste, Save Money": Testing a novel intervention to reduce household food waste. *Environ. Behav.* **2021**, *53*, 151–183. [CrossRef]
54. Ravandi, B.; Jovanovic, N. Impact of plate size on food waste: Agent-based simulation of food consumption. *Resour. Conserv. Recy.* **2019**, *149*, 550–565. [CrossRef]
55. Zhang, P.; Zhang, D.; Cheng, S. The Effect of Consumer Perception on Food Waste Behavior of Urban Households in China. *Sustainability* **2020**, *12*, 5676. [CrossRef]
56. Naicker, A.; Shrestha, A.; Joshi, C.; Willett, W.; Spiegelman, D. Workplace cafeteria and other multicomponent interventions to promote healthy eating among adults: A systematic review. *Prev. Med. Rep.* **2021**, *22*, 101333. [CrossRef] [PubMed]
57. Bandoni, D.H.; Sarno, F.; Jaime, P.C. Impact of an intervention on the availability and consumption of fruits and vegetables in the workplace. *Public Health Nutr.* **2011**, *14*, 975–981. [CrossRef] [PubMed]

 *sustainability* 

Article

Mapping Sustainable Diets: A Comparison of Sustainability References in Dietary Guidelines of Swiss Food Governance Actors

Ronja Teschner [1,*], Jessica Ruppen [2], Basil Bornemann [2,3], Rony Emmenegger [2] and Lucía Aguirre Sánchez [4,5]

[1] Institute for European Global Studies, University of Basel, Riehenstrasse 154, 4058 Basel, Switzerland
[2] Sustainability Research Group, Department of Social Sciences, University of Basel, Bernoullistrasse 14/16, 4056 Basel, Switzerland; jessica.ruppen@stud.unibas.ch (J.R.); basil.bornemann@unibas.ch (B.B.); rony.emmenegger@unibas.ch (R.E.)
[3] Social Transitions Research Group, Department of Social Sciences, University of Basel, Rheinsprung 24, 4051 Basel, Switzerland
[4] Institute of Communication and Public Policy (ICPP), Università della Svizzera Italiana, Via G. Buffi 13, 6900 Lugano, Switzerland; lucia.aguirre.sanchez@usi.ch
[5] Institute of Public Health (IPH), Università della Svizzera Italiana, Via G. Buffi 13, 6900 Lugano, Switzerland
* Correspondence: ronja.teschner@stud.unibas.ch

Abstract: With the growing recognition of the food system for a transformation toward sustainability, there is a need for future guidance on food consumption and policy. In particular, dietary guidelines (DGs) have received increasing attention as potential tools for enabling transformative change. This paper analyzes how and to what extent different state and non-state actors in Switzerland incorporate sustainability aspects in their dietary guidelines. It examines how these DGs account for different dimensions at the basis of sustainability thinking, including the classic environmental, economic, and social dimensions as well as issues of health and governance. Our analysis shows the explicit inclusion of sustainability aspects in all DGs of the chosen actors in Switzerland, addressing at least one sustainability category predominantly. Through the analysis of the different stakeholders, different areas of focus become apparent, with each stakeholder covering specific niches of sustainability. On this basis, the transformative role of non-state actors in developing the concept of sustainable diets is discussed.

Keywords: sustainable diet; sustainable dietary guidelines; qualitative content analysis; sustainable food systems; food governance

Citation: Teschner, R.; Ruppen, J.; Bornemann, B.; Emmenegger, R.; Sánchez, L.A. Mapping Sustainable Diets: A Comparison of Sustainability References in Dietary Guidelines of Swiss Food Governance Actors. *Sustainability* **2021**, *13*, 12076. https://doi.org/10.3390/su132112076

Academic Editors: Ada Margarida Correia Nunes Da Rocha and Belmira Almeida Ferreira Neto

Received: 13 July 2021
Accepted: 26 October 2021
Published: 1 November 2021

1. Introduction

Confronted with anthropogenic challenges, humanity urgently needs to begin operating within planetary boundaries—nine biological and physical thresholds that define the "safe operating space" for humanity [1]. By 2015, four of the planetary boundaries (climate change, biosphere integrity, biogeochemical flows, and land-system change) had already been exceeded or are at risk [2]. The current food system is key to this: Food system dynamics have adverse consequences on planetary and human health [3] and are responsible for 21–37% of total anthropogenic greenhouse gas (GHG) emissions [4]. Agriculture, in particular, is not only a significant contributor to climate change but also the greatest driver of transgressions of other planetary boundaries: biosphere integrity and biochemical flows (related to human-induced changes in global nitrogen and phosphorus cycles), along with land-system use, and freshwater use [5]. The unsustainability of the food system is also critical [6] as the world will face increasing food quality and food security challenges in the coming decades [7]. Consequently, improving human and planetary health, while ensuring food security, requires a shift to more sustainable food systems [8].

The challenge of a sustainability-oriented food system transformation concerns all phases of the food value chain from production through distribution to consumption [9]

and involves the engagement of multiple state and non-state actors [10–13]. Food consumption and, in particular, dietary patterns are increasingly seen as key levers for such a transformation and are increasingly moving into the focus of political and scientific attention [8,14,15]. For example, the EAT-Lancet Commission concludes that improvements in food production can reduce agricultural GHG emissions only by about 10%, while dietary shifts display a reduction potential of up to 80% [16]. According to the IPCC (Intergovernmental Panel on Climate Change), there is a climate change mitigation potential of up to 8.55 Gt CO2e in 2050 from dietary change and reduction of food waste, while only a maximum of 4.6 Gt CO2e can be achieved by supply-side interventions in agriculture [17]. Beyond being only of significance for the environment, changes in nutritional patterns can equally benefit human health and well-being.

While there is broad agreement on the need for more sustainable diets [3,6,8,18–20], the comprehensive definition of sustainable dietary recommendations, that are operational at the consumer level, is still at an early stage [21,22]. The often-cited definitions of sustainable diets are still very general, and not operational at the consumer level. FAO defines sustainable diets as those "with low environmental impacts which contribute to food and nutrition security and to healthy life for present and future generations [...] protective and respectful of biodiversity and ecosystems, culturally acceptable, accessible, economically fair and affordable; nutritionally adequate, safe and healthy; while optimizing natural and human resources" [23] (p. 83). This aspirational definition leaves room for different understandings in research and policy of what constitutes a "sustainable diet", resulting in varying approaches and recommendations for food system transformations. On the one hand, this has to do with the complex embeddedness of diets within the food system, the contestedness of the concept of sustainability itself or the way it is operationalized in food system research [7]. On the other hand, it reflects the increasing politicization of food and, among other things, the associated expansion of the food governance landscape. In addition to governments, other actors from health, business, and civil society are increasingly involved in shaping food governance. They not only play active roles in specific policy processes, but also engage in less tangible ways by developing and disseminating food-related norms and knowledge [24].

With the overall goal of elucidating the societal debate on "sustainable diets", this paper aims to empirically capture the different understandings of the concept held by different actors. Drawing on the example of the Swiss food governance landscape, the paper examines how different food governance actors refer to sustainability in their "dietary guidelines" (DGs). These are broadly defined as norms and knowledge about good nutrition practices that are publicly communicated by food governance actors in the form of recommendations or reports. In our focus are the core statements of the DGs of five actors–the Swiss government, Nestlé, the World Wildlife Fund (WWF), EAT-Lancet Commission, and Schweizer Verband Volksdienst (SV) Group Switzerland. We ask: How do different stakeholders in the Swiss food governance landscape conceptualize sustainable diets in their dietary guidelines? To answer the question, we employ a combination of qualitative and quantitative tools to capture content and to comparatively map it along five sustainability dimensions.

With this study, we make a twofold contribution to the study of sustainable diets. First, by broadening the view beyond official government DGs toward informal ones, we sharpen the understanding of differences and commonalities between interpretations of sustainable diets in a pluralized food supply landscape. Second, we assess sustainable diets from a multi-dimensional understanding, that adds health and governance to the three traditional dimensions of sustainability, namely environmental, social and economic. With this approach, we attempt to map the complexity of sustainability considerations in food systems.

Our argument unfolds as follows. In the next section, we briefly position the object of our analysis, dietary guidelines, by discussing their potential in sustainability-oriented food system governance. Section three presents our research materials and the methodological approach we applied for studying sustainability in dietary guidelines. In the Results section, we outline the stakeholders' sustainability considerations at the category level, and then discuss the codes within the categories in more detail, highlighting similarities and differences between the stakeholders' approaches. We then discuss the implications of our findings for the governance of sustainability-oriented transformations of the food system. We conclude with perspectives for future research on the role of dietary guidelines for food system transformations.

2. Background: Dietary Guidelines in a Changing Food Governance Landscape

The literature about dietary guidelines (DGs), also called nutritional guidelines or food-based dietary guidelines (FBDGs), often refers to the official dietary recommendations released by country governments. While we build on these previous developments, we also will argue for expanding the research focus of dietary recommendations to those provided by nongovernmental food governance actors.

Dietary guidelines were originally created with the purpose of providing recommendations from the government to the population on what constitutes a healthy diet. DGs are the basis of health policy and nutrition education, aimed at promoting population health and preventing diet-related diseases [25]. Gradually, the guidelines have moved from nutrient intake recommendations to food-based recommendations, designed to be easily understood, often visual [25]. The idea of food-based dietary guidelines (FBDGs) was born at the Joint FAO/WHO consultation in 1995, with the aim of making dietary recommendations more accessible to the general public who think in terms of foods instead of nutrients [25]. Thereafter, FBDGs has become the common term when referring to country official DGs.

In recent years, the potential of DGs to address the multiple challenges of sustainability in food systems has attracted increasing interest [18,19]. A growing body of literature continues to call for expanding the scope of DGs to sustainability-oriented dietary recommendations as a potential tool to address the unsustainability of the food systems and eating habits [10,18,19,26,27]. In fact, several countries have incorporated sustainability aspects in their DGs [16,18]—beyond the original focus on health and nutrition only. However, the "sustainabilization" of dietary guidelines has faced some reluctance, and the number of countries that have done so remains limited to date [18,28,29]. Furthermore, findings suggest that policies to mitigate climate change and related international climate agreements are inconsistent with the official dietary recommendations of most countries [27].

In terms of impacts, empirical evidence has shown that adherence to official DGs is low in many countries, including Switzerland [30–32]. Switzerland, along with other countries such as Argentina, Australia, Greece, Honduras, Portugal, and the United Kingdom, is among the 28% of countries that fulfill none of the recommendations of their FBDGs. In view of the limited steering effect of DGs on the food consumption behavior of most individuals, it cannot be expected that the inclusion of sustainability aspects in the guidelines will automatically lead to sustainability-oriented change in population diets.

However, the role of DGs for food system transformation goes beyond their capacity to steer consumer behavior directly. On the one hand, DGs can influence government investment as food policies and programs are often required to be guided by official dietary guidelines. DGs have been found to influence policy and program implementation in different sectors and settings, from educational campaigns and food procurement for hospitals, schools' menus, and vending machines, to food security and agricultural programs aimed at encouraging farmers to grow foods recommended in official DGs [26]. At their full potential, they guide both the public and policy makers to develop health and agricultural policies and interventions, public procurement standards and regulations, food marketing and advertising, and labelling [26,28]. On the other hand, there is some evidence about the signaling function of a food policy change towards sustainability. When policy change is communicated to the public, for instance, it increases consumer acceptance of the eating behavior that the new policy aims to promote [33]. Dietary guidelines have therefore proven as a key component of food policy—despite their limitations in regard to their direct impact on consumption behavior.

In light of recent developments in the food governance landscape, however, the focus on the role of official dietary guidelines for a sustainable transformation of the food system seems too narrow [34]. While food has long been considered an apolitical issue managed in closed circles by administrative experts and interest groups, we are witnessing a wave of politicization of food over the past two decades [35], accompanied by a growing pluralization of the food governance landscape [10–12]. An increasing number of non-state actors are attempting to shape food governance at all stages of the food system, from production to consumption [3]. In addition to already established major food companies and related organized interest groups [10], actors such as environmental NGOs, food movements, consumer networks, and research networks have entered the scene [10,36]. These actors play an increasingly important role not only in "hard" policy-making processes related to food issues, but also in "soft" shaping of food-related practices. They create and disseminate knowledge and norms on good food practices and behaviors into societal discourses and governance arrangements [37]. In doing so, they offer new potentials for strengthening sustainability aspects due to their heterogeneous interests and positioning in the food field. Particularly with regard to emerging discourses such as "sustainable diets", these actors can be expected to try to play a shaping role by contributing their own visions and ideas. To address this pluralization of food governance, we broaden the understanding of dietary guidelines to include all types of norms and knowledge about what, when, and how to eat that are given in recommendations and reports by different actors. Expanding the focus to include dietary recommendations from nongovernmental actors offers the potential to take a fresh look at the definition of "sustainable diets" and open up new entry points for shaping sustainability-oriented change in dietary habits and the food system as a whole.

3. Materials and Methods

3.1. Selection of Stakeholders and Dietary Guidelines

This study investigates the extent to which the DGs promoted by various actors in Switzerland relate to sustainability (see Appendix A for an overview of the data). In a first step, we identified relevant actors in the Swiss food landscape by using purposive sampling to capture a selection of information-rich cases from key theoretical constructs of food system governance. Switzerland is an interesting example for examining sustainability-oriented food system transformation given its clear commitment to the 2030 Agenda [38,39] and its international outreach. The country is the home of important international organizations shaping the food system, several large international food companies and one of the largest civil society conservation organizations. In the process, we attempted to cover a multitude of sectors within food system governance. Selection within each sector was based on the following criteria: First, we selected the stakeholder with the largest sphere of influence within a sector. After identification, it was examined whether the stakeholder provides dietary guidelines or recommendations in English or German that are accessible to the general public and are not older than 2015, except for the official FBDG of Switzerland, the current version of which was published in 2011. The selected guidelines contain a range of information and recommendations on nutrition and food, such as reference intake values and dietary suggestions. This indicates that stakeholders are involved in the creation and dissemination of norms and knowledge about food/eating behaviors. Two researchers independently reviewed the websites of the stakeholders. When a search function was available, we searched for the following keywords: "diets", "diet", "dietary guidelines", "recommendation", "dietary recommendation", "guide", "plate", and "food". Inconsistencies in inclusion were resolved by consensus, and an exchange with the various stakeholders took place in the form of interviews and/or correspondence on the selection of appropriate and representative guidelines. In cases where multiple recommendations from a stakeholder met the inclusion criteria, all of these guidance documents were included in the analysis to provide the most comprehensive picture possible. On the basis of the preliminary data collection and data analysis results, we reviewed and revised our data and used the newly discovered information to make future guideline selections [40].

The final selection includes the Swiss government, Switzerland's largest nature conservation organization WWF, Switzerland's largest community catering company SV Group, the largest food and restaurant company worldwide Nestlé (according to a 2019 ranking), and the international research organization EAT-Lancet Commission that serves as a reference point for sustainability-driven dietary guidelines. Table 1 provides an overview of the guidelines origin, year of publication, language, and the sector to which the selected stakeholders belong.

Table 1. Selected dietary guidelines of stakeholders categorized by the food sector.

Stakeholder	Sector	Guideline	Language	Year
Swiss Government	Government	Eating well and staying Healthy Swiss Nutrition Policy 2017–2024	English	2017
		Swiss Food Pyramid	English	2011
		Der optimale Teller	German	2018
Nestlé Switzerland/ International	Private Sector (Food Company)	Nestlé's Net Zero Roadmap	English	2020
		Nestlé in der Schweiz	German	2019
		The Balanced Plate–day by day	English	2017
WWF Switzerland/ International	Civil Society	Bending the Curve: The Restorative Power of Planet-Based Diets (WWF International)	English	2020
		Factsheet–Umweltgerechtes Essen–der Erde zuliebe (WWF Schweiz)	German	2019
SV Group Switzerland	Private Sector (Community Catering)	Nachhaltigkeitsbericht (extended online version)	German	2020
		SV Restaurant Kundenbroschüre	German	2018
EAT-Lancet Commission	International Organization	Diets for a better Future: Rebooting and Reimagining Healthy and Sustainable Food Systems in the G20	English	2020
		Healthy Diets from Sustainable Food Systems–Food Planet Health	English	2020

3.2. *Measuring Sustainability in Dietary Guidelines*

We followed a three-step approach for the generation of the code book (Table 2) to assess the DGs for its sensitivity in different dimensions of sustainability. First, we identified various interpretations and definitions of sustainable diets in relevant literature, official reports, guidelines, and various forms of documents from private, academic, and public institutions on nutrition, food systems, and sustainability. Search criteria included publications up to 2021. The following keywords inspired our search: "food system(s)", "sustainable diet(s)", "sustainability" or "sustainable", "food-based dietary guideline(s)", "dietary guideline(s)", "nutritional guideline(s)".

In the next step, we used inductive reasoning to develop overarching categories for sustainable diets. This categorization enabled us to summarize the data to illustrate the most critical points within the texts [40]. The inductive proceeding revealed two other categories, governance and health, which are closely related to the food system [7,41–44] —in addition to the generally accepted categories of economics, social, and environment which were deduced from the sustainability literature. The first three categories represent Brundtland's triangular model for sustainable development that integrates social, economic, and environmental dimensions [44]. Despite illuminating the essence of sustainability thinking, they fail to address the other aspects—health and governance—that are central to diet since food consumption has direct health impacts and is embedded in a complex governance landscape [15].

We then divided these five overarching categories into subcategories [45], which are referred to as codes in this analysis. Codes were created during the process in line with the inductive study design [46–48]. As this is an interpretative act, the coding process was repeated several times equally by both researchers—the two first authors—until saturat ion was reached [49]. By use of the software Dedoose (v. 8.3.41; University of California, Los Angeles, CA, USA), all pertinent excerpts were marked, labeled with codes, and assigned scores. The excerpts were generated in the form of words, phrases, and sentences [49], and the codes applied were based on frequency. We did not differentiate between word forms, e.g., "educate" and "education" were coded the same. In addition, we coded implicit and explicit mentions of codes in the text passages. This is illustrated in Appendix B, where the code is applied first to the statement and then the code's level of presence is scored. Quantification by scoring was used as an instrumental step towards visualization of the different profiles (Table 3). Quantification condensed the data into a visually instructive form and rendered the content of guidelines comparable in the context of new players entering the food landscape. Ultimately, each category was combined into a single value, resulting in spider-webs.

The analytical framework and the methods proposed can be easily applied as a screening tool to assess the sustainability quality of DGs in other contexts beyond Switzerland.

Table 2. The code book for sustainable diet assessment.

Category	1. Environment	2. Social	3. Economics	4. Health	5. Governance
Definition	Denotes living within the carrying capacity of supporting ecosystems while meeting human needs [50].	Driving forward social progress for all with socio-economic conditions that are fair and affordable; nutritionally adequate, safe, culturally acceptable, and accessible while empowering animal welfare and gender equality [4].	Practice that reinforces social and environmental objectives considered in relation to trade, industry, and the creation of wealth [4].	Comprises the essential food groups for growth and good health as well as being of complete physical, mental, and social well-being [4]	Denotes the totality of instruments and mechanisms available to steer a society collectively [51].
Code	1.1 Climate Change	2.1 Community	3.1 Affordability	4.1 Well-being	5.1 Certifications and Standards
	1.2 Biodiversity	2.2 Culture	3.2 Cost	4.2 Fruits and Vegetables	5.2 Transparency
	1.3 Land use	2.3 Pleasure	3.3 Labor Rights	4.3 Animal-based Protein	5.3 Regulation
	1.4 Water use	2.4 Animal Welfare	3.4 Sustainable Production Patterns	4.4 Plant-based Protein	5.4 Food Security
	1.5 Soil	2.5 Ethical Buying	3.5 Technology and Innovation	4.5 Tubers or Starchy Vegetables	5.5 Justice
	1.6 Animal Agriculture	2.6 Gender Equality		4.6 Whole Grains	5.6 Education
	1.7 Origin			4.7 Liquid (Unsweetened Drinks)	5.7 Directives
	1.8 Food Waste			4.8 Sweets, Salty, Snacks and Alcoholic Drinks	5.8 Science
	1.9 Energy use			4.9 Dairy Products	
	1.10. Aquatic Ecosystem				

Table 3. Scoring system.

Scale	Definition
0 = not mentioned	Absence of the code.
1 = briefly mentioned	The code is only vaguely mentioned in the text, with only a word or short phrase referring to the feature.
2 = well expressed	The statement consists of a clear clarification of the code that is explained or elaborated within the excerpt and is substantiated with figures, graphs, facts or details.

4. Results

4.1. Overview of the Sustainability Profiles of Dietary Guidelines

The following section unravels how sustainability is captured in different dietary recommendations by different stakeholders. We use these sustainability references as a proxy for understanding sustainable diets. Our method does not assign distinct weights to our proposed five dimensions of sustainability. However, this does not necessarily imply a normative stand about the need to include each aspect of diet sustainability in equal proportions within DGs. Therefore, our visualizations should be interpreted as descriptive tools to illustrate how stakeholders compare to each other and not as a normative tool to show how they compare to an ideal, balanced version of sustainable DGs. The majority of stakeholders communicate a one- or two-dimensional view of sustainability (Figure 1). There are distinct aspects of sustainability associated with each stakeholder's DGs, each with niche-specific considerations. The Swiss government places a strong focus on the health aspect of dietary recommendations (Figure 2), which accounts for 62% of the total coded content. Next in line is governance, at 27%, which leaves little room for the other three categories, all below 10%. The government does not focus on environmental considerations within their recommendations. In contrast, Nestlé's guidelines are dominated by the category environment, amounting to 54%; the following categories are economics with 20% of coded content and governance with 17% (Figure 3). Half of the WWF guidelines' content focuses on the environmental implications of dietary choices. Governance accounts for close to a quarter of the coded content, while health has slightly lower coverage with 18%. The economics and social categories are briefly addressed with short substantiation (Figure 4).

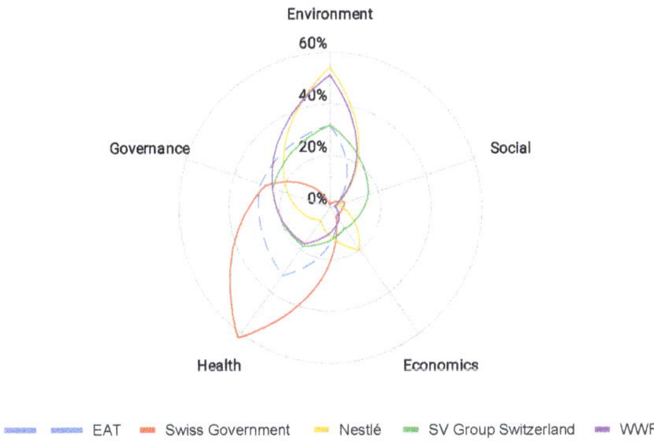

Figure 1. Profiles overview.

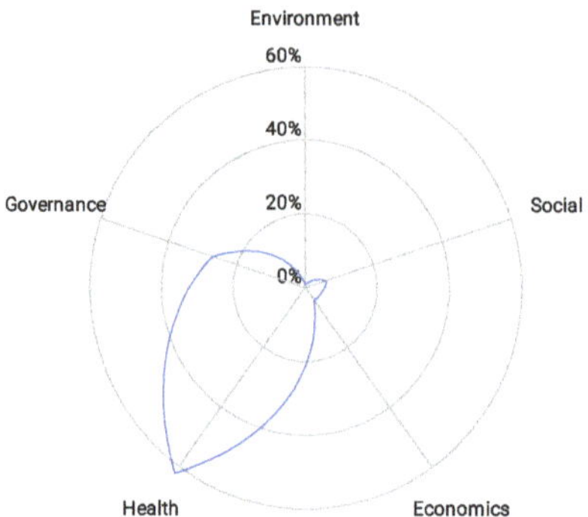

Figure 2. Swiss government profile.

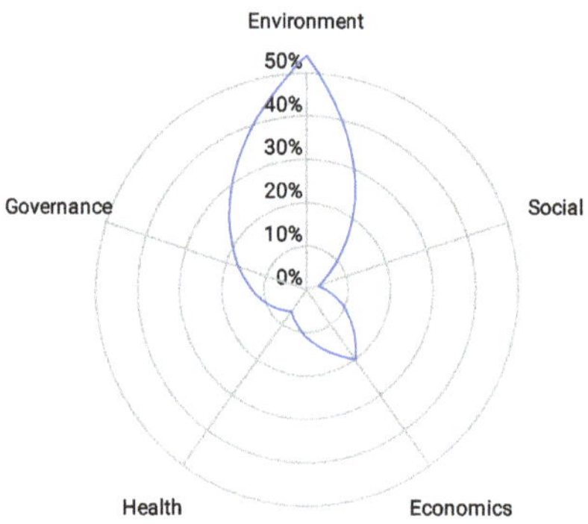

Figure 3. Nestlé profile.

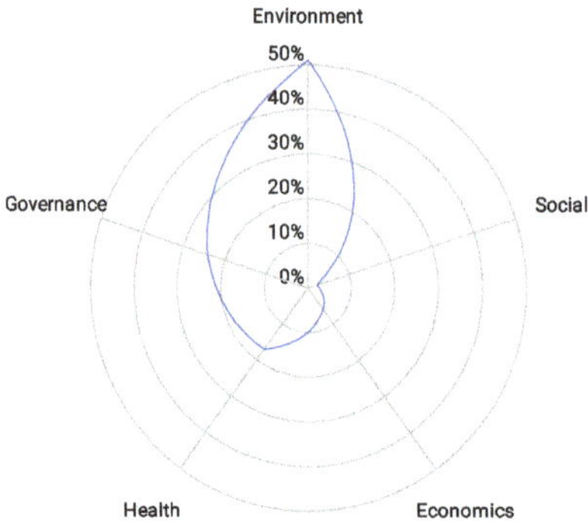

Figure 4. WWF profile.

SV Group Switzerland primarily focuses on the environment category with 32%, accompanied by governance with 24% and health with 18%. Overall, SV Group's profile is fairly balanced, covering all five categories with at least 10%, which differentiates it from the other stakeholders (Figure 5). Finally, the analysis of EAT-Lancet Commission's guidelines show that the categories environment and health are equally covered by 32% of the coded content. Governance follows them slightly behind with 29%. Social and economics categories together account for only 7% of the content (Figure 6).

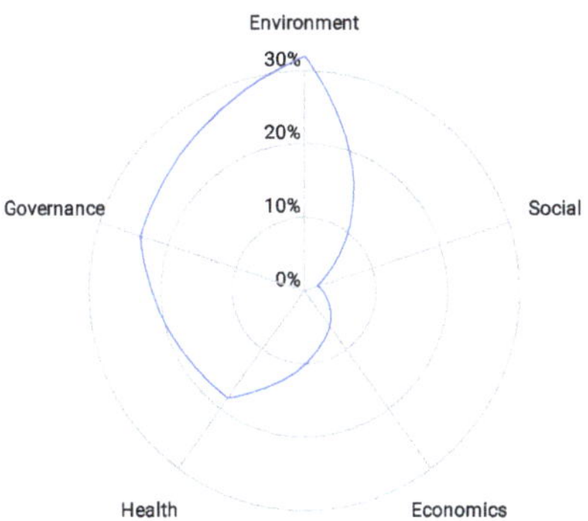

Figure 5. SV Group profile.

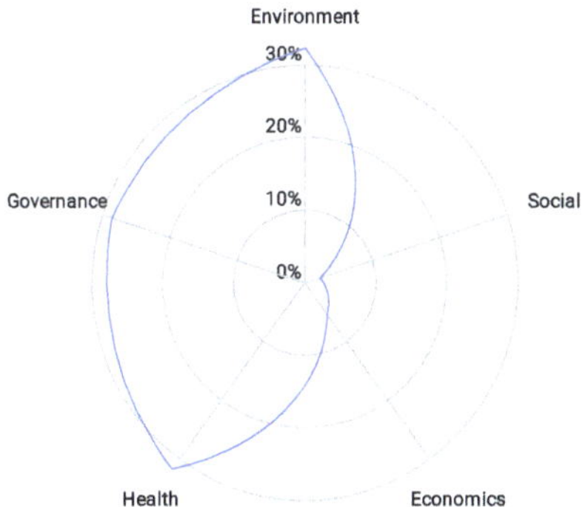

Figure 6. EAT-Lancet Commission profile.

4.2. Similarities and Differences of Sustainability References in the Dietary Guidelines

Overall, dietary recommendations are predominantly framed in reference to health, environmental, and governance considerations, while relatively little attention is paid to social or economic aspects (Figure 7).

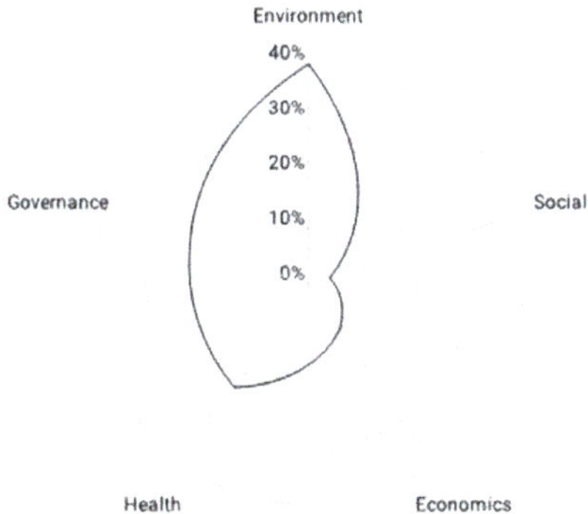

Figure 7. Sustainable diet framework, accumulated score of all documents reviewed (n = 12).

The analysis reveals that the environment category is the top coded category (40%) of the guidelines overall (Figure 7). The focus of stakeholders in the environmental category depends on the food sector they operate in, resulting in them emphasizing different aspects in their sustainability approach in terms of environment (Table 4). The stakeholders cover a multitude of codes, except for the government. While the Swiss government and SV Group Switzerland emphasize consumer-side recommendations, such as choosing regional and seasonal food, Nestlé and WWF focus on production aspects such as land use. As Nestlé is a food production company and WWF a nature conservation organization, this focus on land use is connected to their respective activities. In the guidelines addressed explicitly to Switzerland, there is a pattern of including the environmental impact of food origin that is absent when aiming for an international audience (e.g., Nestle's Net Zero Roadmap or EAT-Lancet Commission's Food Planet Health). Guidelines directed at an international audience focus on the supply chain and logistics of food distribution. All the stakeholders stress the food system's role regarding climate change besides the Swiss government. The actors communicate climate change as a target variable mainly to reduce GHG, while other environmental codes are largely communicated as an influencing variable to achieve this target.

Table 4. Three most frequently coded codes within each category ordered by stakeholder and their approach to the topic within the analyzed guidelines.

Stakeholder	Environment	Social	Economics	Health	Governance
Swiss Government	Origin (67%): Regional and seasonal food. Food Waste (33%): Avoid food waste.	Pleasure (73%): Enjoyment of eating. Community (13%): Social contact. Ethical Buying (7%): Fair trade products.	Cost (50%): Health care costs related to diet and wellness. Affordability (30%): Affordable food. Technology and Innovation (20%): Promote innovation.	Well-being (18%): Ensure physical and mental well-being. Plant-based Protein (15%): Specific alternatives to animal products are offered. Fruits and Vegetables (13%): Recommended to eat vegetarian several days a week.	Directives (25%): Networking nutritional stakeholders, utilizing synergy and coordinating activities. Education (25%): Strengthen nutrition literacy (put knowledge into practice). Regulation (22%): Political focus on improvement of health and well-being.
Nestlé	Climate Change (33%): Reach net-zero by 2050. Land Use (16%): Regenerative agriculture and reforestation. Origin (16%): Source sustainable ingredients, supply chain perspective.	Gender Equality (41%): 30% women in top management positions by 2022. Community (24%): Local farming communities. Animal Welfare (12%): Improving animal welfare.	Sustainable Production Patterns (52%): Cleaner logistics. Technology and Innovation (40%): R&D investments. Cost (5%): Economically viable practices.	Dairy Products (49%): Climate friendly milk pilot project. Sweets, Salty and Alcoholic Drinks (11%): Brief practical guideline on how to eat. Fruits and Vegetables (8%): Brief practical guidelines how to eat.	Certifications and Standards (28%): Accountability towards consumers. Transparency (25%): Transparent to consumers. Regulation (15%): Call for appropriate ground rules from the government side.

Table 4. *Cont.*

Stakeholder	Environment	Social	Economics	Health	Governance
WWF	Land Use (22%): All action items have direct link to land use. Climate Change (19%): Role of the food system in connection to climate change. Biodiversity (13%): Connection of the food system to biodiversity loss.	Culture (58%): National context needs to be considered. Animal Welfare (33%): Species-appropriate animal husbandry. Ethical Buying (8%): Support of fair trade.	Sustainable Production Patterns (77%): Food production. Technology and Innovation (13%): Technological progress as part of the solution. Cost (5%): Briefly mentioned.	Animal-based Protein (28%): Reduction of animal-based proteins to achieve planet-based diet. Well-being (26%): Planet-based diet benefits human health. Dairy Products (19%): Reduction of dairy products to achieve planet-based diet.	Directives (35%): Translate global recommendations for healthier and more sustainable diets to individual countries. Science (21%): Use scientific basis for decision making. Food Security (19%): Feed humanity on existing cropland.
SV Group Switzerland	Origin (40%): Regional and seasonal buying play a significant role. Climate Change (20%): Overarching umbrella in their guidelines (ONE TWO WE Program). Energy Use (9%): Energy saving practices and alternative energy sources.	Animal Welfare (48%): Concrete measures and examples for improving animal welfare (BTS, RAUS). Pleasure (20%): Sustainable produced food needs to taste good and be balanced. Community (14%): Eating brings people together.	Sustainable Production patterns (33%): Resource bundling along the value chain. Technology and Innovation (33%): Process optimization with sound analyses. Labor Rights (15%): Fair trade products and support of labor rights in developing countries are important.	Animal-based Protein (29%): Reduction of meat consumption in order to reduce GHG. Fruits and Vegetables (29%): Increased fruits and vegetables consumption, vegan and vegetarian menus. Dairy Products (17%): Reduction of dairy products.	Certifications and Standards (38%): Labels play a central role. Directives (21%): SDGs as important guidance. Education (14%): Education as essential tool for dietary change.
EAT-Lancet Commission	Climate Change (36%): Overarching umbrella for environmental indicators. Food Waste (13%): At least halve food losses and waste. Land Use (12%): Reorient agricultural priorities.	Culture (89%): Critical not to neglect the reality of cultural diversity and regional differences. Animal Welfare (11%): Explicitly mentioned that this issue is foregone in their guidelines.	Sustainable Production Patterns (57%): Intensify food production sustainably, increase high quality output. Technology and Innovation (39%): Fertilizer and water use efficiency, recycling of phosphorus. Affordability (4%): Little specification.	Animal-based Protein (18%): Reduced meat intake to achieve planetary health diet. Well-being (17%): Optimize health within environmental limits. Fruits and Vegetables (16%): Increased fruits and vegetable intake to achieve planetary health diet.	Directives (41%): FBDGs as central element for changing diets and the global food system; SDGs, Paris Agreement. Science (22%): Development of first universal scientific goals for healthy and sustainable diet. Food Security (20%): Planetary health diet as framework to feed nearly ten billion by 2050.

Within the social category, the stakeholders tend to focus on a single feature rather than addressing diversity. Animal welfare as a social concern is strongly addressed, especially by representatives of the private sector and the nature conservation organization. The analysis shows that the economics category focuses more on technological progress (e.g., sustainable production patterns or technology and innovation) than on employee relations.

However, in the private sector, the topic of labor rights is thematized alongside certification and standards. There is either a focus on the monetary cost of food or on affordability to afford an item, but not both aspects at once, although they are closely linked. Among all parties, only the Swiss government discusses health costs concerning diets and well-being. Table 5 presents previously discussed similarities and differences arising from different categories within the reviewed dietary guidelines.

Table 5. Similarities and differences between the various stakeholders' dietary guidelines.

Category	Similarities	Differences
Environment	All stakeholders, except the Swiss government, emphasize the role of food systems in relation to climate change. Climate change stated as target variable, influencing variables are land use, biodiversity and soil to achieve GHG reduction target. Swiss government and SV Group Switzerland place more emphasis on consumer-side recommendations, such as choosing regional and seasonal food. Nestlé and WWF converge around production-related aspects such as land use.	Guidelines aimed at an international or domestic audience differ in terms of food origin aspects. Seasonality and regionality are highlighted by guidelines specific to the Switzerland, supply chain aspects by international guidelines. Environmental aspects are almost absent in Swiss government guidelines.
Social	There is little coverage overall. When included, there is a focus on one single feature, rarely touch on several social aspects. With exception of the Swiss government, all the other stakeholders recognize animal welfare as a social issue within their dietary guidelines.	Swiss government and SV Group Switzerland emphasize the pleasure of eating and sharing food.
Economics	There is little coverage overall. All stakeholders converge on technological innovation. Focus more on technical production side (e.g., sustainable production patterns, technology and innovation) rather than consumer or worker realities.	Cost and affordability aspects are varied. Either cost or affordability considerations are usually at the forefront, not both simultaneously. Swiss government is the only stakeholder that focuses on economic aspects at the consumer end (health care costs and affordability) while all the others, including EAT-Lancet Commission, focus more on the production side. Private sector (Nestlé and SV Group Switzerland) includes labor rights consistent with certifications and standards.
Health	Reduction of animal-based protein, especially meat. Increase fruits and vegetables as well as plant-based protein.	Decisive reasons for reducing the consumption of animal-based protein are health and/or environmental aspects. Food groups are highlighted differently depending on their sphere of influence within the food system. WWF and SV Group Switzerland, similarly to the EAT-Lancet Commission international benchmark, suggest reduction in animal-based proteins. The Swiss government does not suggest reduction but rather present plant-based alternatives. Nestlé makes no mention to reduction of animal-based proteins.
Governance	Consistently addressed through all profiles. The governance approach focuses on either directives or regulations.	Food security is raised only by international organization (EAT-Lancet Commission) and civil society (WWF), while neglected by the others. Transparency only recognized by private sector (Nestlé and SV Group Switzerland), link to certifications and standards. Only the Swiss government and the SV Group Switzerland mention education as a tool for dietary change.

The health category presents a heterogeneous picture, but among the most frequently communicated codes within the category, four food groups are mentioned particularly often, as is well-being (Table 4). The focus on specific food groups is indicative of the food sector in which stakeholders are active. For example, the private company Nestlé shows a strong focus on dairy products (49%), reflecting this also in its efforts to draw attention to its climate-friendly milk pilot project. As a civil society organization, WWF focuses on reducing animal-based proteins (28%) as a means of promoting a "planet-based diet" [52]. Most profiles emphasize the adverse effects of animal-based foods on individual well-being and climate change. All actors, besides Nestlé, emphasize the intake of more plant- based foods and fewer animal-based foods regularly. In particular, EAT-Lancet Commission, WWF, and SV Group Switzerland are explicit about the required transition from animal-based to plant-based foods. A key component of this shift is a firm emphasis on the consumer demand side, but the reasons for this emphasis are divided. In addition to its health benefits, EAT-Lancet Commission, WWF, and SV Group Switzerland point out its environmental benefits. Despite this acknowledgment, recommendations differ on what constitutes a reduced intake of animal products.

The governance category is consistently addressed through all the profiles (Figure 1), indicating that the importance of this category is recognized by the food system representatives. The stakeholders set a different focus in terms of governing acts: directives or regulations. While EAT-Lancet Commission, WWF, SV Group Switzerland, and the Swiss government choose to highlight international policies such as the SDGs, Nestlé chooses to call for policy action to transform industries with mandatory rules. A contrasting point appears in the inclusion of transparency. The Swiss government, WWF, and the EAT-Lancet Commission have not indicated this feature, whereas private sector representatives Nestlé and SV Group Switzerland point it out in their statements. International organizations and civil society emphasize food security and the importance of feeding almost ten billion people by 2050 on existing farmland. The two organizations, EAT-Lancet Commission and WWF, thus present a broader perspective than one that focuses exclusively on Switzerland. Through strengthening nutrition literacy, education is recognized only by the government and the catering company as an essential tool for dietary change. Similar to the previous categories, the covered aspects depend on the sector of the food system in which the representative stakeholder operates.

5. Discussion

DGs have received increasing attention among state and non-state actors in recent years as potential tools for addressing sustainability in the food system [10,27,53,54]. Our findings on actors in the Swiss food governance landscape warrant this attention. We found that all considered actors include sustainability aspects in their DGs, addressing at least one sustainability dimension—with health, environment and governance being the most predominant. The different DGs do account for the notion of sustainability in diverse ways, assigning different weight to their five pillars and attaching diverse meanings to it. These differences reflect the complexity and ambiguity of sustainability considerations in food systems research in general [4] and of understanding what constitutes a sustainable diet in particular. The pluralistic map of the food governance landscape outlined in this paper provides an overview of different paths to sustainable diets where potential compatibilities and tensions arise. While different stakeholders cover niche-specific aspects of sustainability, we identified recurrent gaps in economic and social sustainability content: The economic sustainability aspect is almost absent, which is surprising given the significance of markets and neoliberal forms of governance for food system transformation at local and global scale [55–57]. Similarly, social aspects of sustainable diets are scarce, what seems problematic given the food system being embedded in and shaped by society in complex ways [6,53]: Consumption choices influence the food system; cultural aspects influence consumer choices. In order to shape the food system in a more sustainable direction, social and economic aspects must therefore be considered [6].

The focus of the Swiss government's recommendations is predominantly on health aspects without embedding them in a broader sustainability context. This confirms the general focus of governments on health concerns and their reluctance to include (extended) sustainability aspects in DGs [27]. Furthermore, it demonstrates a significant inconsistency between the government's DGs and its broader sustainability policy inspired by the SDGs and the Paris Agreement. DGs from non-state actors in contrast go beyond the classic thematization of health aspects and address other sustainability dimensions as well. However, actors focus mainly on aspects that reflect the position they occupy in the food governance system [18]: The private sector player Nestlé is the largest dairy company in the world [58] and is heavily involved in the climate impact of dairy products in Switzerland. The catering company SV Group Switzerland focuses on consumer participation and individual agency by providing transparent information about its supply chain. WWF as a civil society organization actively promotes nature conservation and creates a framework that emphasizes this goal, a "planet-based diet" [52]. As an exception, the international organization EAT-Lancet Commission provides rather comprehensive scientific goals for healthy diets from sustainable food systems that will feed nearly 10 billion people by 2050. The references of these state and non-state actors to individual sustainability dimensions require further analysis as it can reflect both a genuine commitment to sustainable transformation as well as a legitimation or marketing strategy.

Differences between the stakeholders' DGs manifest not only in the weighting of different sustainability dimensions, but also in the way these dimensions are framed. For example, the social dimension, which is only marginally considered overall, is addressed in different ways by the various actors: Nestlé brings forward gender equality (41%), which is quite significant as gender equality is a foundation for progress in achieving multiple factors towards sustainable development [59]. In contrast, SV Group Switzerland rather elaborates on animal welfare (48%), strongly related to animal health and therefore of substantial interest to farmers and their productivity [6]. There are also divergent recommendations relative to common topics, such as reducing animal-based products. Some recommendations still include a relatively high intake of animal protein compared to the internationally accepted threshold of 25–58 g/day [14] for consumption of animal products, weighing their environmental impact. For example, Swiss government guidelines recommend an intake of 100–120 g/day [60] of animal protein or an alternative protein source.

Apart from these differences, the stakeholders share common features. For example, all actors communicate an increase in the intake of fruits and vegetables and plant-based protein and most a reduction of animal-based protein, despite varying benchmarks. This reduction or increased intake is cited from a health and environmental perspective, indicating a belief in positive synergy between the two dimensions.

Based on a generic five-dimensional understanding of sustainability, we show in our analysis how different actors occupy this common framework in different ways. Combining the numerous individual aspects put forward by the actors, the potential for a comprehensive understanding of sustainable diets becomes apparent. At the same time, however, the synopsis also reveals potential differences and tensions. While addressing all five sustainability dimensions in one way or another will be pivotal for a sustainable transformation of the food system, it remains debatable to what degree coherence between different DGs in regard to their understanding of sustainability is a necessity to this. On the one hand, similar references to sustainability dimensions in the guidelines of different actors offer a potential for the formation of actor coalitions. Indeed, the multifaceted challenges of a sustainability-oriented food system transformation require a broader fundamental understanding of human health in the context of planetary health. It forms the core of what Patterson et al. called a shared "transformative agenda" for a sustainable food system [61] (p. 4). It is known from other fields that actor coalitions form around similar cognitive and normative orientations [62]. Assuming that DGs reflect the normative and cognitive belief systems, mapping sustainability can thus reveal opportunities for cooperation, launch mutual exchange and promote learning among actors.

On the other hand, a certain degree of incoherence in dietary recommendations might be inevitable or even necessary for a sustainable transformation to advance. As discussed above, the reviewed DGs reflect different normative understandings of what constitutes a sustainable diet and thus address different target groups. In more general terms, it might be the very complexity of the food system that demands for the coexistence and cooperation of various approaches as no one-size-fits-all will provide the 'solution'. Allowing incoherence to exist, or even proliferate also means to account for an increasingly differentiated food governance landscape in food system transformation towards sustainability. In this vein, converging and diverging DGs can all potentially contribute to incremental change at different nodes within a complex food system. It entails to acknowledge the different actors with their respective priorities and frames in the design of sustainability-oriented transformations [16]. The interdisciplinary operationalization of sustainable diets and the concept of sustainability then require constant negotiation and debate [61]. The framework and empirical analysis we outline in this paper thus contribute to such a debate.

The selected guidelines represent only a snapshot; we acknowledge that the results may differ if other guidelines were considered. However, given the number of guidelines from these stakeholders, it was necessary to place topics outside the scope of the analysis.

6. Conclusions

This paper contributes to the discussion on sustainability-oriented transformations of the food system in three specific ways. First, it provides the methodological and analytical tools for studying and comparing different understandings of sustainable diet, for instance by focusing on how different actors make reference to sustainability in their dietary guidelines. We have done so by moving beyond their original focus of health in favor of a multi-dimensional understanding of what constitutes a sustainable diet. Second, it offers insights into an empirical example, DGs in Switzerland, revealing how different actors in the food system refer to sustainability in their dietary guidelines and making apparent context specific differences and commonalities. To do so allows to identify systematic gaps in the sustainability content of dietary guidelines and to lay the basis for the respective political negotiations about it. Third, it provides a basis for further theoretical reflection on the role of dietary guidelines for a sustainability-oriented transformation of the food system in the context of an increasingly diversified food governance. Multi-stakeholder engagement in food system governance offers a promising new area of focus for bringing about profound changes in sustainable diets and consumer and producer behaviors. An array of possibilities are opened up for non-state actors, emphasizing their transformative role in changing the dialogue around sustainable diets. Given that the food system's unsustainability cannot be solved by the government, market relations, or consumers alone, we acknowledge the importance of multi-stakeholder involvement in defining sustainable diets. This in turn requires opening up space for cooperation around converging and diverging understandings of what might constitute a sustainable diet. To that end, this paper offers a multi-dimensional sustainability framework for developing "*a strategy of incremental change with a transformative agenda*" [61] (p. 4) for diet sustainability, the food system and its sustainable transformation. However, at what specific nodes within the food system the harmonization of different actors' DGs is necessary for capitalizing on potential synergies and counteract trade-offs will require further analytical scrutiny. It requires integrative research that goes beyond the study of content presented in this paper, for instance on how DGs influence behavior change and on how different DGs interact in this process. It requires context-specific empirical analysis of the synergies and tradeoffs between different DGs and how they influence consumer and producer behavior. Finally, it also requires an analysis of the way consumption and production are mediated by converging and diverging notions of what might constitute a sustainable diet.

Author Contributions: Conceptualization, L.A.S., B.B., R.E., J.R. and R.T.; methodology, J.R. and R.T.; validation, L.A.S., B.B., R.E., J.R. and R.T.; formal analysis, J.R and R.T.; investigation, J.R. and R.T.; data curation, J.R. and R.T.; writing—original draft preparation, R.T.; writing—review and editing, L.A.S., B.B., R.E. and J.R.; visualization, J.R. and R.T.; supervision, B.B. and R.E.; project administration, R.T.; funding acquisition, B.B., R.E., J.R. and R.T. All authors have read and agreed to the published version of the manuscript.

Funding: This project has received funding from the Swiss Academies of Arts and Sciences: ID 315, 2020–2021, the Sustainability Research Group of the University of Basel, the Publication Fund of the University of Basel for Open Access, and the European Union's Horizon 2020 research and innovation programme under the Marie Skłodowska-Curie grant agreement No 801076, through the SSPH+ Global PhD Fellowship Programme in Public Health Sciences (GlobalP3HS) of the Swiss School of Public Health.

Institutional Review Board Statement: Not applicable.

Informed Consent Statement: Not applicable.

Data Availability Statement: The data presented in this study are available in (Appendix A).

Conflicts of Interest: The authors declare no conflict of interest.

Appendix A

Table A1. Qualitative content analysis results overview.

Categories/Codes	EAT			Swiss Government			Nestlé			SV Group Switzerland			WWF Schweiz			Total		
	Count	Sum	Percent	Count	Sum	Percent	Count	Sum	Percent	Count	Sum	Percentage	Count	Sum	Percentage	Count	Sum	Percentage
Economics	20	23	5.3%	7	10	3.8%	84	119	19.6%	23	27	10.0%	34	39	5.9%	168	218	9.7%
Affordability	1	1	4.3%	3	3	30.0%	0	0	0.0%	1	1	3.7%	1	1	2.6%	6	6	2.8%
Cost	0	0	0.0%	2	5	50.0%	5	6	5.0%	4	4	14.8%	2	2	5.1%	13	17	7.80%
Labor Rights	0	0	0.0%	0	0	0.0%	3	3	2.5%	3	4	14.8%	1	1	2.6%	7	8	3.7%
Sustainable Production Patterns	11	13	56.5%	0	0	0.0%	43	62	52.1%	8	9	33.3%	26	30	76.9%	88	114	52.3%
Technology and Innovation	8	9	39.1%	2	2	20.0%	33	48	40.3%	7	9	33.3%	4	5	12.8%	54	73	33.5%
Environment	98	141	32.3%	2	3	1.1%	241	330	54.5%	60	85	31.5%	239	340	51.3%	640	899	40.1%
Animal Agriculture	6	8	5.7%	0	0	0.0%	16	19	5.8%	3	4	4.7%	23	32	9.4%	48	63	7.0%
Aquatic Ecosystem	3	3	2.1%	0	0	0.0%	3	4	1.2%	4	7	8.2%	16	23	6.8%	26	37	4.1%
Biodiversity	12	16	11.3%	0	0	0.0%	8	8	2.4%	2	3	3.5%	33	44	12.9%	55	71	7.9%
Climate Change	31	51	36.2%	0	0	0.0%	80	108	32.7%	12	17	20.0%	46	64	18.8%	169	240	26.7%
Food Waste	15	19	13.5%	1	1	33.3%	3	4	1.2%	3	4	4.7%	17	19	5.6%	39	47	5.2%
Land Use	12	17	12.1%	0	0	0.0%	35	52	15.8%	3	4	4.7%	47	75	22.1%	97	148	16.5%
Origin	2	2	1.4%	2	2	66.7%	38	54	16.4%	25	34	40.0%	9	14	4.1%	76	106	11.8%
Renewable Energy	1	1	0.7%	0	0	0.0%	37	49	14.8%	5	8	9.4%	5	7	2.1%	48	65	7.2%
Soil	10	16	11.3%	0	0	0.0%	14	22	6.7%	3	4	4.7%	31	43	12.6%	58	85	9.5%
Water Use	6	8	5.7%	0	0	0.0%	7	10	3.0%	0	0	0.0%	12	19	5.6%	25	37	4.1%
Governance	93	125	28.6%	58	72	27.3%	79	103	17.0%	48	66	24.4%	113	156	23.5%	391	522	23.3%
Certifications and Standards	0	0	0.0%	1	1	1.4%	20	29	28.2%	18	25	37.9%	2	3	1.9%	41	58	11.1%
Directives	36	51	40.8%	12	18	25.0%	3	4	3.9%	12	14	21.2%	34	54	34.6%	97	141	27.0%
Education	3	5	4.0%	15	18	25.0%	8	12	11.7%	6	9	13.6%	3	3	1.9%	35	47	9.0%
Food Security	21	25	20.0%	2	3	4.2%	1	1	1.0%	0	0	0.0%	24	30	19.2%	48	59	11.3%
Justice	6	10	8.0%	1	1	1.4%	3	3	2.9%	2	4	6.1%	11	17	10.9%	23	35	6.7%
Regulation	6	7	5.6%	14	16	22.2%	13	15	14.6%	1	2	3.0%	14	17	10.9%	48	57	10.9%

Table A1. *Cont.*

Categories/Codes	EAT			Swiss Government			Nestlé			SV Group Switzerland			WWF Schweiz			Total		
	Count	Sum	Percent	Count	Sum	Percent	Count	Sum	Percent	Count	Sum	Percentage	Count	Sum	Percentage	Count	Sum	Percentage
Science	21	27	21.6%	13	15	20.8%	9	13	12.6%	4	4	6.1%	25	32	20.5%	72	91	17.4%
Transparency	0	0	0.0%	0	0	0.0%	22	26	25.2%	5	8	12.1%	0	0	0.0%	27	34	6.5%
Health	100	139	31.8%	127	164	62.1%	24	37	6.1%	37	48	17.8%	86	116	17.5%	374	504	22.5%
Animal-based Protein	20	25	18.0%	12	15	9.1%	1	2	5.4%	11	14	29.2%	24	33	28.4%	68	89	17.7%
Dairy Products	12	18	12.9%	15	20	12.2%	12	18	48.6%	7	8	16.7%	16	22	19.0%	62	86	17.1%
Fruits and Vegetables	15	22	15.8%	17	21	12.8%	2	3	8.1%	10	14	29.2%	9	12	10.3%	53	72	14.3%
Liquids (Unsweetened Drinks)	2	3	2.2%	8	12	7.3%	1	2	5.4%	2	2	4.2%	1	2	1.7%	14	21	4.2%
Plant-based Protein	11	16	11.5%	16	24	14.6%	1	2	5.4%	3	5	10.4%	7	10	8.6%	38	57	11.3%
Sweets, Salty Snacks and Alcoholic Drinks	11	13	9.4%	19	22	13.4%	2	4	10.8%	1	1	2.1%	2	2	1.7%	35	42	8.3%
Tubers or starchy vegetables	5	8	5.8%	5	8	4.9%	1	2	5.4%	1	2	4.2%	0	0	0.0%	12	20	4.0%
Well-being	16	23	16.5%	24	29	17.7%	3	3	8.1%	2	2	4.2%	24	30	25.9%	69	87	17.3%
Whole grains	8	11	7.9%	11	13	7.9%	1	1	2.7%	0	0	0.0%	3	5	4.3%	23	30	6.0%
Social	7	9	2.1%	12	15	5.7%	14	17	2.8%	32	44	16.3%	10	12	1.8%	75	97	4.3%
Animal Welfare	1	1	11.1%	1	1	6.7%	2	2	11.8%	14	21	47.7%	3	4	33.3%	21	29	29.9%
Community	0	0	0.0%	2	2	13.3%	5	4	23.5%	5	6	13.6%	0	0	0.0%	12	12	12.4%
Culture	6	8	88.9%	0	0	0.0%	0	0	0.0%	1	1	2.3%	6	7	58.3%	13	16	16.5%
Ethical Buying	0	0	0.0%	1	1	6.7%	1	2	11.8%	3	5	11.4%	1	1	8.3%	6	9	9.3%
Gender Equality	0	0	0.0%	0	0	0.0%	4	7	41.2%	1	2	4.5%	0	0	0.0%	5	9	9.3%
Pleasure	0	0	0.0%	8	11	73.3%	2	2	11.8%	8	9	20.5%	0	0	0.0%	18	22	22.7%

Appendix B

Table A2. Selected examples to illustrate the process from raw data to weighted results.

Excerpt	Category/Applied Code	Code Definition	Scaling	Justification
"In this section we use these results to draw insights on the global implications of current food consumption patterns in G20 countries and how the 5 Gt CO2eq food budget may need to be more equitably distributed to achieve healthy diets for all."	Environment/Climate Change	Global warming, GHG emissions, Carbon Storage Ecological Footprint, Carbon Budget, Temperature, Precipitation	1	based on the following text passage: " . . . 5 Gt CO2eq food budget . . . " Carbon Budget. A single word or short phrase that refers to the feature and is not further explained or elaborated within the excerpt.
	Governance/Justice	Democratic Values, Intergenerational Justice, Food Distribution, Fair Economic Conditions, Equal Opportunities	1	based on the following text passage: "...equitably distributed to achieve healthy diets for all." Equal Opportunities and Food Distribution. A single word or short phrase that refers to the feature and is not further explained or elaborated within the excerpt.
"Mealtimes are not just about the intake of energy and nutrients; they are also about pleasure, relaxation and social contact. Taking time, switching off and eating and drinking in peace and quiet help to promote the enjoyment of eating."	Social/Community	Alone/Together, Sharing, Eating Modes, Farming/Local Communities Support, Team Spirit, Cohesion Exchange	1	based on the following text passage: "... social contact." Social Contact elaborates on togetherness and eating modes. A single word or short phrase that refers to the feature and is not further explained or elaborated within the excerpt.
	Social/Pleasure	Taste, Aesthetics, Mindful Eating, Time, Comfort, Cordiality, Enjoyment of Eating	2	based on the following text passage: "...pleasure, relaxation and social contact. Taking time, switching off and eating and drinking in peace and quiet help to promote the enjoyment of eating." Within the excerpt, the statement is substantiated. Rationale for the significance of the characteristic is elaborated and the excerpt refers to concrete actions.

References

1. Rockström, J.; Steffen, W.; Noone, K.; Persson, Å.; Chapin, F.S.; Lambin, E.F.; Lenton, T.M.; Scheffer, M.; Folke, C.; Schellnhuber, H.J.; et al. A Safe Operating Space for Humanity. *Nature* **2009**, *461*, 472–475. [CrossRef] [PubMed]
2. Steffen, W.; Richardson, K.; Rockstrom, J.; Cornell, S.E.; Fetzer, I.; Bennett, E.M.; Biggs, R.; Carpenter, S.R.; de Vries, W.; de Wit, C.A.; et al. Planetary Boundaries: Guiding Human Development on a Changing Planet. *Science* **2015**, *347*, 1259855. [CrossRef] [PubMed]
3. Nutrition and Food Systems. *A Report by the High Level Panel of Experts on Food Security and Nutrition of the Committee on World Food Security*; HLPE: Rome, Italy, 2017.
4. *Transforming Food Systems for Affordable Healthy Diets*; FAO: Rome, Italy, 2020.

5. Campbell, B.M.; Beare, D.J.; Bennett, E.M.; Hall-Spencer, J.M.; Ingram, J.S.I.; Jaramillo, F.; Ortiz, R.; Ramankutty, N.; Sayer, J.A.; Shindell, D. Agriculture Production as a Major Driver of the Earth System Exceeding Planetary Boundaries. *E&S* **2017**, *22*, art8. [CrossRef]
6. Mason, P.; Lang, T. *Sustainable Diets: How Ecological Nutrition Can Transform Consumption and the Food System*; Routledge: London, UK, 2017; ISBN 978-0-415-74470-6.
7. Béné, C.; Oosterveer, P.; Lamotte, L.; Brouwer, I.D.; de Haan, S.; Prager, S.D.; Talsma, E.F.; Khoury, C.K. When Food Systems Meet Sustainability – Current Narratives and Implications for Actions. *World Dev.* **2019**, *113*, 116–130. [CrossRef]
8. Tobi, R.C.A.; Harris, F.; Rana, R.; Brown, K.A.; Quaife, M.; Green, R. Sustainable Diet Dimensions. Comparing Consumer Preference for Nutrition, Environmental and Social Responsibility Food Labelling: A Systematic Review. *Sustainability* **2019**, *11*, 6575. [CrossRef]
9. Ingram, J. A Food Systems Approach to Researching Food Security and Its Interactions with Global Environmental Change. *Food Sec.* **2011**, *3*, 417–431. [CrossRef]
10. Ingram, J.S.I.; Ericksen, P.; Liverman, D.M. *Food Security and Global Environmental Change*; Earthscan: London, UK; Washington, DC, USA, 2010; ISBN 978-1-84971-127-2.
11. Verbruggen, P.; Havinga, T. (Eds.) *Hybridization of Food Governance: Trends, Types and Results*; Edward Elgar Publishing: Cheltenham, UK, 2017; ISBN 978-1-78536-169-2.
12. Schilpzand, R.; Liverman, D.; Tecklin, D.; Gordon, R.; Pereira, L.; Saxl, M.; Wiebe, K. Governance beyond the State: Non-State Actors and Food Systems. In *Food Security and Global Environmental Change*; Earthscan: London, UK; Washington, DC, USA, 2010; pp. 272–300. ISBN 978-1-84971-127-2.
13. Messerli, P.; Murniningtyas, E.; Eloundou-Enyegue, P.; Foli, E.G.; Furman, E.; Glassman, A.; Hernández Licona, G.; Kim, E.M.; Lutz, W.; Moatti, J.P.; et al. *Sustainable Development Report 2019: The Future Is Now – Science for Achieving Sustainable Development*; United Nations: New York, NY, USA, 2019.
14. The Planetary Health Diet. Available online: https://eatforum.org/learn-and-discover/the-planetary-health-diet/ (accessed on 10 September 2020).
15. Lang, T.; Barling, D. Nutrition and Sustainability: An Emerging Food Policy Discourse. *Proc. Nutr. Soc.* **2013**, *72*, 1–12. [CrossRef]
16. Willett, W.; Rockström, J.; Loken, B.; Springmann, M.; Lang, T.; Vermeulen, S.; Garnett, T.; Tilman, D.; DeClerck, F.; Wood, A.; et al. Food in the Anthropocene: The EAT–Lancet Commission on Healthy Diets from Sustainable Food Systems. *Lancet* **2019**, *393*, 447–492. [CrossRef]
17. *Climate Change 2014: Mitigation of Climate Change; Working Group III Contribution to the Fifth Assessment Report of the Intergovernmental Panel on Climate Change*; Climate change 2014; Cambridge University Press: New York, NY, USA, 2014; ISBN 978-1-107-65481-5.
18. Lang, T.; Mason, P. Sustainable Diet Policy Development: Implications of Multi-Criteria and Other Approaches, 2008–2017. *Proc. Nutr. Soc.* **2018**, *77*, 331–346. [CrossRef]
19. Santaoja, M.; Jauho, M. Institutional Ambiguity and Ontological Politics in Integrating Sustainability into Finnish Dietary Guidelines. *Sustainability* **2020**, *12*, 5330. [CrossRef]
20. Burlingame, B. *Sustainable Diets and Biodiversity - Directions and Solutions for Policy Research and Action Proceedings of the International Scientific Symposium Biodiversity and Sustainable Diets United Against Hunger*; FAO: Rome, Italy, 2012; ISBN 978-92-5-107288-2.
21. Béné, C.; Fanzo, J.; Prager, S.D.; Achicanoy, H.A.; Mapes, B.R.; Alvarez Toro, P.; Bonilla Cedrez, C. Global Drivers of Food System (Un)Sustainability: A Multi-Country Correlation Analysis. *PLoS ONE* **2020**, *15*, e0231071. [CrossRef]
22. Mertens, E.; van't Veer, P.; Hiddink, G.J.; Steijns, J.M.; Kuijsten, A. Operationalising the Health Aspects of Sustainable Diets: A Review. *Public Health Nutr.* **2017**, *20*, 739–757. [CrossRef]
23. Burlingame, B.; Dernini, S. *Sustainable Diets and Biodiversity: Directions and Solutions for Policy, Research and Action*; FAO: Rome, Italy, 2012; ISBN 978-92-5-107288-2.
24. Díaz-Méndez, C.; Lozano-Cabedo, C. Food Governance and Healthy Diet an Analysis of the Conflicting Relationships among the Actors of the Agri-Food System. *Trends Food Sci. Technol.* **2020**, *105*, 449–453. [CrossRef]
25. *Preparation and Use of Food-Based Dietary Guidelines: Report of a Joint FAO/WHO Consultation*; WHO Technical Report Series; WHO: Geneva, Switzerland, 1998; ISBN 978-92-4-120880-2.
26. Wijesinha-Bettoni, R.; Khosravi, A.; Ramos, A.I.; Sherman, J.; Hernandez-Garbanzo, Y.; Molina, V.; Vargas, M.; Hachem, F. A Snapshot of Food-Based Dietary Guidelines Implementation in Selected Countries. *Glob. Food Secur.* **2021**, *29*, 100533. [CrossRef]
27. Springmann, M.; Spajic, L.; Clark, M.A.; Poore, J.; Herforth, A.; Webb, P.; Rayner, M.; Scarborough, P. The Healthiness and Sustainability of National and Global Food Based Dietary Guidelines: Modelling Study. *BMJ* **2020**, m2322. [CrossRef]
28. Fischer, C.G.; Garnett, T. *Plates, Pyramids, and Planets: Developments in National Healthy and Sustainable Dietary Guidelines: A State of Play Assessment*; FAO: Rome, Italy, 2016.
29. Willett, W.C.; Hu, F.B.; Rimm, E.B.; Stampfer, M.J. Building Better Guidelines for Healthy and Sustainable Diets. *Am. J. Clin. Nutr.* **2021**, *114*, 401–404. [CrossRef] [PubMed]
30. Suggs, L.S.; Della Bella, S.; Marques-Vidal, P. Low Adherence of Swiss Children to National Dietary Guidelines. *Prev. Med. Rep.* **2016**, *3*, 244–249. [CrossRef]
31. de Abreu, D.; Guessous, I.; Gaspoz, J.-M.; Marques-Vidal, P. Compliance with the Swiss Society for Nutrition's Dietary Recommendations in the Population of Geneva, Switzerland: A 10-Year Trend Study (1999–2009). *J. Acad. Nutr. Diet.* **2014**, *114*, 774–780. [CrossRef]

32. de Abreu, D.; Guessous, I.; Vaucher, J.; Preisig, M.; Waeber, G.; Vollenweider, P.; Marques-Vidal, P. Low Compliance with Dietary Recommendations for Food Intake among Adults. *Clin. Nutr.* **2013**, *32*, 783–788. [CrossRef] [PubMed]
33. Graça, J.; Cardoso, S.G.; Augusto, F.R.; Nunes, N.C. Green Light for Climate-Friendly Food Transitions? Communicating Legal Innovation Increases Consumer Support for Meat Curtailment Policies. *Environ. Commun.* **2020**, *14*, 1047–1060. [CrossRef]
34. Havinga, T.; van Waarden, F.; Casey, D. (Eds.) *The Changing Landscape of Food Governance: Public and Private Encounters*; Edward Elgar Publishing: Cheltenham, UK, 2015; ISBN 978-1-78471-540-3.
35. Nestle, M.; Pollan, M. *Food Politics: How the Food Industry Influences Nutrition and Health*; University of California Press: Berkeley, CA, USA, 2013; ISBN 978-0-520-27596-6.
36. Clapp, J.; Fuchs, D.A. (Eds.) *Corporate Power in Global Agrifood Governance*; MIT Press: Cambridge, MA, USA, 2009; ISBN 978-0-262-01275-1.
37. Fuchs, D.; Kalfagianni, A. The Causes and Consequences of Private Food Governance. *Bus. Polit.* **2010**, *12*, 1–34. [CrossRef]
38. Confideration, S. *Switzerland Implements the 2030 Agenda for Sustainable Development Switzerland's Country Report 2018*; Federal Department of Foreign Affairs FDFA: Bern, Switzerland, 2018.
39. 2030 Sustainable Development Strategy. Available online: https://www.eda.admin.ch/agenda2030/en/home/strategie/strategie-nachhaltige-entwicklung.html (accessed on 10 May 2021).
40. Drisko, J.W.; Maschi, T. *Content Analysis*; Oxford University Press: New York, NY, USA, 2016; ISBN 978-0-19-021549-1.
41. Béné, C.; Prager, S.D.; Achicanoy, H.A.E.; Toro, P.A.; Lamotte, L.; Bonilla, C.; Mapes, B.R. Global Map and Indicators of Food System Sustainability. *Sci Data* **2019**, *6*, 279. [CrossRef] [PubMed]
42. Scott, P. Global Panel on Agriculture and Food Systems for Nutrition: Food Systems and Diets: Facing the Challenges of the 21st Century. *Food Sec.* **2017**, *9*, 653–654. [CrossRef]
43. Ericksen, P.J. Conceptualizing Food Systems for Global Environmental Change Research. *Glob. Environ. Chang.* **2008**, *18*, 234–245. [CrossRef]
44. *World Commission on Environment and Development (WCED) Report of the World Commission on Environment and Development: Our Common Future*; United Nations: Oxford, UK, 1987.
45. Bengtsson, M. How to Plan and Perform a Qualitative Study Using Content Analysis. *Nurs. Open* **2016**, *2*, 8–14. [CrossRef]
46. Woods, N.F.; Catanzaro, M. *Nursing: Research Theory and Practice. M. Catanzaro Using Qualitative Analytical Techniques*; The CV Mosby Company: St.Louis, MO, USA, 1988.
47. Richards, L.; Morse, J.M. *Readme First for a User's Guide to Qualitative Methods*, 3rd ed.; SAGE: Los Angeles, CA, USA, 2013; ISBN 978-1-4129-9806-2.
48. Downe-Wamboldt, B. Content Analysis: Method, Applications, and Issues. *Health Care Women Int.* **1992**, *13*, 313–321. [CrossRef] [PubMed]
49. Saldaña, J. *The Coding Manual for Qualitative Researchers*, 2nd ed.; SAGE: Los Angeles, CA, USA, 2013; ISBN 978-1-4462-4736-5.
50. *Caring for the Earth. A Strategy for Sustainable Living*; IUCN/UNEP/WWF: Gland, Switzerland, 1991.
51. Bäckstrand, K. Multi-Stakeholder Partnerships for Sustainable Development: Rethinking Legitimacy, Accountability and Effectiveness. *Eur. Env.* **2006**, *16*, 290–306. [CrossRef]
52. Loken, B.; Opperman, J.; Orr, S.; Fleckenstein, M.; Halevy, S.; McFeely, P.; Park, S.; Weber, C. *Bending the Curve: The Restorative Power of Planet-Based Diets*; WWF: Gland, Switzerland, 2020.
53. Johnston, J.L.; Fanzo, J.C.; Cogill, B. Understanding Sustainable Diets: A Descriptive Analysis of the Determinants and Processes That Influence Diets and Their Impact on Health, Food Security, and Environmental Sustainability. *Adv. Nutr.* **2014**, *5*, 418–429. [CrossRef] [PubMed]
54. van Bers, C.; Delaney, A.; Eakin, H.; Cramer, L.; Purdon, M.; Oberlack, C.; Evans, T.; Pahl-Wostl, C.; Eriksen, S.; Jones, L.; et al. Advancing the Research Agenda on Food Systems Governance and Transformation. *Curr. Opin. Environ. Sustain.* **2019**, *39*, 94–102. [CrossRef]
55. Laforge, J.M.L.; Anderson, C.R.; McLachlan, S.M. Governments, Grassroots, and the Struggle for Local Food Systems: Containing, Coopting, Contesting and Collaborating. *Agric Hum Values* **2017**, *34*, 663–681. [CrossRef]
56. Lawrence, G.; Smith, K. *Neoliberal Globalization and Beyond: Food, Farming, and the Environment*; Legun, K., Keller, J., Bell, M., Carolan, M., Eds.; Cambridge University Press: Cambridge, UK, 2020; pp. 411–428. ISBN 978-1-108-55455-8.
57. Clapp, J.; Moseley, W.G. This Food Crisis Is Different: COVID-19 and the Fragility of the Neoliberal Food Security Order. *J. Peasant. Stud.* **2020**, *47*, 1393–1417. [CrossRef]
58. Nestlé Working with Dairy Farmers. Available online: https://www.nestle.com/brands/dairy/dairycsv (accessed on 23 September 2021).
59. Ahmed, S.; Byker Shanks, C. *Supporting Sustainable Development Goals Through Sustainable Diets*; Leal Filho, W., Wall, T., Azul, A.M., Brandli, L., Özuyar, P.G., Eds.; Springer International Publishing: Cham, Germany, 2020; pp. 688–699. ISBN 978-3-319-95680-0.
60. Swiss Society for Nutrition SSN, Federal Food Safety and Veterinary Office FSVO. Swiss Food Pyramid. 2011, p. 5. Available online: https://www.sge-ssn.ch/media/sge_pyramid_E_basic_20161.pdf (accessed on 13 July 2020).
61. Patterson, J.; Schulz, K.; Vervoort, J.; van der Hel, S.; Widerberg, O.; Adler, C.; Hurlbert, M.; Anderton, K.; Sethi, M.; Barau, A. Exploring the Governance and Politics of Transformations towards Sustainability. *Environ. Innov. Soc. Transit.* **2017**, *24*, 1–16. [CrossRef]
62. Hajer, M.A. *The Politics of Environmental Discourse*; Oxford University Press: Oxford, UK, 1997; ISBN 978-0-19-829333-0.

sustainability

MDPI

Article

Valorisation of Organic Waste By-Products Using Black Soldier Fly (*Hermetia illucens*) as a Bio-Convertor

Kieran Magee [1], Joe Halstead [2], Richard Small [3] and Iain Young [1,*]

[1] Institute of Life Course and Medical Sciences, University of Liverpool, Liverpool L1 8JX, UK; kmagee@liverpool.ac.uk
[2] AgriGrub Ltd., NIAB Innovation Hub, Hasse Road, Soham, Cambridgeshire CB7 5UW, UK; joe@agrigrub.co.uk
[3] Inspro Ltd. Brook House, Mead Road, Cranleigh, Surrey GU6 7BG, UK; richard@sns-eu.com
* Correspondence: isyoung@liverpool.ac.uk

Abstract: One third of food produced globally is wasted. Disposal of this waste is costly and is an example of poor resource management in the face of elevated environmental concerns and increasing food demand. Providing this waste as feedstock for black soldier fly (*Hermetia illucens*) larvae (BSFL) has the potential for bio-conversion and valorisation by production of useful feed materials and fertilisers. We raised BSFL under optimal conditions (28 °C and 70% relative humidity) on seven UK pre-consumer food waste-stream materials: fish trimmings, sugar-beet pulp, bakery waste, fruit and vegetable waste, cheese waste, fish feed waste and brewer's grains and yeast. The nutritional quality of the resulting BSFL meals and frass fertiliser were then analysed. In all cases, the volume of waste was reduced (37–79%) and meals containing high quality protein and lipid sources (44.1 ± 4.57% and 35.4 ± 4.12%, respectively) and frass with an NPK of 4.9-2.6-1.7 were produced. This shows the potential value of BSFL as a bio-convertor for the effective management of food waste.

Keywords: black soldier fly larvae; *Hermetia illucens*; bio-convertor; nutrient recovery; aquaculture feed; organic waste

Citation: Magee, K.; Halstead, J.; Small, R.; Young, I. Valorisation of Organic Waste By-Products Using Black Soldier Fly (*Hermetia illucens*) as a Bio-Convertor. *Sustainability* **2021**, *13*, 8345. https://doi.org/10.3390/su13158345

Academic Editors: Ada Margarida Correia Nunes Da Rocha and Belmira Almeida Ferreira Neto

Received: 28 June 2021
Accepted: 23 July 2021
Published: 27 July 2021

1. Introduction

It is estimated that the human population will exceed 9 billion by 2050. An increase in food production of around 50% will be needed to meet their needs [1]. Despite this rising demand, one third of all food produced globally is lost or wasted, equating to approximately 1.3 billion tonnes per year [2]. In the UK alone, 10.2 million tonnes of food waste was generated in 2015, of which 7.1 million tonnes was household waste and the remaining 3.1 million tonnes was from the post farm-gate supply chain [3]. The 'waste hierarchy' [4] illustrates destinations for waste ranked by environmental impact: prevention and minimisation (of waste generation such as redistribution), reuse (for other purposes, including use as animal feed and biomaterial processing), recycling (including composting and anaerobic digestion), energy recovery (including incineration for heat generation) and, finally, disposal. This hierarchy has been incorporated into UK law though the Waste (England and Wales) Regulations 2011, the Waste Regulations (Northern Ireland) 2011 and the Waste (Scotland) Regulations 2012 [5]. According to the waste hierarchy, prevention and redistribution have the lowest environmental impact; however, once food has started to spoil, recycling becomes the next best option. Anaerobic digestion (AD) is a go-to technology favoured globally for recycling food and other organic waste into bioenergy. However, these systems suffer from poor stability and low efficiency, due to the characteristics of food waste [6].

Some food waste can be recycled to make animal feed (within certain confines of the law). Bakery or confectionery products, providing they do not contain and or have not been in contact with meat, fish, or shellfish, can be used. Food or catering waste from kitchens which process meat, vegetarian kitchens which handle dairy products, restaurants and commercial kitchens producing vegan food and international catering waste cannot be used [7]. Animal by-products (ABPs) are subject to greater restrictions to maintain safe food supply chains and appropriate management of high-risk materials. ABPs are divided into three categories, categories one and two being high risk materials, while category three are low risk; category three materials can be processed into farm animal or pet feeds, among other products [8].

There is a growing interest in the use of insects as natural bio-convertors of organic waste valorising the waste by consuming the waste, incorporating it into their bodies and, in the process, converting it into valuable products. Life cycle assessment (LCA) has shown that insect bio-conversion is efficient and environmentally sustainable; direct greenhouse gas (GHG) emissions produced by insects are 47 times lower and the resulting global warming potential (GWP) is half that of open air composting [9]. Production of insects as a food also uses comparatively little space (reducing land use), but, usually, has a relatively high energy use (and GWP) for heating, to achieve a suitable culture environment and for drying the insects after harvest. This high energy demand may be exacerbated by the need to transport the food source to an insect production facility [10]. Overall, energy consumption can be reduced by using waste heat from other industrial processes, or using AD or incineration of other low category waste to heat the system [11]. Insect products can replace less sustainable products, such as fishmeal or soy products in aquaculture feeds; if this product replacement is also accounted for during LCA assessment, then the GWP of insect production decreases further [9,10].

Insects have previously been used in organic waste management strategies, recovering nutrients in the form of the constituent parts of the insect yielding high quality protein, lipid and chitin [12–15]. Insects, or their derivative proteins and fats, are utilized as food for humans [16,17] and in animal feeds [18,19]. Further, as a lipid-rich source, they can also be used for the production of biodiesel [10]. In this paper we focus primarily on the value of the products as constituents of animal feed.

Of the many insect species that have been studied [16,17], the black soldier fly (BSF) (*Hermetia illucens*) stands out. It has many characteristics that make it particularly attractive for commercial scale production in the UK. It is a species of true fly (Diptera) of the family Stratiomyidae. It originates from the Americas, although they are now more widespread in tropical and temperate regions [20–23]. They do not tolerate colder climates, such as those found in north-western Europe [24]. Therefore, if any escape from culture facilities, they are unlikely to survive the winter and become an invasive species. The BSF larvae (BSFL) are capable of consuming a wider range of organic materials than other fly species [25,26]. The adult stage is not a vector for human, animal, or plant pathogens. It does not possess mouth apparatus, so cannot bite [25,27,28]. BSF have an excellent nutritional profile, high in protein with a high quality amino acid profile and high levels of lipid, including economically and nutritionally valuable fatty acids [25]. *Hermetia illucens* was included in the seven insect species listed in the Commission Regulation (EU) 2017/893 [29] as safe for production for food use. This regulation permits the use of processed animal proteins (PAPs) derived from those insect species for aquaculture feeds, pet feeds and fur animal feeds in the EU. Insects grown for the production of processed insect proteins (PIPs) that are fed to other farmed animals are categorized as 'farmed animals' (Article 3(6) of Regulation (EC) No 1069/2009; [30]). As such, they are subject to regulations in the use of feed materials used to grow them [31]. The fact that they can be used in aquaculture feeds, their high-quality nutritional profile and their utility for bioconversion of waste strongly suggest that there will be an increase in the demand for BSFL and BSFL products.

Production of BSFL PIPs involves hatching BSFL, then growing them on an organic feed material, until they reach an appropriate life stage. They are then separated from the remains of the feed and larval residue (known as frass) and harvested. BSFL products have several potential uses, the primary use, that we discuss here, is the production of meal. Liu and Chen [32] concluded the early pre-pupae is the most appropriate life stage to harvest for meal production. This stage was, therefore, used during this study. BSFL meal can be further processed to concentrate protein while extracting the lipids for other uses as high-quality feed ingredients. In addition, lipids can also be used for biodiesel production [33] and chitin can be extracted from the meal for many uses, including food, pharmaceuticals, textiles, waste water treatment and cosmetics [34]. The BSFL lipid content is of particular interest because of the high content (approximately 28.6 ± 8.6% of insect mass) and because it is rich in useful fatty acids [25]. Spranghers and Ottoboni [14] showed that the fatty acid profile of BSFL is highly influenced by their diet, highlighting an opportunity for manipulation of the product via diet.

The 'frass' is a by-product that consists of faecal matter, residual growing substrate and shed exoskeletons from previous instars. It has value as a good quality, slow release organic fertiliser, with higher NPK values than other animal by-products recognized as fertilizers, such as composted poultry litter and worm castings [35]. Frass can also be processed via AD for further energy recovery, as it possesses suitable characteristics [36]. The anaerobic biodegradability fraction (fd) of BSFL frass is equal to that of food waste (89%); however, it has higher bio-methane potential (502 ± 9 mL CH_4/g VS) than food waste (449 ± 53 mL CH4/g VS) [36]. Food waste also causes two main problems for AD, poor stability, due to volatile fatty acids and low organic loading rates and effectively low efficiency [37,38], caused by high levels of easily biodegradable suspended solids [6]. However, significantly, utilising BSFL to bio-convert food waste into high quality feed ingredients can be classed as "prevention"; therefore, it is preferred over AD as a method for processing organic food waste in the waste hierarchy [4,39].

The increased interest in the use of BSFL as an organic waste management tool and a source of raw materials for the manufacture of animal feed has led to a better understanding of how nutrient density and feed substrate quality can influence the development and growth of BSFL [40,41]. BSFL have been shown to achieve a good feed conversion ratio (FCR), ranging between 1.4 and 2.6, when fed food waste materials [42]. Diets consisting of high protein and high lipid achieve the best results. The nutritional profile of the end larval material is also affected and, while protein levels do vary, the lipid levels and profile are more highly affected [14,43]. In this study, we look at the impact of seven potential organic waste streams (Table 1) from pre-consumer and manufacturing situations on BSFL bio-concentration of nutrients and, primarily, fatty acids. We evaluate which of our organic waste materials are most suitable for use in modulating and manipulating fatty acid profiles of BSFL meal and draw conclusions about the valorisation of organic waste by-products via BSFL treatment.

In response to the change in EU law [29] allowing use of insect meals and the growing interest in this area, it is very likely that these meals will become highly valued as aquaculture feed ingredients. Because fish lack the enzymes to completely synthesize polyunsaturated fatty acids (PUFA), or highly unsaturated fatty acids (HUFA) of the n-3 and n-6 series de novo [44], these must be provided preformed via the diet, making them essential fatty acids (EFAs): linoleic acid (18:2n-6), α-linolenic acid (18:3n-3), arachidonic acid (20:4n-6), Eicosapentaenoic acid (EPA, 20:5n-3) and Docosahexaenoic acid (DHA, 22:6n-3) [45]. This study pays close attention to these EFAs and their potential as feed ingredients.

Table 1. Pre-consumer and manufacturing organic waste streams identified for study.

Organic Waste	Source	Reason Waste Was Chosen for Investigation	Waste Category and Disposal Method	Median Gate Fezzze (GBP/tonne) [46]
Fish trimmings	Collected from local fish monger	Waste generated at fish processing facilities. Seeking to track long chain fatty acids in BSFL.	Fallen stock and digestive tracks—category II. Parts of stock unconsumed—category III.	Material recovery facilities (MRFs): all wastes, GBP 25, contracts from 2018 are GBP 35. In vessel composting (IVC): mixed food and green, GBP 50; all feedstock types, GBP 46. AD: all gates fees, GBP 27. Energy recovery: GBP 89. Landfill: non-hazardous waste, including landfill tax (standard rate for 2017/18 is GBP 88.95/tonne), GBP 113.
Sugar beet pulp	British sugar	Highly produced by sugar industry and meat-free.	Covered under fruit and vegetable waste.	
Bakery waste	Local bakery	Available in high volumes due to short shelf life and meat free	Non-animal by-product approved, depackaged and shred.	
Fruit and vegetable waste	Household waste (representative of supermarket waste)	Available in high volumes and meat-free	Non-animal by-product approved, depackaged and shred.	
Cheese waste	Harvey & Brockless (H&B) Cheese in London	Available in high volumes, meat-free and high in fat. Investigating how BSFL respond to high fat material.	Covered under dairy products. Treated as bakery and fruit and vegetable, depackaged and shred.	
Industrial fish feed waste	Skretting feed manufacturing facility	By product of aquaculture feed industry	As for fish trimmings.	
Brewer's grains and yeast	Firebird Brewery	Available in high volumes from brewing industry and meat-free.	Often used as animal feed or disposed via landfill [47].	

2. Materials and Methods

2.1. Outcomes

2.1.1. Growth Performance of BSFL Fed Identified Organic Materials

BSFL are produced under optimal conditions, fed on our identified feed materials and converted into BSFL meals. Growth performance is assessed to explore BSFL meal production and waste reduction potential of each organic material.

2.1.2. BSFL Bioconcentration and Modification of Fatty Acid Profile

Samples of feed materials and BSFL meals were collected and analysed for nutritional quality—protein, lipid and fatty acid profiles. Data were analysed to assess bioconcentration of nutrients during production of BSFL meals, in order to identify suitable organic by-products for nutrient recovery by BSFL.

2.1.3. Valorisation of Organic Waste By-Products via BSFL Treatment

The value of the BSFL outputs, meals and frass, was estimated to assess valorisation of the identified organic waste streams via BSFL treatment.

2.2. Processing Organic Waste Materials

Water was added to the sugar beet, bakery waste, cheese waste and fish feed waste to achieve 70% water content prior to feeding, while the fish trimmings, fruit and vegetable waste and brewer's grains and yeast already contained a high enough water content. All feedstock materials were homogenized, prior to feeding, in order to optimize processing by BSFL. Samples (100 g) of each material were frozen at −20 °C and sent to Nottingham University for proximate analyses and fatty acid analyses.

Material energy content was determined using a Parr 6300 bomb calorimeter connected to a Parr 6520 water recirculation system. One-gram Benzoic acid tablets standardized for bomb calorimetry (26.454 MJ/kg, Parr Instrument Co, item No: 3415) were used as standards. Material protein content was analysed using a Thermo Scientific FlashEA® 1112 N/Protein Analyzer in conjunction with the EAGER software. The lipid content of each sample was analysed using rapid Soxhlet extraction, using a Gerhardt Soxtherm. The extracted lipid samples were further analysed to determine the fatty acid profile of each sample by applying a direct method for fatty acid methyl ester (FAME) synthesis, in conjunction with GC analyses (Perkin Elmer Clarus 500 Gas Chromatograph), utilizing a Varian capillary column CP-Sil 88 for FAME; column length, 100 m, column width, 0.25 mm. Gas flow for air was 450 mL/min and hydrogen was 45 mL/min; the temperature set point was 250 °C. Ash was determined using the AOAC official method 942.05 [48]. Fibre content was analysed using the Gerhardt Fibrebag method.

2.3. BSFL Production

Each of the experimental organic waste stream materials were fed to five replicate groups of 25 larvae, for a total of 35 groups of BSFL. Larvae were grown under environmentally controlled conditions (28 °C and 70% relative humidity) and kept in the dark during production. BSFL growth and performance were assessed through feed conversion ratio (FCR), specific growth rate (SGR) and larval growth rate (LGR). Efficiency of conversion of ingested food (ECI) was assessed. Waste material reduction was assessed through waste reduction index (WRI) and substrate reduction (SR). All using the following equations:

$$FCR = TFI\ (kg) \div Weight\ gain\ (kg)$$

where *TFI* (total feed intake) = total feed given and *Weight gain* = weight at end of study period–weight at start of study period [49].

$$SGR\ (\%) = 100 \times (\ln W_2 - \ln W_1) \times (t_2 - t_1)^{-1}$$

where ln = natural log, W_1 = initial weight, W_2 = final weight, t_1 = starting time point (day one) and t_2 = end time point (final day number) [50].

$$LGR\ (g/day) = (W_2 - W_1)/number\ of\ days$$

where W_1 = initial larval weight (g), W_2 = final larval weight (g) [51].

$$ECI = B/(W - R)$$

where B = total biomass (larvae) (g), W = total amount of feed provided (g) and R = remaining substrate (g) [51].

$$WRI = (W - R/W)/days\ of\ trial\ (d) \times 100$$

where W = total amount of feed provided and R = remaining substrate [51].

$$SR = W - R/W \times 100$$

where W = total amount of feed provided and R = remaining substrate [51].

2.4. Nutritional Analyses of BSFL Pre-Pupae Fed Each Organic Material

BSFL groups were harvested at the pre-pupae stage and frozen at −20 °C. They were dried in a drying cabinet at 60 °C for 4 days, then ground into BSFL meal using a benchtop hand grinder. Samples of each meal were sent to Skretting UK for analysis—crude protein, crude lipid, amino acid profile and fatty acid profile. The BSFL meal nutrient content for crude protein, crude lipid and fatty acids were analysed as the feed materials above. Amino acid profiles were determined using hydrolyses and an amino acid analyser. These data were combined with Skretting UK's undisclosed data regarding protein digestibility of BSFL meal for value estimation as an aquaculture feed ingredient.

2.5. Data Analyses

A representative frass sample was collected during production of the *BSFL* and analysed by NMR laboratories for nitrogen (N), phosphorus (P), potassium (K) (NPK) and magnesium (Mg) to assess the quality of the waste product for use as a fertiliser.

Nutrient bioconcentration by *BSFL* from each of the organic waste materials was investigated by generating an apparent bioconcentration factor (*aBCF*) for each nutrient, i.e., crude protein, crude lipid, fatty acids and fatty acid groups, which are present in both feed materials and *BSFL* meals, calculated as follows:

$$aBCFi = \frac{(FAi/TDFA)BSFL\ meal}{(FAi/TDFA)Diet} \quad \text{or} \quad aBCFi = \frac{(Nutrienti)BSFL\ meal}{(Nutrienti)Diet}$$

where i = specific *FA* (g/100 g DM), or the sum of a group of *FA* (SFA, MUFA, MUFA trans, PUFA and branched *FA*) and *TDFA* = total detected fatty acids (g/100 g DM), or nutrient (g/100 g DM of crude protein or crude lipid) [43].

2.6. Value Estimation of BSFL Outputs

Each BSFL meal was inputted into Skretting's aquaculture feed formulation software programme; this programme assigned a value to each BSFL meal (as an ingredient), based on their nutritional qualities compared to the quality of all the other available feed ingredients and their current market prices (correct as of February 2019). The potential value of frass can be estimated based on N, P and K content along with current costs of those nutrients available through other marketed fertilisers, as described by Kissel and Risse [52].

2.7. Statistical Analysis

Tests for differences were carried out with 95% confidence levels ($p \leq 0.05$) between each test substance. The Kolmogorov–Smirnov tests for normality were carried out, with one-way ANOVA tests, followed by post hoc Tukey's tests, used for parametric data, and Mann–Whitney U tests, for non-parametric data.

3. Results

3.1. BSFL Growth, Performance and Substrate Reduction

Growth and performance of the BSFL varied significantly depending on the feedstock they were fed (Table 2). Sugar beet pulp and cheese waste were used the least efficiently by BSFL, attaining the lowest ECI, LGR and SGR, subsequently reaching the highest FCR. Bakery waste achieved the greatest performance (FCR and SGR), while fish feed waste was the most efficiently used (ECI). The BSFL consumed more cheese waste (WRI) than any other feedstock. The greatest reduction in feedstock substrate (SR) was seen with fruit and vegetable waste. BSFL mortality rate when fed fish feed waste (48.8%) was significantly higher ($p < 0.05$) than when fed all other feedstocks (4–12%).

3.2. BSFL Nutrient Bioconcentration

3.2.1. Organic Waste Material and BSFL Meal Profiles

The variety of organic waste materials used display a range in protein (8.4–54.0 g/100 g DM) and lipid (0.4–57.3 g/100 g DM) levels (full nutritional profile of the BSFL meal provided in Appendix A). Each material also displays varied fatty acid profiles (Table 3).

Table 2. Substrate reduction alongside growth and performance of BSFL when raised on the identified organic feedstocks.

Diet	Feed Conversion Rate (FCR)	Specific Growth Rate (SGR)	Larval Growth Rate (LGR) (mg/day)	Efficiency of Conversion of the Ingested Food (ECI)	Waste Reduction Index (WRI) (g/day)	Substrate Reduction (SR) (%)
Fish trimmings	5.98 ± 2.77^{ac}	16.92 ± 3.36^{a}	9.25 ± 4.94^{a}	0.32 ± 0.08^{a}	28.17 ± 5.26^{ad}	-54.49 ± 8.59^{a}
Sugar beet pulp	20.54 ± 8.68^{b}	9.95 ± 1.84^{b}	2.19 ± 0.54^{b}	0.11 ± 0.04^{b}	14.79 ± 0.68^{b}	-60.98 ± 8.31^{ac}
Bakery waste	4.84 ± 1.45^{a}	17.66 ± 1.68^{a}	9 ± 2.83^{a}	0.35 ± 0.08^{a}	22.66 ± 1.26^{ae}	-70.26 ± 9.49^{bc}
Fruit and vegetable waste	7.97 ± 1.1^{ac}	16.08 ± 1.21^{a}	6.45 ± 0.79^{ac}	0.15 ± 0.01^{b}	28.02 ± 2.21^{ad}	-79.28 ± 6.18^{b}
Cheese waste	12.92 ± 2.06^{c}	9.55 ± 0.99^{b}	2.71 ± 0.19^{c}	0.11 ± 0.03^{b}	45.17 ± 4.98^{c}	-63.86 ± 5.63^{ac}
Fish feed waste	6.42 ± 1.25^{ac}	17.39 ± 1.43^{a}	16.05 ± 1.9^{d}	0.55 ± 0.07^{c}	31 ± 1.42^{d}	-37.27 ± 5.4^{d}
Brewer's grain and yeast	6.78 ± 1.12^{ac}	16.59 ± 1.19^{a}	6.99 ± 0.38^{ac}	0.31 ± 0.04^{a}	21.19 ± 3.32^{be}	-52.04 ± 5.02^{a}

LGR, ECI, WRI and SR were calculated on a DM basis. Organic materials which do not share a common letter in each column are significantly different $p < 0.05$.

Table 3. (**a**) Nutritional profile of waste stream feed materials, proximate. (**b**) Fatty acid profiles of organic waste stream feed materials.

(a)

Parameter	Fish Trimmings	Sugar Beet Pulp	Bakery Waste	Fruit and Vegetable Waste	Cheese Waste	Fish Feed Waste	Brewer's Grains and Yeast
Dry matter (DM) (%)	31.07	52.65	58.63	12.47	53.54	93.93	21.66
Crude protein (g/100 g DM)	42.42	8.62	18.22	8.42	31.71	54.02	49.95
Crude fat (g/100 g DM)	36.47	0.36	2.66	1.68	57.27	10.40	6.56
Fibre (g/100 g DM)	0.00	4.21	0.65	0.10	0.22	1.63	0.88
Ash (g/100 g DM)	5.22	4.22	1.97	0.66	3.35	6.51	1.03
Energy (MJ/kg)	7.5	8.47	11.11	2.02	16.33	20.89	4.41

(b)

Fatty Acids (g/100 g DM)		Fish Trimmings	Sugar Beet Pulp	Bakery Waste	Fruit and Vegetable Waste	Cheese Waste	Fish Feed Waste	Brewer's Grains and Yeast
Caproic acid	C6:0	1.6	0.0	0.1	0.1	6.4	0.3	0.1
Caprylic acid	C8:0	0.01	0.01	0.01	0.00	0.31	0.01	0.00
Capric acid	C10:0	0.00	0.00	0.00	0.00	0.76	0.00	0.00
Undecanoic acid	C11:0	0.00	0.00	0.00	0.00	0.18	0.00	0.00
Lauric acid	C12:0	0.02	0.01	0.01	0.00	1.02	0.02	0.01
Tridecanoic acid	C13:0	3.03	0.05	0.77	0.55	0.88	1.46	3.06
Myristic acid	C14:0	0.95	0.01	0.01	0.01	3.74	0.63	0.01
Myristoleic acid	C14:1n-5	0.03	0.00	0.00	0.01	0.64	0.01	0.00
Pentadecanoic acid	C15:0	0.08	0.00	0.00	0.00	0.41	0.04	0.00
cis-10 pentadecanoic acid	C15:1	0.65	0.00	0.11	0.09	0.29	0.40	0.31
Palmitic acid	C16:0	2.62	0.08	0.29	0.05	10.75	1.55	0.35
Palmitoleic acid	C16:1n-7	0.71	0.00	0.00	0.00	0.74	0.42	0.01
cis-10 heptadecanoic acid	C17:1	0.03	0.00	0.00	0.00	0.03	0.01	0.01
Stearic acid	C18:0	0.50	0.02	0.04	0.01	3.45	0.28	0.03
Elaidic acid, Oleic acid	C18:1n-9	2.70	0.03	0.62	0.03	7.57	2.49	0.22
Linoleic acid	C18:2n-6	0.95	0.03	0.45	0.07	0.54	1.38	0.66
α-linolenic acid	C18:3n-3	0.20	0.00	0.06	0.01	0.16	0.38	0.07
Gamma-linolenic acid (GLA)	C18:3n-6	0.01	0.00	0.00	0.00	0.01	0.01	0.00
Arachidic acid	C20:0	0.04	0.00	0.01	0.00	0.06	0.05	0.00
Gondoic acid	C20:1n-9	4.42	0.00	0.03	0.00	0.03	0.98	0.07
Eicosadienoic acid	C20:2n-6	0.12	0.00	0.01	0.00	0.02	0.05	0.00
cis-11,14,17 eicosatrienoic acid	C20:3n-3	0.12	0.00	0.00	0.00	0.00	0.03	0.00
cis-8,11,14 eicosatrienoic acid	C20:3n-6	0.03	0.00	0.00	0.00	0.03	0.01	0.01
Arachidonic acid	C20:4n-6	0.03	0.00	0.00	0.00	0.03	0.04	0.00
Eicosapentaenoic acid (EPA)	C20:5n-3	0.24	0.00	0.00	0.00	0.01	0.56	0.01
Heneicosanoic acid	C21:0	0.02	0.00	0.00	0.00	0.03	0.01	0.03
Behenic acid	C22:0	0.02	0.00	0.00	0.00	0.04	0.06	0.01
Erucic acid	C22:1n-9	0.58	0.00	0.01	0.00	0.01	0.12	0.01
cis-13,16-docosadienoic acid	C22:2	0.07	0.00	0.00	0.00	0.02	0.01	0.01
Docosahexaenoic acid (DHA)	C22:6n-3	0.22	0.00	0.00	0.00	0.00	0.61	0.00
tricosanoic acid	C23:0	0.01	0.00	0.00	0.00	0.02	0.00	0.00
lignoceric acid	C24:0	0.03	0.00	0.00	0.01	0.03	0.01	0.01
nervonic acid	C24:1	0.60	0.00	0.00	0.00	0.00	0.08	0.01
Sum Sat FA		8.95	0.23	1.23	0.77	28.12	4.43	3.66
Sum unsaturated FA		11.73	0.06	1.30	0.22	10.13	7.56	1.39
Sum monoenes		9.73	0.03	0.78	0.14	9.30	4.50	0.64
Sum n-6 FA		1.14	0.03	0.46	0.07	0.63	1.48	0.68
Sum n-3 FA		0.79	0.00	0.06	0.01	0.17	1.57	0.07
Unsat/Saturated		1.31	0.28	1.06	0.29	0.36	1.71	0.38
n-6/n-3		1.44	13.47	7.17	6.09	3.66	0.94	9.29
n-3/n-6		0.69	0.07	0.14	0.16	0.27	1.06	0.11

The nutrient profiles of the BSFL meals (Table 4) were influenced by the different organic feed materials. However, there was less variation between the BSFL meals than there was between the organic waste stream materials, with average crude protein levels of 44.1% (±4.57) and lipid levels of 35.4% (±4.12), compared to 30.48% (±19.11) and 16.48%

(±21.86). Leucine, aspartic acid and glutamic acid were the three most prevalent amino acids across all the BSFL meals, with tyrosine being exceptionally higher in BSFL fed cheese waste, fish feed waste and brewer's grain and yeast. The fatty acids most prevalent across all the BSFL meals included Lauric acid, Myristic acid, Palmitic acid, Palmitoleic acid, Oleic acid and linoleic acid. BSFL fed fish trimmings also contained raised levels of EPA. BSFL fed brewer's grains contained the highest level of n6 fatty acids with a high level of n3 fatty acids, containing EFA's linoleic and α-linolenic acid, alongside a good level of EPA and small amounts of DHA, with only the fish trimmings and fish feed waste fed BSFL meals possessing higher levels of EPA and DHA.

Table 4. (**a**) Nutritional profiles of BSFL meals, proximate and amino acid analysis. (**b**) Nutritional profiles of BSFL meals, fatty acid analysis.

(a)

Parameter (g/100 g DM)		BSFL Meals						
		Fish Trimmings	**Sugar Beet Pulp**	**Bakery Waste**	**Fruit and Vegetable Waste**	**Cheese Waste**	**Fish Feed Waste**	**Brewer's Grains and Yeast**
Proximate	Crude protein	46.62	43.15	43.07	36.03	43.18	45.70	51.05
	Crude fat	35.05	35.49	37.63	40.30	36.94	35.22	27.00
Amino acids								
Essential	Arginine	2.06	1.79	1.84	1.43	1.95	2.14	1.96
	Histidine	1.43	1.34	1.31	1.01	1.30	1.44	1.34
	Isoleucine	1.97	1.84	1.76	1.45	1.82	1.99	2.11
	Leucine	3.85	3.58	2.79	2.28	2.85	3.15	3.30
	Lysine	2.41	2.19	2.41	1.97	2.38	2.53	3.15
	Methionine	0.87	0.78	0.79	0.63	0.82	0.80	0.85
	Cystine	0.29	0.27	0.33	0.29	0.23	0.22	0.37
	Phenylalanine	1.73	1.69	1.81	1.51	1.85	2.05	2.44
	Tyrosine	2.03	1.91	1.96	1.48	5.28	5.73	5.44
	Threonine	1.68	1.56	1.61	1.28	1.64	1.80	1.92
	Valine	2.67	2.52	2.62	2.11	2.64	2.91	3.04
Non-essential	Alanine	2.94	2.92	2.81	2.36	2.76	2.91	4.46
	Aspartic acid	3.86	3.76	3.89	3.10	3.69	4.12	4.05
	Glutamic acid	4.47	4.26	4.37	3.48	4.40	4.40	5.13
	Glycine	2.59	2.35	2.35	1.84	2.37	2.61	2.65
	Proline	2.65	2.47	2.32	1.91	2.88	2.97	3.35
	Serine	1.83	1.73	1.70	1.33	1.69	1.83	1.94
Sum of AA		39.31	36.95	36.68	29.47	40.56	43.61	47.51

Tryptophan was not tested for.

(b)

Parameter (g/100 g DM)		BSFL Meals						
		Fish Trimmings	**Sugar Beet Pulp**	**Bakery Waste**	**Fruit and Vegetable Waste**	**Cheese Waste**	**Fish Feed Waste**	**Brewer's Grains and Yeast**
Fatty acids								
Caprylic acid	C8:0	<LOD	<LOD	<LOD	<LOD	0.01	<LOD	<LOD
Capric acid	C10:0	0.34	0.35	0.34	0.28	0.37	0.48	0.23
Lauric acid	C12:0	12.59	20.37	19.80	15.39	12.35	16.71	7.18
Myristic acid	C14:0	2.33	3.69	3.88	3.39	3.59	2.97	1.83
Myristelaidic acid	C14:1n-5	0.06	0.09	0.08	0.28	0.31	0.08	0.06
Pentadecanoic acid	C15:0	0.13	<LOD	0.04	0.08	0.22	0.06	0.08
Palmitic acid	C16:0	4.71	4.29	4.74	5.72	7.39	3.61	4.60
Palmitoleic acid	C16:1n-7	2.23	0.97	1.05	2.06	1.80	1.18	1.34
	C16:2n-6	0.04	<LOD	<LOD	<LOD	<LOD	0.02	<LOD
Stearic acid	C18:0	0.55	0.55	0.68	0.73	0.95	0.44	0.69
	C18:1n-5	<LOD	<LOD	<LOD	<LOD	<LOD	0.02	0.01
Elaidic acid, Oleic acid	C18:1n-9	5.56	3.02	3.99	8.87	6.68	4.13	4.23
cis-vaccenic acid	C18:1n-7	0.42	<LOD	0.11	0.12	0.23	0.26	0.38
	C18:2n-4	<LOD	<LOD	<LOD	<LOD	0.04	0.01	<LOD
Linoleic acid	**C18:2n-6**	**2.29**	**1.28**	**2.11**	**1.65**	**1.27**	**2.00**	**3.57**
α-linolenic acid	**C18:3n-3**	**0.33**	**0.18**	**0.30**	**0.36**	**0.23**	**0.28**	**0.42**
Gamma-linolenic acid (GLA)	C18:3n-6	0.03	<LOD	<LOD	<LOD	0.02	0.01	0.01
Stearidonic acid (SDA)	C18:4n-3	0.23	<LOD	<LOD	<LOD	0.01	0.22	0.17
Arachidic acid	C20:0	<LOD	0.04	<LOD	<LOD	0.03	0.03	0.06
	C20:1n-8	0.09	<LOD	<LOD	<LOD	<LOD	0.24	0.20
Gadoleic acid	C20:1n-11	0.44	<LOD	<LOD	<LOD	0.11	0.34	0.30
Eicosadienoic acid	C20:2n-6	<LOD	<LOD	<LOD	<LOD	0.07	0.07	0.06

Table 4. *Cont.*

(b)

Parameter (g/100 g DM)		BSFL Meals						
		Fish Trimmings	Sugar Beet Pulp	Bakery Waste	Fruit and Vegetable Waste	Cheese Waste	Fish Feed Waste	Brewer's Grains and Yeast
Arachidonic acid	**C20:4n-6**	**0.14**	<LOD	<LOD	**0.04**	**0.05**	**0.03**	**0.02**
Eicosapentaenoic acid (EPA)	**C20:5n-3**	**1.20**	**0.09**	**0.04**	**0.08**	**0.18**	**0.78**	**0.53**
Behenic acid	C22:0	<LOD	<LOD	<LOD	<LOD	0.03	0.02	0.04
Cetoleic acid	C22:1n-11	<LOD	<LOD	<LOD	<LOD	<LOD	0.12	0.07
Docosahexaenoic acid (DHA)	**C22:6n-3**	**0.32**	<LOD	<LOD	<LOD	<LOD	**0.06**	**0.07**
Sum Sat FA		20.64	29.28	29.47	25.59	24.93	24.33	14.70
Sum unsaturated FA		13.37	5.64	7.68	13.46	10.99	9.85	11.43
Sum monoenes		8.79	4.08	5.23	11.32	9.12	6.38	6.58
Sum n-6 FA		2.50	1.28	2.11	1.69	1.41	2.13	3.66
Sum n-3 FA		2.08	0.27	0.34	0.44	0.42	1.34	1.20
Unsat/Saturated		0.65	0.19	0.26	0.53	0.44	0.40	0.78
n-6/n-3		1.20	4.70	6.22	3.82	3.35	1.59	3.06
n-3/n-6		0.83	0.21	0.16	0.26	0.30	0.63	0.33
Unknown		2.95	1.61	0.9	1.4	2.74	2.93	3.24

<LOD = below level of detection. Essential fatty acids highlighted in bold.

3.2.2. BSFL Bioconcentration of Nutrients

According to the method used to calculate aBCF, values higher than unity are considered to indicate nutrient concentration. These results show that BSFL bioconcentrate nutrients from most organic food materials very well. Lauric acid is the fatty acid which BSFL accumulate the greatest across all feed materials (Table 5). BSFL fed cheese waste is the only meal where lauric acid is not the most bioconcentrated FA; EPA is higher. BSFL achieve the greatest bioconcentration of the overall desired EFAs from fish trimmings, followed by brewer's grains and cheese waste.

Table 5. Apparent bioconcentration factor (aBCF) of nutritional parameters tested for and present in both feed materials and BSFL meals, achieved by BSFL during production feeding on each waste stream material.

Parameter (%DM)		BSFL Apparent Bioconcentration Factor (aBCF)						
		Fish Trimmings	Sugar Beet Pulp	Bakery Waste	Fruit and Vegetable Waste	Cheese Waste	Fish Feed Waste	Brewer's Grains and Yeast
Crude protein		1.1	5.0	2.4	4.3	1.4	0.8	1.0
Crude fat		1.0	98.4	14.1	23.9	0.6	3.4	4.1
Fatty acids								
Caprylic acid	C8:0	0	0	0	0	0.1	0	0
Capric acid	C10:0	216.4	0	43.4	6.4	0.7	57.4	0
Lauric acid	C12:0	561.9	21.1	114.6	157.9	18.8	230.2	329.0
Myristic acid	C14:0	2.6	4.9	29.1	15.2	1.5	1.4	57.2
Pentadecanoic acid	C15:0	1.7	0	1.4	2.7	0.8	0.4	8.2
Palmitic acid	C16:0	1.9	0.5	1.1	4.7	1.1	0.7	3.1
Palmitoleic acid	C16:1n-7	3.3	9.2	17.7	42.5	3.8	0.8	64.2
Stearic acid	C18:0	1.1	0.3	1.1	2.1	0.4	0.5	5.5
Elaidic acid, Oleic acid	C18:1n-9	2.1	1.1	0.5	10.6	1.4	0.5	4.6
Linoleic acid	C18:2n-6	2.5	0.5	0.3	1.0	3.6	0.4	1.3
α-linolenic acid	C18:3n-3	1.7	0.9	0.3	1.3	2.3	0.2	1.6
Gamma-linolenic acid (GLA)	C18:3n-6	2.4	0	0	0	4.8	0.2	0
Arachidic acid	C20:0	0.0	0.4	0	0	0.7	0.2	3.1
Eicosadienoic acid	C20:2n-6	0.0	0	0	0	5.2	0.5	3.0
Arachidonic acid	C20:4n-6	4.7	0	0	2.6	2.5	0.3	4.7
Eicosapentaenoic acid (EPA)	C20:5n-3	5.1	0	4.9	14.1	19.8	0.4	19.7
Behenic acid	C22:0	0	0	0	0	1.3	0.1	2.0
Docosahexaenoic acid (DHA)	C22:6n-3	1.5	0	0	0	0	0.0	0
Sum Sat FA		2.4	1.3	1.7	1.4	1.4	1.6	1.0
Sum unsaturated FA		1.2	0.9	0.4	2.5	1.7	0.4	2.0
Sum monoenes		0.9	1.2	0.5	3.4	1.5	0.4	2.5
Sum n-6 FA		2.3	0.5	0.3	1.0	3.5	0.4	1.3
Sum n-3 FA		2.7	1.4	0.4	1.6	3.8	0.3	4.0

3.3. *Valorisation of BSFL Products*

3.3.1. Value of BSFL Meals as Aquaculture Feed Ingredients

The predicted value of each BSFL meal, as feed ingredients within the aquaculture feed market, based on nutritional quality, is as follows:

- BSFL fed fish trimmings = GBP 824 per tonne
- BSFL fed sugar beet pulp = GBP 743 per tonne
- BSFL fed bakery waste = GBP 792 per tonne
- BSFL fed fruit and vegetable = GBP 792 per tonne
- BSFL fed cheese waste = GBP 787 per tonne
- BSFL fed fish feed waste = GBP 822 per tonne
- BSFL fed brewer's grains and yeast = GBP 819 per tonne

(correct as of February 2019).

3.3.2. Quality and Value of BSFL Frass

Analyses of the BSFL frass revealed a magnesium level of 0.26% (2589 mg/kg) and an NPK of 4.9-2.6-1.7. Therefore, it would take approximately 5 tonnes of dry material or 8 tonnes of wet material (62.4% DM) to reach a maximum fertilizer application rate of 250 kg/ha of nitrogen. The BSFL frass contains high NPK levels compared to other manure fertilisers (Table 6). Utilising other fertiliser prices and NPK content (Table 7), the BSFL frass has been estimated to be worth a value of GBP 57.12/tonne. When the value of other manure fertilisers is calculated using the midpoint values for NPK taken from Table 5, the BSFL frass compares very favourable, achieving the highest value (Table 8).

Table 6. Comparison of BSFL frass NPK values with other manure fertilizers [53].

Fertiliser	Nitrogen (N) %	Phosphorus (P) %	Potassium (K) %
BSFL frass	4.9	2.6	1.7
Cow manure	0.5–2	0.2–0.7	0.4–2
Horse manure	0.7–1.5	0.2–0.7	0.6–0.8
Pig manure	0.4–2	0.5–1	0.4–1.2
Poultry manure	1.5–6	1–4	0.5–3
Sheep manure	2.2–3.6	0.3–0.6	0.7–1.7
Rabbit manure	3–4.8	1.5–2.8	1–1.3

Table 7. Cost of each nutrient (N, P and K) based on other fertiliser prices.

Fertiliser	Cost (GBP/Tonne)	Kg of Nutrient Per Tonne	Cost of Nutrient (GBP/Tonne)	Average Cost of Nutrient (GBP/kg)
Ammonium nitrite (34.5% N)	258	345	0.75	0.67
Granular Urea-standard specification (46% N)	272	460	0.59	
Muriate of Potash (MOP) (60% K_20)	283	600	0.47	0.47
Diammonium Phosphate (DAP) (46% P_2O_5)	350	460	0.76	0.71
Triple Super Phosphate (TSP) (46% P_2O_5)	302	460	0.66	

Prices correct as of October 2019 [54].

Table 8. Estimated value of BSFL frass compared to other manure fertilisers based on midpoint NPK content, taken from Table 5, and value of nutrients, taken from Table 6.

Fertiliser	Cost of Nutrient (GBP/Tonne)			Total Cost (GBP/Tonne)
	Nitrogen (N)	Phosphorus (P)	Potassium (K)	
BSFL frass	32.8	12.2	12.1	57.1
Cow manure	8.4	2.1	8.5	19.0
Horse manure	7.4	2.1	5.0	14.5
Pig manure	8.0	3.5	5.7	17.2
Poultry manure	25.1	11.8	12.4	49.3
Sheep manure	19.4	2.1	8.5	30.1
Rabbit manure	26.1	10.1	8.2	44.4

4. Discussion

The results achieved here clearly concur with that of other studies [32,55]—the nutritional profile of BSFL is highly influenced by their diet.

Protein is frequently the most expensive component of agricultural diets, especially in aquaculture diets [56]. The BSFL meals produced here are high in crude protein (>43%), except when produced using fruit and vegetable waste (36%). This level reaches as high as 51% when produced using brewer's grains. These meals are rich in the amino acids leucine, aspartic acid and glutamic acid and very rich in tyrosine, when produced with cheese waste, fish feed waste and brewer's grains, on an % DM basis. Fishmeal is considered a very high-quality protein source for aquaculture diets [57]. Compared to an average 65% (70.7% DM) seen in fishmeal [58], the quality of this BSFL meal is also high, with a well-balanced amino acid profile. However, BSFL meal contains relatively lower levels of the three common limiting amino acids, arginine, lysine and methionine, although levels of the latter two are still good; however, it is richer in histidine, isoleucine, phenylalanine, tyrosine, valine, alanine and proline, on a percentage protein basis (Table A1a). Soybean meal is the most commonly used plant protein in aquaculture feeds, despite its nutritional restrictions [59–61]. The BSFL meals are overall richer in the amino acids alanine, glycine, histidine, methionine, proline and valine, compared to high protein soybean meal [62], on a percentage protein basis (Table A1a).

The fatty acid profile is the nutritional aspect of the BSFL meal most affected by diet. The most prevalent fatty acids across all BSFL meals are lauric acid, oleic acid, palmitic acid and linoleic acid, both on a % DM (Table 3b) and % total fatty acid basis (Table A1b), although they do vary considerably, depending on the BSFL food source.

Lauric acid, the most prevalent fatty acid found in BSF meal, is credited with antimicrobial, antiviral and antifungal properties [63–65]. It has been shown to reduce *Campylobacter* spp. in broilers [66]. The other most abundant fatty acids found here also have many uses, including use in food [67], as emulsifiers in soap [68], as emollients in cosmetics [69] and as excipients in pharmaceuticals [67]. A high level of linoleic acid, which also has been credited with antimicrobial properties [70], along with the other essential FAs, is desirable in agriculture and aquaculture feed ingredients for many species.

The bioconcentration data we present indicate how efficiently each nutrient is recovered from the organic materials by BSFL treatment. The bioconcentration of each nutritional element varies with each organic feed stock. BSFL fed on fruit and vegetable waste had high EPA conversion rates; however, the final amount of EPA remained low in the BSFL meal, because the EPA levels were low in the fruit and vegetable waste material. The highest conversion rates for the essential fatty acids is seen in BSFL fed on fish trimmings, cheese waste and brewer's grains. BSFL fed fish trimmings and brewer's grains also had the highest levels of essential FAs. These food waste feed stocks, therefore, are likely the most viable that we tested for production of feed ingredients for use in aquaculture or agriculture feeds. The BSFL that were fed fish feed waste also had high levels of these essential FAs; however, the bioconcentration factor was low (viz. there was no apparent concentration during BSFL treatment), so the high levels of essential FAs were due to the high levels of these nutrients in the fish feed waste material.

These results provide evidence that manipulation of the fatty acid profile is achievable via diet. The BSFL meal that was produced, beyond use in feeds, has several possibilities for further refinement, such as lipid extraction. The high lipid content of BSFL meal, once extracted, would be suitable feedstock for biodiesel production [33]; BSFL that were fed fruit and vegetable waste generated the highest lipid levels. This meal also has the lowest crude protein level achieved here. While a detailed examination is beyond the scope of this study, we could speculate that lipid extraction would also provide an improved protein meal, as well as the richest biodiesel lipid feedstock.

Depending on the target nutrient or product and target use, we have identified several industrial and pre-consumer organic by-products that could be processed via BSFL treatment and provided data to show which of these generate high-value nutrients for feed production or other added value products, instead of simply sending this "food waste" for AD, composting, landfill, or incineration.

We have shown that all of our selected organic waste materials can be utilised as feedstock by BSFL; however, our growth and performance data indicate which are most suitable to accomplish higher levels of production of BSFL products. In all cases, waste reduction (DM basis) was achieved. Therefore, BSFL treatment could be a viable method of reducing all these pre-consumer organic waste materials, generating lower volumes of more valuable and accessible materials.

BSFL treatment, as well as being more sustainable according to the waste hierarchy, clearly provides opportunities for increased valorisation of organic by-products, especially as the frass produced from BSFL treatment is more suitable than many organic food materials for AD treatment, or, as discussed, has value as fertiliser. The value calculated here is purely based on NPK content compared to the costs of other manure fertilisers. However, when looking to source BSFL frass fertilisers, they attract a considerably higher value than that of its NPK content would suggest; Ecothrive charge is a frass soil conditioner selling at GBP 29.95 per 3.5 kg, which scales to GBP 8557.14 per tonne (correct as of April 2020 [71]), a vastly improved value from the estimated GBP 57 per tonne. BSFL frass qualities which contribute to this high value include slow release of nutrients, support of soil microbiota [72] and promotion of plant health and growth; BSFL frass has also indicated insecticidal properties against wireworm [35].

5. Conclusions

This study shows how seven organic waste by-products influence the nutritional profile of BSFL. We identify that fish feed trimmings, along with brewer's grains and yeast, are ideal organic waste materials for BSFL treatment in order to generate high quality BSFL meals which would be suitable for inclusion in aquaculture or agriculture feeds. We have also identified that fruit and vegetable waste is a potential candidate for BSFL treatment, followed by lipid extraction, for recovery and production of lipid feedstock for biodiesel due to the higher lipid content.

BSFL treatment is a viable option for recovering and recycling organic waste by-products, especially if it is traceable to source. It provides opportunity for the valorisation of these organic waste products as constituents of animal feed, providing a more environmentally friendly alternative route than landfill or AD.

Author Contributions: Conceptualization, I.Y. and R.S.; methodology, K.M., J.H.; validation, J.H.; formal analysis, K.M.; investigation, K.M. and J.H.; resources, R.S., J.H. and I.Y.; data curation, K.M.; writing—original draft preparation, K.M.; writing—review and editing, K.M., I.Y.; visualization, K.M.; supervision, K.M. and I.Y.; project administration, I.Y. and R.S.; funding acquisition, I.Y. and R.S. All authors have read and agreed to the published version of the manuscript.

Funding: This research was funded by Innovate UK, grant number 7714.

Institutional Review Board Statement: Not applicable.

Informed Consent Statement: Not Applicable.

Data Availability Statement: Not Applicable.

Acknowledgments: We would like to acknowledge the following for their contributions to our research: Tim Parr and Jon Stubberfield at the School of Biosciences, University of Nottingham, Sutton Bonington Campus, Loughborough for providing analysis, in kind, of the seven organic waste stream materials we trialled. Skretting UK for provision of the waste fish feed material, for providing nutritional analyses of the BSFL meals and for utilising their feed formulation software to associate a value of the BSFL meals as aquaculture feed ingredients based on their quality. Sarah Gaunt and SPG innovation Ltd. were pivotal in navigation of the funding landscape to help us deliver proof of concept projects for our innovative technologies.

Conflicts of Interest: Mr Richard Small is the owner/CEO of Inspro Ltd.—a company that aims to become a supplier of BSFL meal. All other authors declare no conflict of interest.

Appendix A

Table A1. (**a**) Nutritional profile of BSFL meals, amino acid given as % protein, compared to an average 65% fishmeal and an average high protein soybean meal. (**b**) Nutritional profile of BSFL meals, fatty acids given as % total fatty acids, compared to an average 65% fishmeal and an average high protein soybean meal.

(**a**)

	Diet Component	Average 65% Protein Fish-meal	High Protein Soy-bean Meal	Fish Trim-mings	Sugar Beet Pulp	Bakery Waste	Fruit and Veg-etable Waste	Cheese Waste	Fish Feed Waste	Brewer's Grains and Yeast
Proximate analyses (% DM)	Crude Protein (% DM)	70.7	55.2	46.62	43.15	43.07	36.03	43.18	45.7	51.05
	Lipid (crude fat) (% DM)	10	1.7	35.05	35.49	37.63	40.3	36.94	35.22	27
Essential amino acids (% protein)	Arginine (Arg)	6.21	7.30	4.42	4.15	4.27	3.97	4.52	4.68	3.84
	Histidine (His)	2.50	2.7	3.07	3.11	3.04	2.80	3.01	3.15	2.62
	Isoleucine (Ile)	4.14	4.6	4.23	4.26	4.09	4.02	4.21	4.35	4.13
	Leucine (Leu)	7.17	7.7	8.26	8.30	6.48	6.33	6.60	6.89	6.46
	Lysine (Lys)	7.50	6.2	5.17	5.08	5.60	5.47	5.51	5.54	6.17
	Methionine (Met)	2.72	1.4	1.87	1.81	1.83	1.75	1.90	1.75	1.67
	Cystine (Cys)	0.86	1.6	0.62	0.63	0.77	0.80	0.53	0.48	0.72
	Phenylalanine (Phe)	3.90	5.1	3.71	3.92	4.20	4.19	4.28	4.49	4.78
	Tyrosine (Tyr)	3.04	3.5	4.35	4.43	4.55	4.11	12.23	12.54	10.66
	Threonine (Thr)	4.14	3.8	3.60	3.62	3.74	3.55	3.80	3.94	3.76
	Tryptophan (Try/Trp)	1.00	1.4	-	-	-	-	-	-	-
	Valine (Val)	4.98	4.8	5.73	5.84	6.08	5.86	6.11	6.37	5.95
Non-essential amino acids (% protein)	Alanine (Ala)	6.29	4.3	6.31	6.77	6.52	6.55	6.39	6.37	8.74
	Aspartic acid (Asp)	9.09	11.3	8.28	8.71	9.03	8.60	8.55	9.02	7.93
	Glutamic acid (Glu)	12.57	17.9	9.59	9.87	10.15	9.66	10.19	9.63	10.05
	Glycine (Gly)	6.65	4.2	5.56	5.45	5.46	5.11	5.49	5.71	5.19
	Proline (Pro)	4.34	5	5.68	5.72	5.39	5.30	6.67	6.50	6.56
	Serine (Ser)	3.89	4.6	3.93	4.01	3.95	3.69	3.91	4.00	3.80

Tryptophan was not analysed.

(**b**)

	Diet Component	Average 65% Protein Fish-meal	High Protein Soy-bean Meal	Fish Trim-mings	Sugar Beet Pulp	Bakery Waste	Fruit and Veg-etable Waste	Cheese Waste	Fish Feed Waste	Brewer's Grains and Yeast
Essential fatty acids (% total fatty acids)	C18:2n-6 (Linoleic acid)	2.1	54	6.53	3.61	5.61	4.09	3.44	5.68	13.22
	C18:3n-3 (α-linolenic acid)	1.9	7.2	0.94	0.51	0.80	0.89	0.62	0.80	1.56
	C20:4n-6 (Arachidonic acid)	2.4		0.40	<LOD	<LOD	0.10	0.14	0.09	0.07
	C20:5n-3 (Eicosapentaenoic acid (EPA))	9		3.42	0.25	0.11	0.20	0.49	2.21	1.96
	C22:6n-3 (Docosahexaenoic acid (DHA))	6.6		0.91	<LOD	<LOD	<LOD	<LOD	0.17	0.26

Table A1. *Cont.*

(b)

Diet Component	Average 65% Protein Fish-meal	High Protein Soy-bean Meal	Fish Trim-mings	Sugar Beet Pulp	Bakery Waste	Fruit and Veg-etable Waste	Cheese Waste	Fish Feed Waste	Brewer's Grains and Yeast
Non-essential fatty acids (% total fatty acids)									
C8:0 (Caprylic acid)			<LOD	<LOD	<LOD	<LOD	0.03	<LOD	<LOD
C10:0 (Capric acid)			0.97	0.99	0.90	0.69	1.00	1.36	0.85
C12:0 (Lauric acid)			35.92	57.40	52.62	38.19	33.43	47.44	26.59
C14:0 (Myristic acid)	6	0.2	6.65	10.40	10.31	8.41	9.72	8.43	6.78
C14:1n-5 (Myristelaidic acid)			0.17	0.25	0.21	0.69	0.84	0.23	0.22
C15:0 (Pentadecanoic acid)			0.37	<LOD	0.11	0.20	0.60	0.17	0.30
C16:0 (Palmitic acid)	17.8	11.2	13.44	12.09	12.60	14.19	20.01	10.25	17.04
C16:1n-7 (Palmitoleic acid)	7.2	0.1	6.36	2.73	2.79	5.11	4.87	3.35	4.96
C16:2n-6			0.11	<LOD	<LOD	<LOD	<LOD	0.06	<LOD
C18:0 (Stearic acid)	3.6	3.8	1.57	1.55	1.81	1.81	2.57	1.25	2.56
C18:1n-5			<LOD	<LOD	<LOD	<LOD	<LOD	0.06	0.04
C18:1n-9 (Elaidic acid, Oleic acid)	12.3	23.1	15.86	8.51	10.60	22.01	18.08	11.73	15.67
C18:1n-7 (cis-vaccenic acid)			1.20	<LOD	0.29	0.30	0.62	0.74	1.41
C18:2n-4			<LOD	<LOD	<LOD	<LOD	0.11	0.03	<LOD
C18:3n-6 (Gamma-linolenic acid (GLA))			0.09	<LOD	<LOD	<LOD	0.05	0.03	0.04
C18:4n-3 (Stearidonic acid (SDA))	1.5		0.66	<LOD	<LOD	<LOD	0.03	0.62	0.63
C20:0 (Arachidic acid)			<LOD	0.11	<LOD	<LOD	0.08	0.09	0.22
C20:1n-8			0.26	<LOD	<LOD	<LOD	<LOD	0.68	0.74
C20:1n-9 (Eicosenoic acid)	6.6								
C20:1n-11 (Gadoleic acid)			1.26	<LOD	<LOD	<LOD	0.30	0.97	1.11
C20:2n-6 (Eicosadienoic acid)			<LOD	<LOD	<LOD	<LOD	0.19	0.20	0.22
C22:0 (Behenic acid)			<LOD	<LOD	<LOD	<LOD	0.08	0.06	0.15
C22:1 n-9 (Erucic acid)	7.7								
C22:1n-11 (Cetoleic acid)			<LOD	<LOD	<LOD	<LOD	<LOD	0.34	0.26
C22:5n-3 (Docosapentaenoic acid (DPA))	2.6								

<LOD = below level of detection.

References

1. FAO. *The Future of Food and Agriculture–Trends and Challenges*; Food and Agriculture Organization of the United Nations: Rome, Italy, 2017.
2. FAO. Key Facts on Food Loss and Waste You Should Know! Save Food: Global Initiative on Food Loss and Waste Reduction. 2019. Available online: http://www.fao.org/save-food/resources/keyfindings/en/ (accessed on 12 March 2019).
3. WRAP. *Courtauld 2025 Signatory Data Report: 2015 and 2016*; Harris, B., Ed.; WRAP: Banbury, UK, 2017.
4. European Commission. *Directive 2008/98/EC of The European Parliament and of The Council of 19 November 2008 on Waste and Repealing Certain Directives*; Official Journal of the European Union: Aberdeen, UK, 2008.
5. House of Commons; Environment, Food and Rural Affairs Committee. *Food Waste in England, Eighth Report of Session 2016–2017*; Environment, Food and Rural Affairs Committee's Website: London, UK, 2017.
6. Ye, M.; Liu, J.; Ma, C.; Li, Y.-Y.; Zou, L.; Qian, G.; Xu, Z.P. Improving the stability and efficiency of anaerobic digestion of food waste using additives: A critical review. *J. Clean. Prod.* **2018**, *192*, 316–326. [CrossRef]
7. DEFRA. Statutory Guidance. Food and Drink Waste Hierarchy: Deal with Surplus and Waste. 2018. Available online: https://www.gov.uk/government/publications/food-and-drink-waste-hierarchy-deal-with-surplus-and-waste/food-and-drink-waste-hierarchy-deal-with-surplus-and-waste (accessed on 25 July 2021).
8. DEFRA and APHA. Animal By-Product Categories, Site Approval, Hygiene and Disposal. 2018. Available online: https://www.gov.uk/guidance/animal-by-product-categories-site-approval-hygiene-and-disposal (accessed on 25 July 2021).
9. Mertenat, A.; Diener, S.; Zurbrügg, C. Black Soldier Fly biowaste treatment–Assessment of global warming potential. *Waste Manag.* **2019**, *84*, 173–181. [CrossRef] [PubMed]
10. Salomone, R.; Saija, G.; Mondello, G.; Giannetto, A.; Fasulo, S.; Savastano, D. Environmental Impact of Food Waste Bioconversion by Insects: Application of life Cycle Assessment to Process Using *Hermetia illucens*. *J. Clean. Prod.* **2017**, *140*, 890–905. [CrossRef]
11. Lalander, C.; Nordberg, A.; Vinneras, B. A comparison in product-value potential in four treatment strategies for food waste and faeces, assessing composting, fly larvae composting and anaerobic digestion. *GCB Bioenergy* **2018**, *10*, 84–91. [CrossRef]

12. Caligiani, A.; Marseglia, A.; Leni, G.; Baldassarre, S.; Maistrello, L.; Dossena, A.; Sforza, S. Composition of black soldier fly prepupae and systematic approaches for extraction and fractionation of proteins, lipids and chitin. *Food Res. Int.* **2018**, *105*, 812–820. [CrossRef] [PubMed]
13. Sosa, D.A.T.; Fogliano, V. Potential of Insect-Derived Ingredients for Food Applications. In *Insect Physiology and Ecology*; Intech Open: London, UK, 2017; pp. 215–231.
14. Spranghers, T.; Ottoboni, M.; Klootwijk, C.; Ovyn, A.; Deboosere, S.; Meulenaer, B.D.; Michiels, J.; Eeckhout, M.; Clercq, P.D.; Smet, S.D. Nutritional composition of black soldier fly (*Hermetia illucens*) prepupae reared on different organic waste substrates. *J. Sci. Food Agric.* **2017**, *97*, 2594–2600. [CrossRef]
15. Wasko, A.; Bulak, P.; Polak-Berecka, M.; Nowak, K.; Polakowski, C.; Bieganowski, A. The first report of the physicochemical structure of chitin isolated from *Hermetia illucens*. *Int. J. Biol. Macromol.* **2016**, *92*, 316–320. [CrossRef]
16. Rumpold, B.A.; Schluter, O.K. Potential and challenges of insects as an innovative source for food and feed production. *Innov. Food Sci. Emerg. Technol.* **2013**, *17*, 1–11. [CrossRef]
17. Rumpold, B.A.; Schluter, O.K. Nutritional composition and safety aspects of edible insects. *Mol. Nutr. Food Res.* **2013**, *57*, 802–823. [CrossRef]
18. Makkar, H.P.S.; Tran, G.; Heuze, V.; Ankers, P. State-of-the-art on use of insects as animal feed. *Anim. Feed Sci. Technol.* **2014**, *197*, 1–33. [CrossRef]
19. Henry, M.; Gasco, L.; Piccolo, G.; Fountoulaki, E. Review on the use of insects in the diet of farmed fish: Past and future. *Anim. Feed Sci. Technol.* **2015**, *203*, 1–22. [CrossRef]
20. Callan, E.M. *Hermetia illucens* (L.) (Dipt., Stratiomyidae), a cosmopolitan American species long established in Australia and New Zealand. *Entomol. Mon. Mag.* **1974**, *109*, 232–234.
21. Kim, J.I. Newly recording two exotic insects species from Korea. *J. Kor. Biota.* **1997**, *2*, 223–225.
22. May, B.M. The occurrence in New Zealand and the life-history of the soldier fly *Hermetia illucens* (L.) (Diptera: Stratiomyidae). *New Zealand J. Sci.* **1961**, *4*, 55–65.
23. Üstüner, T.; Hasbenli, A.; Rozkosny, R. The first record of *Hermetia illucens* (Linnaeus, 1758) (Diptera, Stratiomyidae) from the Near East. *Stud. Dipterol.* **2003**, *10*, 181–185.
24. Spranghers, T.; Noyez, A.; Schildermans, K.; Clercq, P.D. Cold Hardiness of the Black Soldier Fly (Diptera: Stratiomyidae). *J. Econ. Entomol.* **2017**, *110*, 1501–1507. [CrossRef]
25. Wang, Y.-S.; Shelomi, M. Review of Black Soldier Fly (*Hermetia illucens*) as Animal Feed and Human Food. *Foods* **2017**, *6*, 91. [CrossRef]
26. Kim, W.; Bae, S.; Park, K.; Lee, S.; Choi, Y.; Han, S.; Koh, Y. Biochemical characterization of digestive enzymes in the black soldier fly, *Hermetia illucens* (Diptera: Stratiomyidae). *J. Asia-Pac. Entomol.* **2011**, *14*, 11–14. [CrossRef]
27. Cicková, H.; Newton, G.L.; Lacy, R.C.; Kozánek, M. The use of fly larvae for organic waste treatment. *Waste Manag.* **2015**, *35*, 68–80. [CrossRef]
28. Sheppard, D.C.; Tomberlin, J.K.; Joyce, J.A.; Kiser, B.C.; Sumner, S.M. Rearing Methods for the Black Soldier Fly (Diptera: Stratiomyidae). *J. Med Entomol.* **2002**, *39*, 695–698. [CrossRef] [PubMed]
29. European Commission. *Commission Regulation (EU) 2017/893 of 24 May 2017 amending Annexes I and IV to Regulation (EC) No 999/2001 of the European Parliament and of the Council and Annexes X, XIV and XV to Commission Regulation (EU) No 142/2011 As Regards the Provisions on Processed Animal Protein*; Official Journal of the European Union: Aberdeen, UK, 2017.
30. European Commission. *Regulation (Ec) No 1069/2009 of The European Parliament AND OF The Council Laying Down Health Rules as Regards Animal By-Products and Derived Products Not Intended for Human Consumption and Repealing Regulation (EC) No 1774/2002 (Animal By-Products Regulation)*; Official Journal of the European Union: Aberdeen, UK, 2009.
31. APHA. *Import of Processed Animal Protein (PAP) Derived from Farmed Insects Not for Human Consumption from Third Countries. Import Information Note (IIN) ABP/45*; The Animal and Plant Health Agency (APHA): Surrey, UK, 2020.
32. Liu, X.; Chen, X.; Wang, H.; Yang, Q.; Rehman, K.u.; Li, W.; Cai, M.; Li, Q.; Mazza, L.; Zhang, J.; et al. Dynamic changes of nutrient composition throughout the entire life cycle of black soldier fly. *PLoS ONE* **2017**, *12*, e0182601. [CrossRef]
33. Wang, C.; Qian, L.; Wang, W.; Wang, T.; Deng, Z.; Yang, F.; Xiong, J.; Feng, W. Exploring the potential of lipids from black soldier fly: New paradigm for biodiesel production (I). *Renew. Energy* **2017**, *111*, 749–756. [CrossRef]
34. Gortari, M.C.; Hours, R.A. Biotechnological processes for chitin recovery out of crustacean waste: A mini-review. *Electron. J. Biotechnol.* **2013**, *16*, 14.
35. Temple, W.D.; Radley, R.; Baker-French, J.; Richardson, F. *Use of Enterra Natural Fertilizer (Black Soldier Fly Larvae Digestate) As a Soil Amendment*; Enterra Feed Corporation: Maple Ridge, BC, Canada, 2013; Available online: https://easyasorganics.com.au/wp-content/uploads/2021/02/I-172_Frass_Research_Final-Report.pdf (accessed on 25 July 2021).
36. Win, S.S.; Ebner, J.H.; Brownell, S.A.; Pagano, S.S.; Cruz-Dilone, P.; Trabold, T.A. Anaerobic digestion of black solider fly larvae (BSFL) biomass as part of an integrated biorefinery. *Renew. Energy* **2018**, *127*, 705–712. [CrossRef]
37. Braguglia, C.M.; Gallipoli, A.; Gianico, A.; Pagliaccia, P. Anaerobic bioconversion of food waste into energy: A critical review. *Bioresour. Technol.* **2018**, *248*, 37–56. [CrossRef]
38. Zhang, C.; Su, H.; Baeyens, J.; Tan, T. Reviewing the anaerobic digestion of food waste for biogas production. *Renew. Sustain. Energy Rev.* **2014**, *38*, 383–392. [CrossRef]
39. WRAP. Why Take Action: Legal/Policy Case. Available online: http://www.wrap.org.uk/content/why-take-action-legalpolicy-case (accessed on 25 July 2021).

40. Bava, L.; Jucker, C.; Gislon, G.; Lupi, D.; Savoldelli, S.; Zucali, M.; Colombini, S. Rearing of *Hermetia illucens* on Different Organic By-Products: Influence on Growth, Waste Reduction, and Environmental Impact. *Animals* **2019**, *9*, 289. [CrossRef] [PubMed]
41. Barragan-Fonseca, K.B.; Dicke, M.; van Loon, J.J.A. Influence of larval density and dietary nutrient concentration on performance, body protein, and fat contents of black soldier fly larvae (*Hermetia illucens*). *Entomol. Exp. Appl.* **2018**, *166*, 761–770. [CrossRef]
42. Oonincx, D.G.A.B.; Broekhoven, S.v.; van Huis, A.; van Loon, J.J.A. Feed Conversion, Survival and Development, and Composition of Four Insect Species on Diets Composed of Food By-Products. *PLoS ONE* **2015**, *10*, e0144601. [CrossRef] [PubMed]
43. Danieli, P.P.; Lussiana, C.; Gasco, L.; Amici, A.; Ronchi, B. The Effects of Diet Formulation on the Yield, Proximate Composition, and Fatty Acid Profile of the Black Soldier Fly (*Hermetia illucens* L.) Prepupae Intended for Animal Feed. *Animals* **2019**, *9*, 178. [CrossRef] [PubMed]
44. Henderson, R.J. Fatty acid metabolism in freshwater fish with particular reference to polyunsaturated fatty acids. *Arch Tierernahr.* **1996**, *49*, 5–22. [CrossRef]
45. Tacon, A.G.J. *The Nutrition and Feeding of Farmed Fish and Shrimp–A Training Manual, 1. The Essential Nutrients*; Food and Agriculture Organization of The United Nations: Brasilia, Brazil, 1987.
46. Dick, H.; Scholes, P. *Comparing the Costs of Alternative Waste Treatment Options*; WRAP: Banbury, UK, 2019.
47. Kerby, C.; Vriesekoop, F. An Overview of the Utilisation of Brewery By-Products as Generated by British Craft Breweries. *Beverages* **2017**, *3*, 24. [CrossRef]
48. Thiex, N.; Novotny, L.; Crawford, A. Determination of ash in animal feed: AOAC official method 942.05 revisited. *J. AOAC Int.* **2012**, *95*, 1392–1397. [CrossRef]
49. NRC. *Nutrient Requirements of Fish and Shrimp*; Animal Nutrition Series; The National Academies Press: Washington, DC, USA, 2011.
50. Korkmaz, A.S.; Cakirogullari, G.C. Effects of partial replacement of fish meal by dried baker's yeast (*Saccharomyces cerevisiae*) on growth performance, feed utilization and digestibility in koi carp (*Cyprinus carpio* L., 1758) fingerlings. *J. Anim. Vet. Adv.* **2011**, *10*, 346–351. [CrossRef]
51. Jucker, C.; Lupi, D.; Moore, C.D.; Leonardi, M.G.; Savoldelli, S. Nutrient Recapture from Insect Farm Waste:Bioconversion with *Hermetia illucens* (L.) (Diptera: Stratiomyidae). *Sustainability* **2020**, *12*, 362. [CrossRef]
52. Kissel, D.E.; Risse, M.; Sonon, L.; Harris, G. *Calculating the Fertilizer Value of Broiler Litter*; The U.S. Department of Agriculture and Counties of the State Cooperating, The University of Georgia and Fort Valley State University: Fort Valley, GA, USA, 2015.
53. Penhallegon, R. *Nitrogen-Phosphorus-Potassium Values Oforganic Fertilizers*; Oregon State University Extension Service: Lane County Office: Eugene, OR, USA, 2003.
54. AHDB. *Monthly Average Prices for October 2019 in GB Fertiliser Price Market Update A.a.H.D.B. (AHDB), Editor*; Agriculture and Horticulture Development Board (AHDB): Warwickshire, CV8 2TL, UK, 2019. Available online: https://ahdb.org.uk/GB-fertiliser-prices (accessed on 25 July 2021).
55. Tschirner, M.; Simon, A. Influence of different growing substrates and processing on the nutrient composition of black soldier fly larvae destined for animal feed. *J. Insects Food Feed* **2015**, *1*, 249–259. [CrossRef]
56. Wilson, R.P. Protein and amino acids. In *Fish Nutrition*, 3rd ed.; Halver, J.E., Hardy, R.W., Eds.; Elsevier Science: San Diego, CA, USA, 2002; pp. 144–179.
57. Miles, R.D.; Chapman, F.A. *The Benefits of Fish Meal in Aquaculture Diets*; Place, N.T., Ed.; The Fisheries and Aquatic Sciences Department, U.S. Department of Agriculture, UF/IFAS Extension, University of Florida: Gainesville, FL, USA, 2018.
58. INRA-CIRAD-AFZ. Fish Meal, Protein 65%. INRA-CIRAD-AFZ Feed Tables 2020. Available online: https://www.feedtables.com/content/fish-meal-protein-65 (accessed on 10 April 2020).
59. Bhat, T.H.; Balkhi, M.H.; Banday, T. Use of soybean products in aquafeeds: A review. In *Soybean in Aquaculture*; University of Agricultural Sciences and Technology of Kashmir: Srinagar, India, 2012.
60. Hasan, A.; Tan, J. *The Current State of Plant-Based Proteins in Aquaculture Feed*; Biomin: Getzersdorf, Austria, 2020. Available online: https://www.biomin.net/science-hub/the-current-state-of-plant-based-proteins-in-aquaculture-feed/ (accessed on 25 July 2021).
61. Cremer, M.C. *Use and Future Prospects for Use of Soy Products in Aquaculture*; USSEC-U.S. Soybean Export Council: Singapore, 2019. Available online: https://ussec.org/resources/future-prospects-soy-products-aquaculture/ (accessed on 25 July 2021).
62. Heuzé, V.; Tran, G.; Kaushik, S. Soybean Meal. Feedipedia 4 March. Available online: http://www.feedipedia.org/node/674 (accessed on 10 April 2020).
63. Dayrit, F.M. The Properties of Lauric Acid and Their Significance in Coconut Oil. *J. Am. Oil Chem. Soc.* **2015**, *92*, 1–15. [CrossRef]
64. Kabara, J.J.; Swieczkowski, D.M.; Conley, A.J.; Truant, J.P. Fatty Acids and Derivatives as Antimicrobial Agents. *Antimicrob. Agents Chemother.* **1972**, *2*, 23–28. [CrossRef] [PubMed]
65. Walters, D.R.; Walker, R.L.; Walker, K.C. Lauric Acid Exhibits Antifungal Activity Against Plant Pathogenic Fungi. *J. Phytopathol.* **2003**, *151*, 228–230. [CrossRef]
66. Zeiger, K.; Popp, J.; Becker, A.; Hankel, J.; Visscher, C.; Klein, G.; Meemken, D. Lauric acid as feed additive–An approach to reducing *Campylobacter* spp. in broiler meat. *PLoS ONE* **2017**, *12*, e0175693. [CrossRef]
67. Hayes, D.G. Fatty Acids Based Surfactants and Their Uses. In *Fatty Acids, Chemistry, Synthesis, and Applications*; Ahmad, M.U., Ed.; Academic Press and AOCS Press: Cambridge, MA, USA, 2017; pp. 355–384.
68. McNaught, A.D.; Wilkinson, A. *Compendium of Chemical Terminology, (the "Gold Book")*, 2nd ed.; Chalk., S.J., Ed.; Blackwell Scientific Publications: Oxford, UK, 1997.

69. Kelm, G.R.; Wickett, R.R. The Role of Fatty Acids in Cosmetic Technology. In *Fatty Acids, Chemistry, Synthesis, and Applications*; Ahmad, M.U., Ed.; Academic Press and AOCS Press: Cambridge, MA, USA, 2017; pp. 385–404.
70. Huang, C.B.; George, B.; Ebersole, J.L. Antimicrobial activity of n-6, n-7 and n-9 fatty acids and their esters for oral microorganisms. *Arch. Oral Biol.* **2010**, *55*, 555–560. [CrossRef] [PubMed]
71. Ecothrive. Ecothrive Charge, Soil Conditioner and Biostimulant. 2020. Available online: https://www.ecothrive.co.uk/catalogue/charge_6/ (accessed on 25 July 2021).
72. Schmitt, E.; de Vries, W. Potential benefits of using *Hermetia illucens* frass as a soil amendment on food production and for environmental impact reduction. *Curr. Opin. Green Sustain. Chem.* **2020**, *25*, 100335.

MDPI

Article

Evaluation of Food Waste at a Portuguese Geriatric Institution

Margarida Liz Martins [1,2,3,4], Ana Sofia Henriques [5] and Ada Rocha [3,5,6,*]

1 Biology and Environment Department, University of Trás-Os-Montes e Alto Douro, 5000-801 Vila Real, Portugal; mliz@utad.pt
2 CBQF—Centro de Biotecnologia e Química Fina—Laboratório Associado, Universidade Católica Portuguesa, 4169-005 Porto, Portugal
3 GreenUPorto—Sustainable Agrifood Production Research Centre, 4485-646 Vairão, Portugal
4 CITAB—Centre for the Research and Technology of Agro-Environmental and Biological Sciences, 5000-801 Vila Real, Portugal
5 Faculty of Food Science and Nutrition, University of Porto, 4150-180 Porto, Portugal; ashenriques24@gmail.com
6 LAQV-Requimte, University of Porto, 4150-180 Porto, Portugal
* Correspondence: adarocha@fcna.up.pt; Tel.: +351-225-074-320

Abstract: Care institutions attending to older adults are responsible for their food supply, which influences their health and quality of life. Food waste at care institutions has been reported to be a matter of great concern, that requires regular monitoring. In this study, we aim to quantify food waste in the food service of an elderly institution, both as leftovers and plate waste. Data collection was performed over 15 consecutive days, at lunch and dinner served to older adults. The aggregate weighing of food was performed before and after distribution, as well as after consumption. Leftovers and plate waste were calculated by the differences in weight. During the study period, 2987 meals were evaluated, corresponding to 1830 kg of food produced, of which only 67% was consumed. For each meal, approximately 610 g of food was produced per older adult, and only about 410 g were consumed, corresponding to 150 g of leftovers and 50 g of plate waste. Food waste represented 36.1% of meals served, composed of 24.1% leftovers and 12.0% plate waste. The wasted meals would be enough to feed 1486 older adults and would correspond to annual losses of approximately €107,112. Leftovers and plate waste were above the limits of acceptability (below 6% and 10%, respectively), indicating excessive food waste. High values of leftovers are related to the food service system and staff, pointing to the need for improvements during the planning and processing of meals. On the other hand, high plate waste values are associated with consumers, indicating the low adequacy of the menu regarding to older adults' habits and preferences.

Keywords: elderly institution; food waste; leftovers; older adults; plate waste

Citation: Liz Martins, M.; Henriques, A.S.; Rocha, A. Evaluation of Food Waste at a Portuguese Geriatric Institution. *Sustainability* **2021**, *13*, 2452. https://doi.org/10.3390/su13052452

Academic Editor: Riccardo Testa

Received: 28 January 2021
Accepted: 22 February 2021
Published: 24 February 2021

1. Introduction

The increasing proportion of older adults over the age of 65 is a reality that has been emerging over the last decades. It is expected that, in the next 50 years, the proportion of the elderly in the population will grow significantly; by one estimate, in 2060, there will be three elderly people for each young person [1]. According to the latest official national data, in Portugal, about 19% of the population is over the age of 65 and about 4.2% is institutionalized in care support institutions [2].

Care institutions attending to older adults are responsible for their food supply, which influences older adults' nutritional status and consequently their health and quality of life. Food and meals are a central issue in the life of old people and play a significant role in elderly institutions [3–5].

The satisfaction of older adults with meals is one of the main factors contribution to reducing malnutrition and optimizing institutional service [6–8]. High levels of plate waste contribute to malnutrition-related complications in institutionalized older adults [7–9].

Inadequate food intake can result from multiple causes associated with the aging process, such as motor and cognitive constraints, losses of gastrointestinal function (dentition, swallowing, digestion, etc.), loss of appetite and decreased sensory abilities, among other physiological processes which may result from pathological situations [9,10]. Several other causes have been described related to institutionalization itself, the loss of family or a spouse, isolation and social marginalization, reduction of purchasing power [10–12], entrenched eating habits, preferences [3,12,13] and the lack of quality of the food service [3,14,15].

Food waste assessment allows dietary intake to be estimated and intervention needs in the food service to be identified. An efficient food service should deliver both good value meals for older adults and high-quality nutrition with minimal waste. The main goal is to promote the acceptability of meals, contributing to an adequate nutritional intake while at the same time minimizing food waste [15]. Waste at a food service is usually associated with inefficiencies in the food production system, and food waste quantification may be used as an indicator of service quality [16].

Food waste is a matter of great concern for governments and institutions as it has financial, environmental, ethical, political and social impacts [17–21]. In food service, it may occur at all stages of the food production system, including storage, meal preparation, cooking and distribution [17,21,22]. Monitoring plate waste (corresponding to food that is served but not consumed) allows for the evaluation of the adequacy of portions in relation to consumers' needs as well as menu acceptability [13,23]. According to several authors, food waste below 10% is considered acceptable [24,25]. Plate waste in hospitals and geriatric institutions has been found to be higher than in other food service settings [7,10,16,20,22,23].

The objective of this study is to evaluate the food waste at an elderly institution determined by leftovers and plate waste and its economic impact, with the aim of improving food service efficiency, as well as contributing to the promotion of older adults' nutritional status.

2. Materials and Methods

This study was developed in a private long-stay geriatric institution with a capacity of 120 beds. The food service works in a cook-and-serve system with a staff of 10 persons.

The typical meal included a soup, a main dish (meat or fish, carbohydrates and vegetable sources) or a diet dish and dessert (fruit or sweet). All meals served at lunch and dinner to 103 older adults during the study period were included. Lunch was served from 12.00 to 13.00 and dinner from 18.00 to 19.00.

Food waste was evaluated by leftovers and plate waste determination, using the aggregated weighing method by weighing all meal components together.

Leftovers correspond to the food that is prepared and cooked but not served to consumers, and it is usually associated with inefficiencies of meal planning. To evaluate leftovers, all containers were weighed empty and after the plating of meals. At the end of the meal, containers were collected and weighed. The amount of leftovers was determined by the weight difference between initial and final values. The percentage of leftovers was calculated by the ratio of leftovers (g) to the food produced (g) [25,26]. According to NHSE Hospitality, values of 6% are considered acceptable [25].

Plate waste refers to food left on the plate by consumers that is discarded. To evaluate plate waste, plates were collected after the meal, non-edible items were removed and food waste was separated into individual garbage bags for the soup, dish, diet dish and dessert. Food served (g) was determined by the difference between the total food produced (g) and the amount of leftovers (g). Food consumed (g) was calculated by the difference between the amount of food served (g) and plate waste (g). The amounts of food served, consumed and wasted per capita were obtained by the ratio between these values and the number of meals served. The percentage of plate waste was calculated by the ratio of food discarded to the food served to older adults [26]. Different guidelines state that values below 10% are considered acceptable [24,25].

The hypothetical number of older adults that could be fed with food that was not wasted was obtained by the ratio between the amount of leftovers (g) or plate waste (g), respectively, and the individual portion.

Meal cost was determined by taking into account the expenses involved with raw materials, labor and resources for meal planning and preparation. The cost associated with plate waste was calculated by the number of older adults that could be fed with plate waste multiplied by the meal cost.

All weighing was performed on a digital scale accurate to the nearest gram (SECA_ model 851, Germany).

Statistical software package IBM SPSS Statistics, version 22.0 and Excel Microsoft Office Program Professional Plus 2010 were used for data analyses. Mean, standard deviations (SD) and maximum and minimum values were used to provide descriptive analysis. The Mann–Whitney test was used to compare plate waste and leftovers according to the protein source (meat or fish) and meal (lunch or dinner), and the Spearman correlation was used to correlate food produced, leftovers and plate waste. The confidence level was set at 95%.

3. Results

3.1. Food Produced, Served, Consumed and Wasted

During the study period, 2987 meals were evaluated, corresponding to about 1830 kg of food produced, of which only 67% was consumed. Approximately 608 kg was wasted, corresponding to about 40.5 kg of food wasted per day. For each meal, approximately 610 g of food was produced for each older adult, while only about 410 g was consumed, corresponding to 150 g of leftovers and 50 g of plate waste.

Food waste represented 36.1% of meals served during the study period, corresponding to 24.1% of leftovers and 12.0% of plate waste. The food produced, served, consumed and wasted according to the meal component is presented in Table 1.

Table 1. Food produced, served, consumed and wasted according to the meal component (n = 2987).

		Food Produced (kg)	Food Served (kg)	Food Consumed (kg)	Leftovers (kg)	Plate Waste (kg)	Leftovers (%)	Plate Waste (%)
Soup (n = 29)	**Mean**	37.3	28.6	26.5	8.6	2.1	23.1	7.4
	SD	3.2	2.6	2.4	1.7	0.4	4.0	1.4
	Maximum	46.7	36.4	33.5	10.5	2.9	27.8	9.3
	Minimum	34.7	36.4	24.6	4.2	1.3	12.2	4.7
Main Dish (n = 29)	**Mean**	24.2	18.7	15.4	5.5	3.3	19.5	17.6
	SD	10.3	5.3	4.6	6.2	1.3	11.3	5.0
	Maximum	65.6	35.0	29.8	30.6	5.8	48.8	27.5
	Minimum	8.8	7.4	6.7	0.7	0.7	3.3	9.2
Diet Dish (n = 20)	**Mean**	3.6	2.4	2.1	1.2	0.4	30.5	14.1
	SD	1.8	1.08	0.9	0.9	0.3	11.2	8.1
	Maximum	7.5	4.5	3.7	3.0	1.0	57.1	30.9
	Minimum	1.56	0.67	0.56	0.31	0.0	14.7	0.0

[1] SD—Standard Deviation.

Soup leftovers ranged between 12.2% and 27.8%, with an average value of 23.1%, and plate waste ranged between 4.7% and 9.3%, with an average value of 7.4% (Table 1).

An average value of 19.5% of leftovers was found for the main dish, with the highest value for a fish dish (48.8%) and the lowest for a meat dish (3.2%). The main dish plate waste ranged between 9.2% and 27.5%, both corresponding to fish dishes, presenting an average value of 17.6% (Table 1).

Regarding the diet dish, 16 different types of dishes were included, and leftovers ranged between 14.7% and 57.1%, with an average value of 30.5%. About 8 kg was wasted after meal consumption, corresponding to 370 g per older adult. Plate waste ranged between 0% and 30.9%, with an average value of 14.1% (Table 1).

Leftovers were above the limit of acceptability for all different meal components (6%). Soup waste values were the only values below the limit of acceptability (10%) (Figure 1).

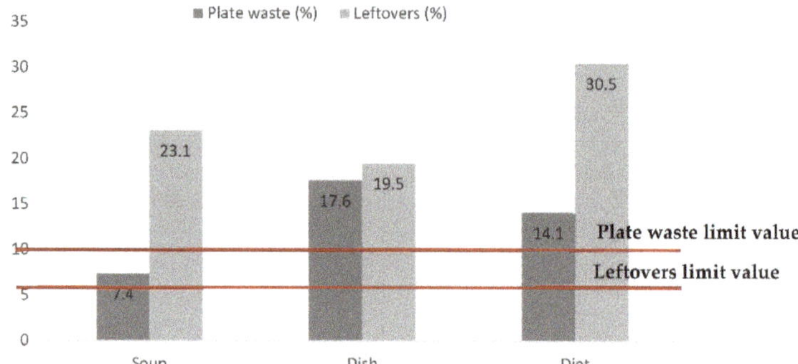

Figure 1. Comparison between plate waste and leftovers by meal component with relation to the limit values of acceptability (n = 2987) [25].

Plate waste was higher for fish dishes (19.0%) than meat dishes (16.6%) (Table 2) for both the main dish and diet dish ($p < 0.05$).

Table 2. Leftovers and plate waste according to the protein source of the main dish (n = 2987).

	Food Produced (kg)	Leftovers (kg)	Leftovers (%)	Plate Waste (kg)	Plate Waste (%)
Meat dishes (n = 24)	338.1	79.9	23.6	43.0	16.6
Fish dishes (n = 21)	330.5	81.1	24.5	47.5	19.0
p-value *	0.75		0.16		0.03

* *p*-value according to the Mann–Whitney test at a confidence level of 95%.

Comparing the food produced and wasted during lunch and dinner, leftovers were higher at dinner (25.9%) than at lunch (22.8%). The same tendency was observed for plate waste, which was higher at dinner (13.4%) than t lunch (10.9%) ($p < 0.05$). Considering the food produced and served per older adult, a higher quantity was observed for lunch than for dinner. Each older adult wasted about 50 g at lunch and 60 g at dinner (Figure 2).

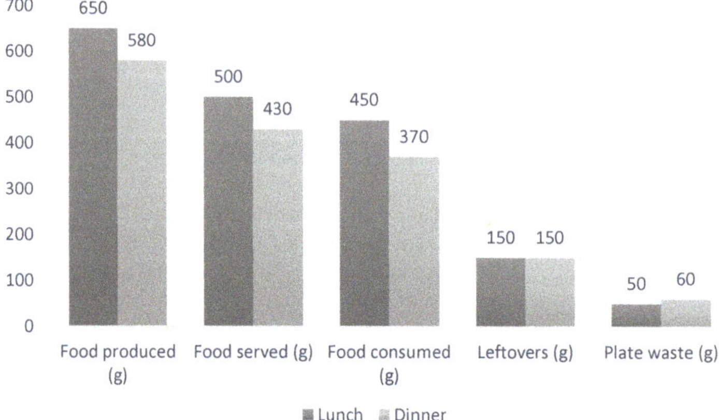

Figure 2. Comparison between food produced, served, consumed, plate waste and leftovers by meal (n = 2987) [25].

3.2. Food Waste—Social and Economic Impact

Considering the mean portion consumed by older adults, the amount of wasted meals would be enough to feed 1486 older adults (Table 3). Taking into account the meal cost, it is possible to estimate that the financial losses resulting from food waste corresponded to approximately €4458 during the study period and resulted in annual losses of approximately €107112 (Table 3).

Table 3. Social and economic impact of food waste.

Food Waste	Total Food Waste (kg)	Food Waste Per Capita (kg)	N° of Older Adults That Could Be Fed	Cost (Euros)		
				Study Period	During One Month	During One Year
Plate waste	**166.1**	**0.05**	406	1219.3	2438.6	29,262.6
Leftovers	441.9	0.15	1081	3243.7	6487.5	77,849.8
Total	608.0	0.20	1487	4463.0	8926.0	107,112.4

4. Discussion

Food waste was determined to be 36% in this study, corresponding to approximately 150 g per capita per meal—far above the limits of acceptability according to previous studies developed in different settings [9,15,16,18,27]. According to other authors, since there is no standard limit for food waste, each food service must monitor food waste as a routine to develop a target and define the limits according to specific characteristics and consumers [16,18,25]. The values found in this study are considered unacceptable since the target population includes long-term residents and the number of older adults is constant; this should allow more accurate estimations of quantities to be produced.

Other authors in an Italian hospital setting reported a food waste value of 41.6% [15].

In a Portuguese case study developed at a hospital, it was observed that plate waste represented 35% of the food served. The authors showed that 0.5% of the Portuguese National Health budget is squandered as food waste considering economic losses [18].

In our study, plate waste was higher at dinner. Opposite results were found in a Spanish study, showing a plate waste of 37.7 ± 29.9% for lunch and 30.4 ± 23.6% for dinner [9].

Williams and Walton reviewed 32 studies in health settings and showed that plate waste values ranged between 9.1% and 42.9% [7].

The amount of food waste may be affected by several factors, such as the poor diversity of the menu, the inadequacy of menus to the food and cultural habits of older adults, inadequate per capita portions and poor appearance of meals [6,10,12,15,27]. Considering that meal portions are determined to satisfy the nutritional needs of older adults, high continuous values of plate waste may compromise their nutritional intake and contribute to undernutrition, reported as an important concern in this age group [4,9,15].

Monitoring food portions that are effectively plated and evaluating consumer's intake and satisfaction may be useful to identify the specific causes of food waste. This may allow the implementation of measures to avoid waste and optimize foodservice efficiency [15,16]. The use of standard recipes for every meal preparation would enable menu standardization, contribute to meal planning and decrease food waste. Food handlers' awareness about the importance of waste control is essential to engage the whole food service team on this effort [15,16].

In this study the average amount of plate waste was 12%—slightly above the 10% recommended. This value was probably related to the target group. Older adults usually have constraints that affect food behavior, such as physiological changes associated with aging. Loss of sensorial capacities, difficulties in chewing and other physical limitations usually compromise food intake, contributing to plate waste [6,9,10,15]. Additionally, psychological factors such as widowhood, leaving home, depression and loss of cognitive capacity may also affect food intake. Changes in the food offer and the modification of routines also affect food intake and the satisfaction of institutionalized older adults. Inadequate portioning, the poor appearance of meals, diet inadequacy with regards to chewing limitations and the absence of assistance during meals are usually associated with higher values of food waste by older adults [6,8–10,15].

According to Sanga, in a hospital setting, lack of appetite was mentioned by 50% of consumers as the main factor affecting the acceptance of meals. The causes for poor appetite are related to aging, medication, poor dentition and loss of sensorial capacity [28]. Additionally, food waste was also related to consumers' lack of awareness of the environmental/social/economic costs of food waste.

Only soup presented a food waste value in the range of acceptability, pointing to a higher satisfaction of older adults with this specific food preparation.

The high food waste values found for diet dishes may be related to consumers rather than to preparation characteristics. This type of meal is mainly served to consumers with diet restrictions, which is frequently associated with a lack of appetite. On the other hand, this kind of preparation is usually less flavored and has low amounts of salt. Satisfaction with food is highly related to sensorial meal characteristics, such as taste, flavor, texture, temperature and smell [6,8]. Considering the poor diversity of diet preparations and their poor sensory characteristics, the higher waste values were expected.

Fish-based dishes presented higher leftovers and plate waste values. Constraints associated to older adults' loss of hand mobility may contribute to the higher plate waste; for example, difficulties cutting and removing bones from fish.

Plate waste can be evaluated by weighing food, a visual estimation of the amount of food remaining on the plate or by a consumer report after the meal. The weighing method is the most accurate method to evaluate food waste, minimizing the bias associated with the observer's subjectivity and to memory and social desires that frequently occur in visual estimation and self-reporting, respectively [26]. Additionally, in these specific target group—older adults with some cognitive disabilities—methodologies that do not interfere with individuals should be preferred.

Taking into account the results from this study, it is important to develop strategies to reduce food waste. Combined efforts, consisting of engaging employees involved in meal preparation and distribution, as well as increasing consumers' awareness about food waste, are required. Additionally, it is important for each institution to monitor their own food waste and identify effective strategies to improve the nutritional intake of their vulnerable users [7].

It is also important to define limits for food waste in different settings in order to establish institutional goals for food handlers that should be evaluated and monitored continuously. This goal will help food services to meet the sustainable development goals: "By 2030, halving per capita global food waste at the retail and consumer levels and reducing food losses along production and supply chains, including post-harvest losses" [29].

In order to promote food acceptability by consumers, it is recommended to standardize processes to ensure the quality of meals. Satisfaction surveys may also be useful to evaluate older adults' satisfaction. Consumer participation in menu planning may also be useful to promote menu adequacy.

5. Conclusions

The value of food waste was 36% in this study, corresponding to approximately 150 g per capita per meal. This amount would be enough to feed 1486 older adults and corresponds to an annual loss of approximately €107,112.

Leftovers and plate waste were above the limits of acceptability (below 6% and 10%, respectively), indicating excessive food waste. High values of leftovers are related to the food service system and staff, pointing to the need for improvements during planning and processing of meals. On the other hand, high plate waste values are associated to consumers, indicating low menu adequacy to older adults' habits and preferences. Control of food waste will deliver significant benefits, including a decrease in the amount of organic residues produced, increases of profits and the improvement of the satisfaction and nutritional status of older adults.

Author Contributions: Methodology, formal analysis, data curation and writing—original draft preparation, M.L.M.; data curation and investigation, A.S.H.; conceptualization, writing—review and editing and supervision, A.R. All authors have read and agreed to the published version of the manuscript.

Funding: This research received no external funding.

Institutional Review Board Statement: Not applicable.

Informed Consent Statement: Not applicable.

Acknowledgments: This research was supported by National Funds by FCT—Portuguese Foundation for Science and Technology—within the scope of UIDB/50016/2020, UIDB/05748/2020, UIDP/05748/2020 and UIDB/04033/2020. The authors thank Phil Lonsdale for the English grammar and structure revision of the manuscript.

Conflicts of Interest: The authors declare no conflict of interest.

References

1. Instituto Nacional de Estatística (INE). Projeções de População Residente em Portugal 2008–2060. 2009. Available online: https://www.ine.pt/ (accessed on 14 March 2020).
2. Instituto Nacional de Estatística (INE). Estatísticas de Portugal. Recenseamento da População e Habitação—Census 2011. 2012. Available online: http://censos.ine.pt (accessed on 4 January 2021).
3. Hansen, K.V. Food and meals in caring institutions—A small dive into research. *Int. J. Health Care Qual. Assur.* **2016**, *29*, 380–406. [CrossRef]
4. Sellier, C. Malnutrition in the elderly, screening and treatment. *Soins Gerontol.* **2018**, *23*, 12–17. [CrossRef] [PubMed]
5. Iuliano, S.; Olden, A.; Woods, J. Meeting the nutritional needs of elderly residents in aged-care: Are we doing enough? *J. Nutr. Health Aging* **2013**, *17*, 503–508. [CrossRef] [PubMed]
6. Lee, K.H.; Mo, J. The Factors Influencing Meal Satisfaction in Older Adults: A Systematic Review and Meta-analysis. *Asian Nurs. Res. (Korean Soc. Nurs. Sci.)* **2019**, *13*, 169–176. [CrossRef]
7. Williams, P.; Walton, K. Plate waste in hospitals and strategies for change. *Eur. J. Clin. Nutr. Met.* **2011**, *6*, 235–241. [CrossRef]
8. McAdams, B.; von Massow, M.; Gallant, M. Food Waste and Quality of Life in Elderly Populations Living in Retirement Living Communities. *J. Hous. Elder.* **2018**, *33*, 1–13. [CrossRef]
9. Simzari, K.; Vahabzadeh, D.; Nouri Saeidlou, S.; Khoshbin, S.; Bektas, Y. Food intake, plate waste and its association with malnutrition in hospitalized patients. *Nutr. Hosp.* **2017**, *34*, 1376–1381. [PubMed]
10. Keller, H.H.; Carrier, N.; Slaughter, S.E.; Lengyel, C.; Steele, C.M.; Duizer, L.; Morrison, J.; Brown, K.S.; Chaudhury, H.; Yoon, M.N.; et al. Prevalence and Determinants of Poor Food Intake of Residents Living in Long-Term Care. *J. Am. Med. Dir. Assoc.* **2017**, *18*, 941–947. [CrossRef]
11. Stroebele-Benschop, N.; Depa, J.; de Castro, J.M. Environmental Strategies to Promote Food Intake in Older Adults: A Narrative Review. *J. Nutr. Gerontol. Geriatr.* **2016**, *35*, 95–112. [CrossRef]
12. Whitelock, E.; Ensaff, H. On Your Own: Older Adults' Food Choice and Dietary Habits. *Nutrients* **2018**, *10*, 413. [CrossRef]
13. Oliveira, D.; Liz Martins, M.; Rocha, A. Food waste index as an indicator of menu adequacy and acceptability in a Portuguese mental health hospital. *Acta Portuguesa Nutrição* **2020**, *20*, 14–18. (In Portuguese)
14. Wilson, E.D.; Garcia, A.C. Environmentally friendly health care food services: A survey of beliefs, behaviours, and attitudes. *Can. J. Diet. Pract. Res.* **2011**, *72*, 117–122. [CrossRef]
15. Schiavone, S.; Pelullo, C.P.; Attena, F. Patient Evaluation of Food Waste in Three Hospitals in Southern Italy. *Int. J. Environ. Res. Public Health* **2019**, *16*, 4330. [CrossRef] [PubMed]
16. Ferreira, M.; Liz Martins, M.; Rocha, A. Food waste as an index of foodservice quality. *Br. Food J.* **2013**, *115*, 1628–1637. [CrossRef]
17. Betz, A.; Buchli, J.; Gobel, C.; Muller, C. Food waste in the Swiss food service industry—Magnitude and potential for reduction. *Waste Manag.* **2015**, *35*, 218–226. [CrossRef]
18. Dias-Ferreira, C.; Santos, T.; Oliveira, V. Hospital food waste and environmental and economic indicators—A Portuguese case study. *Waste Manag.* **2015**, *46*, 146–154. [CrossRef] [PubMed]
19. Food and Agriculture Organization of the United Nations. *The State of Food and Agriculture—Moving forward on Food Loss and Waste Reduction*; FAO: Rome, Italy, 2019.
20. Garcia-Herrero, L.; De Menna, F.; Vittuari, M. Food waste at school. The environmental and cost impact of a canteen meal. *Waste Manag.* **2019**, *100*, 249–258. [CrossRef]

21. Ishangulyyev, R.; Kim, S.; Lee, S.H. Understanding Food Loss and Waste-Why Are We Losing and Wasting Food? *Foods* **2019**, *8*, 297. [CrossRef] [PubMed]
22. Eriksson, M.; Persson Osowski, C.; Malefors, C.; Bjorkman, J.; Eriksson, E. Quantification of food waste in public catering—A case study from a Swedish municipality. *Waste Manag.* **2017**, *61*, 415–422. [CrossRef]
23. Liz Martins, M.; Rodrigues, S.S.; Cunha, L.M.; Rocha, A. School lunch nutritional adequacy: What is served, consumed and wasted. *Public Health Nutr.* **2020**, 1–9. [CrossRef] [PubMed]
24. Conselho Federal de Nutricionistas. *Resolução CFN N° 380/2005*; CFN: Brasília, Brazil, 2005.
25. NHSE Hospitality. *Managing Food Waste in the NHS*; Department of Health: Leeds, UK, 2005.
26. Carr, D.; Levins, J. Plate Waste Studies. Available online: http://www.nfsmi.org/Information/recipes4.pdf (accessed on 20 October 2020).
27. Liz Martins, M.; Rodrigues, S.S.P.; Cunha, L.M.; Rocha, A. Factors influencing food waste during lunch of fourth-grade school children. *Waste Manag.* **2020**, *113*, 439–446. [CrossRef] [PubMed]
28. Stanga, Z.; Zurfluh, Y.; Roselli, M.; Sterchi, A.B.; Tanner, B.; Knecht, G. Hospital food: A survey of patients' perceptions. *Clin. Nutr.* **2003**, *22*, 241–246. [CrossRef]
29. United Nations. *Resolution Adopted by the General Assembly 70/1. Transforming Our World: The 2030 Agenda for Sustainable Development*; UN: New York, NY, USA, 2015.

Article

Evaluation of Determinants of Food Waste in Family Households in the Greater Porto Area Based on Self-Reported Consumption Practices

Taíse Portugal [1], Susana Freitas [2], Luís Miguel Cunha [3,*] and Ada Margarida Correia Nunes Rocha [4,*]

1 Faculty of Nutrition and Food Sciences, University of Porto, 4150 180 Porto, Portugal; up201602964@fc.up.pt
2 Departmento de Educação, Comunicação e Marketing, Lipor, 4435-746 Baguim do Monte, Portugal; susana.freitas@lipor.pt
3 GreenUPorto, Department of Geosciences, Environment and Planning, Faculty of Sciences, University of Porto, 4485-646 Vila do Conde, Portugal
4 LAQV-REQUIMTE/GreenUPorto, Faculty of Nutrition and Food Sciences, University of Porto, 4150 180 Porto, Portugal
* Correspondence: lmcunha@fc.up.pt (L.M.C.); adarocha@fcna.up.pt (A.M.C.N.R.)

Received: 30 September 2020; Accepted: 20 October 2020; Published: 22 October 2020

Abstract: Despite food waste occurring along the entire food supply chain, a significant proportion occurs in domestic settings. Large quantities of domestic food waste have been attributable to consumer behaviors during buying, cooking, consumption, and disposal. The main objective of this research was to understand the major determinants of household food waste from families in the north of Portugal. A convenience sample was used, which was drawn from households in the Greater Porto Area. Data were collected through a self-reported questionnaire that included three groups of structured questions related to perceived behavior and attitudes towards food consumption, leftover usage, and food waste. Exploratory data analysis was used to identify underlying dimensions. No relationships were found between socio-demographic data and food waste, buying behavior, or destination/use of leftovers. The majority of the participants reported a high level of planning of their grocery shopping. Fruits and vegetables presented the highest frequency of consumption, followed by sources of carbohydrates and sources of proteins. The storage of cooked food from different food groups presented a single factor, grouping the majority of the individual food leftovers, going from fruits and vegetables to sources of carbohydrates and proteins. The reported levels of wastage of the different food products were grouped into three dimensions: waste of vegetables, waste of protein sources, and waste of sources of carbohydrates. Waste of precooked foods emerged as an independent item, and it was the individual item with the highest frequency. The families studied reported a positive attitude concerning buying, consumption, and wastage, revealing a particular awareness of food waste and its social and environmental impact.

Keywords: household food waste; planning routines; shopping routines; food practices

1. Introduction

The postmodern society is a consumer society where 'having' becomes more important than 'being' [1,2]. In both rich and poor countries, statistics indicate that waste increases together with the increase in consumption [3]. The modern society faces a social drama as a result of the dimension of food waste, which has severe impacts on the economy and environment, while millions of people are starving all over the world [4].

According to the Food and Agriculture Organization (FAO), approximately one-third of the food produced for human consumption in the world is lost, with about 1300 million tons being lost or wasted [5]. Focusing on the European Union (EU), this figure totals 76 kg per person per year, representing approximately 45% of the total food waste in the entire supply chain, excluding agricultural production. In view of this situation, the European Commission has established the target of reducing food waste by one-half by 2020 throughout the EU [6].

The European Parliament defined food waste as "all food, defined as: any substance or product, whether processed, partially processed or unprocessed, intended to be, or reasonably expected to be ingested by humans, that has become waste" [7].

The reasons for food waste are variable, including inefficient storage and transportation practices, adoption of very tight expiration dates, and promotions that encourage people to buy greater quantities [7,8].

Despite food waste occurring along the entire food supply chain, significant proportions occur in domestic settings [9]. Household food waste is largely uncontrolled despite numerous initiatives that have been implemented to reduce it. Large quantities of domestic food waste have been attributable to consumer behaviors during buying, cooking, consumption, and disposal [10].

In 2015, in a survey developed in Portugal by the consumer defense association *Defesa do Consumidor* (DECO) with 1725 consumers, high amounts of food waste were found. More than 50% of respondents reported throwing away food with expired dates. The main reasons referred to were related to the difficulty of understanding labels [11].

Food waste is a very widespread phenomenon that is also found in families who are generally aware of this problem and make resolutions to avoid this kind of behavior. In Portugal, limited information is available concerning household food waste.

The main objective of this research was to better understand the major determinants of household food waste from families living in eight municipalities in the north of Portugal, near Lipor, which is the Intermunicipal Waste Management Service. It was also intended to characterize the food waste in terms of reasons to waste, most frequently wasted foods, and buying patterns, as well as to identify the most common destinations of leftovers. This association manages about 500 thousand tons of urban residues produced by about 1 million inhabitants yearly [12].

2. Materials and Methods

2.1. Sample Description

A convenience sample was used, drawn from households in the area surrounding Lipor, corresponding to eight municipalities in the Greater Porto Area (Espinho, Gondomar, Maia, Matosinhos, Porto, Póvoa de Varzim, Valongo, Vila do Conde) that were registered in Lipor's database, including people registered and living nearby and that usually attended courses and activities of the organization. Data were exported for verifications during September 2017. Every month, new members were included. The selected database included 27,830 entries; nevertheless, only 10,484 had an email contact. From those, only 8785 were validated to receive the questionnaire due to incomplete questionnaire filling and limitations arising from the European Regulation on Data Protection. Later, the academic community of the University of Porto was also included to enlarge the recruitment base and compensate for the lack of participation.

2.2. Data Collection

Data were collected through a self-reported questionnaire. The questionnaire included 22 questions and was organized into three groups (see Appendix A for the attitudinal and behavioral questions).

The group "Shopping and consumption" included four questions to identify buying and consumption patterns of the household. This group included four questions, each with several items: one multiple choice, describing the place of purchase; and three frequency questions, with answers given using seven-point scales, including shopping behaviors (seven items evaluated on an anchored scale, with answers ranging from (1) Never to (7) On every purchase), an abbreviated food frequency questionnaire (13 items answered on a fully described scale, with answers ranging from (1) Never or less then 1 time/month to (7) More than once a day), and a food leftovers frequency questionnaire (identical set of 13 items evaluated on an anchored scale, with answers ranging from (1) Never to (7) On every consumption moment). The second group, "Food and food waste habits", included four questions to evaluate additional behaviors related to leftovers and food waste, including the usual destination of leftovers (five items evaluated on an anchored scale, with answers ranging from (1) Never to (7) On every consumption moment), a food waste frequency questionnaire (identical set of 13 items evaluated on an anchored scale, with answers ranging from (1) Never to (7) On every consumption moment), and the reasons for food waste (five items evaluated on an anchored scale, with answers ranging from (1) Totally disagree to (7) Totally agree). The third group included sociodemographic data.

The first draft of the questionnaire was developed and tested on a group of individuals of different ages and education levels who worked at Lipor or in the Faculty of Food Sciences and Nutrition, University of Porto, to evaluate the clarity and understanding of the questionnaire. The final questionnaire, entitled "Evaluation of consumption habits", was applied using the software "Google Forms".

In the first phase, the link for this questionnaire was sent by email to 8000 people between the 9th and the 31st of January 2018. Questionnaires were addressed to family members with greater responsibility in meal preparation. Due to the low number of answers, it was sent by dynamic email to the members of the University of Porto academia, and was available between the 10th April and the 1st of May 2018.

Free and informed consent was obtained from all participants after the study was explained, and both anonymity and confidentiality of the data were guaranteed. This study respected all the ethical principles and recommendations of the Helsinki Declaration and the World Health Organization (WHO).

2.3. Data Analysis

The theoretical distributions of the variables were analyzed using means, deviations, the histogram of distribution, and the Shapiro–Wilk test.

Exploratory factor analysis (EFA) was performed over each set of questions. In the EFA, valid items were extracted, considering only those with factorial loads above 0.5. The EFA was performed with varimax rotation. A Kaiser–Meyer–Olkin (KMO) value greater than 0.60 was used to verify the adequacy. Consistency of the built factors was measured using Cronbach's alpha.

Factors emerging from the food waste frequency data were adjusted using generalized linear models (GLM) to determine which variables were associated with the food waste factors, with multiple linear models being developed. The independent variables in each model were those variables that presented a Pearson correlation coefficient greater than 0.30 and differences in Student's t-test. Homoscedasticity and model fit were evaluated by residual analysis and the Chi-squared test.

The tests were conducted using the statistical program IBM SPSS Statistics v.24.

3. Results and Discussion

A total of 438 fully completed questionnaires were collected. Most respondents were women (Table 1), which was expected, since, in Portugal, the majority of household gatekeepers are women. A high predominance of respondents were aged between 30 and 49 years (38.4%), representing the most prevalent age group in Portugal according to the available data [13]. The majority of respondents have higher education (74.7%), which does not represent the reality of the Portuguese population according to National Statistics, which state that in this age group, only around 27% of the population has a higher degree level [14].

Table 1. Socio-demographic characterization of the respondents (n = 438).

Characteristics	n	%
Sex		
Female	336	76.7
Male	102	23.3
Age group		
18–29 years old	160	36.5
30–49 years old	168	38.4
≥ 50 years old	110	25.1
Level of education		
Without a higher degree	111	25.3
With a higher degree	327	74.7
Level of education of main family provider		
Without a higher degree	147	33.8
With a higher degree	288	66.2
Household per capita income		
<2018's MNS *	206	50.7
≥2018's MNS	200	49.3
Has car		
Yes	338	77.9
No	96	22.1
Type of home		
Own	327	75.3
Rented	93	24.6

* MNS: Minimum national salary for Portugal.

Nevertheless, 50.7% of the families have an average per capita income below the minimum national salary for Portugal of 580 EUR. The family income values were used based on the national minimum wage for the year 2018. Regarding the type of housing, 75.3% said they had their own housing and 77.9% of the respondents had a car (Table 1).

No relationships were found between socio-demographic data of our respondents and food waste, buying behavior, or destination of leftovers ($p > 0.01$). On the contrary, Baptista et al. [15] found a relationship between social characteristics and household waste in Portugal: Families with children waste more food; nevertheless, a tendency for reducing food waste with age was found [15]. Similarly, according to the research developed by Evans 2012 [16], having children is associated with larger amounts of waste in total.

In a study on the attitudes of Greek household members regarding food waste generation [17], a positive attitude towards food waste prevention was revealed, driven by financial necessity and a high degree of misconception of food labeling. It was found that the higher the educational level, the better the reported behavior towards food waste prevention, as a straightforward outcome of the correct comprehension of food labels. In addition, semi-urban and middle-income households make better use of the refrigerator and have better management of leftovers by cooking them, respectively.

In our study, 82% of households make purchases at large department stores (hypermarkets, supermarkets). Only 1% of respondents usually buy at organic stores, which is in line with data from various studies that have been reported, highlighting that food is mainly bought from major supermarket chains and that only a small percentage buys from smaller stores [10,18,19].

The characterization of buying behavior was divided into three dimensions, though only one of the dimensions presented adequate consistency ($\alpha > 0.5$), corresponding to "Planning". The majority of the participants reported a high level of planning of their grocery shopping (Table 2).

Table 2. Buying behavior of respondents (n = 438) evaluated over a seven-point anchored scale *.

1.*	**Behavior at buying occasions (explained variance, Cronbach's alpha (α)), KMO = 0.608**	**Mean ± SD**	**Loadings**
	Factor 1—Planning (var: 28.3%; α: 0.746)	**5.5 ± 1.1**	
	Buying with shopping list	5.2 ± 1.6	0.898
	Buying without shopping list (reversed scale)	6.0 ± 1.1	0.874
	Evaluation of contents of refrigerator and pantry before shopping	5.4 ± 1.5	0.623
	Factor 2—Promotions (var: 17.7%; α: 0.224)	**4.1 ± 1.2**	
	Preference for a specific product or brand	4.3 ± 1.5	0.749
	Purchases based on vouchers and promotions	3.9 ± 1.7	0.635
	Factor 3—Convenience (var: 17.6%; α: 0.389)	**2.3 ± 1.0**	
	Purchase of take-away meals	2.4 ± 1.3	0.862
	Get trendy foods and meals	2.1 ± 1.2	0.647
2.	**Food buying frequency (explained variance, Cronbach's alpha (α)), KMO = 0.713.**	**Mean ± SD**	**Loadings**
	Factor 1—Fruits and vegetables (var: 21.3%; α: 0.714;)	**5.4 ± 1.0**	
	Vegetables and salads	5.7 ± 1.4	0.787
	Pulses (dried and fresh)	4.6 ± 1.5	0.715
	Fruits	6.2 ± 1.2	0.691
	Soup	5.3 ± 1.6	0.661
	Factor 2—Carbohydrate suppliers (var: 19.6%; α: 0.657)	**5.1 ± 1.0**	
	Rice/pasta	5.3 ± 1.3	0.809
	Potatoes	4.4 ± 1.4	0.755
	Bread	5.7 ± 1.4	0.648
	Factor 3—Sources of proteins (var: 14.6%; α: 0.548)	**4.6 ± 0.9**	
	Eggs	3.9 ± 1.1	0.655
	Seafood (fish and shellfish)	4.1 ± 1.2	0.651
	Meat (poultry, pork, and beef)	4.8 ± 1.5	0.616
	Dairy products (milk, yogurt, and cheese)	5.6 ± 1.7	0.516
	Precooked food	**2.1 ± 1.0**	
3.	**Frequency of storage of cooked food (explained variance, Cronbach's alpha (α)), KMO = 0.915**	**Mean ± SD**	**Loadings**
	Factor 1—Keep leftovers (var: 55.9%; α: 0.915)	**5.5 ± 1.4**	
	Pasta/Rice	5.9 ± 1.5	0.807
	Bread	6.0 ± 1.6	0.793
	Pulses (dried and fresh)	5.5 ± 1.9	0.780
	Seafood (fish and shellfish)	5.2 ± 2.0	0.773
	Meat (poultry, pork, and beef)	5.7 ± 1.8	0.765
	Soup	6.2 ± 1.4	0.740
	Fruit	5.4 ± 1.2	0.740
	Potatoes	5.0 ± 2.1	0.739
	Vegetables and salads	5.2 ± 2.1	0.737
	Eggs	4.6 ± 2.4	0.675
	Dairy products (milk, yogurt, and cheese)	5.4 ± 2.2	0.663
	Precooked food	**3.3 ± 2.4**	
4.	**Behavior concerning leftover usage (explained variance, Cronbach's alpha (α)), KMO = 0.624**	**Mean ± SD**	**Loadings**
	Factor 1—Use of leftovers (var 54.6%; α: 0.576)	**4.9 ± 1.2**	
	To prepare new culinary items	4.7 ± 1.6	0.783
	Freeze for another occasion	4.4 ± 1.8	0.736
	Eat on the next day	5.7 ± 1.2	0.699
	Throw away	**2.2 ± 1.2**	
	Animal feed	**2.6 ± 2.0**	
5.	**Factors related with food waste (explained variance, Cronbach's alpha (α)), KMO = 0.792**	**Mean ± SD**	**Loadings**
	Factor 1—No purchasing planning (var: 41.7%; α: 0.792)	**2.3 ± 1.4**	
	Excess purchase due to promotion	2.1 ±1.6	0.882
	No shopping list	2.1 ± 1.6	0.875
	No control of stored items at home	2.7 ± 1.9	0.659
	Factor 2—No planning during cooking (var: 30. 9%; α: 0.546)	**3.5 ± 1.7**	
	Excess food is made for meal	3.7 ± 2.2	0.850
	Foods with short shelf life	3.2 ± 1.9	0.727

* Anchored scale: (1) with answers ranging from 1 (Never) to 7 (On every purchase); (2) a fully described scale, with answers ranging from 1 (Never or less than 1 time/month) to 7 (More than once a day); (3 and 4) with answers ranging from 1 (Never) to 7 (On every consumption moment); (5) with answers ranging from 1 (Totally disagree) to 7 (Totally agree).

Careful planning of grocery shopping was reported by several others as an effective strategy to prevent over-purchasing and, ultimately, food waste [10,18,20]. Planning includes using a shopping list, performing meal planning, and checking the refrigerator and store room before shopping.

Consumed foods were grouped into three types: fruits and vegetables, sources of carbohydrates, and sources of proteins, the first presenting the highest frequency of consumption, followed by sources of carbohydrates and sources of proteins. Additionally, the participants reported low frequency of consumption of precooked food.

The storage of cooked food presented a single factor, grouping the majority of the individual food leftovers going from fruits and vegetables to sources of carbohydrates and proteins.

In general, participants presented a high level of leftover storage with the exception of precooked food. The observed behavior of storing leftovers is in agreement with data published by Schanes et al. in 2018, who reported a stronger motivation to reduce food waste, pointing to financial concerns rather than environmental or social concerns [21]. On the other hand, the behavior regarding precooked food may indicate a higher level of the confidence that consumers have in home-cooked items compared to out-of-home prepared food. Households frequently have difficulties in assessing the durability of leftovers and tend to be concerned about their safety. Concerns about foodborne illnesses coupled with the desire to eat fresh foods are decisive reasons for wasting foods [22–24].

As already reported by others, health concerns are usually associated with the increase of the amount of waste of highly perishable food, such as meat, fish, and dairy products, due to the knowledge of the increased risks and consequences of consumption of such products if spoiled [25].

Generally, there was a high frequency of use of leftovers, particularly to be eaten on the following day, in line with the high frequency of use of meals taken from home, mainly for lunchtime meals (Direção Geral de Saúde, 2020) [26], with very low frequency of disposal of leftovers and use for pet feed. In this study, pet ownership was not evaluated, so it was not possible to correlate these data.

On the contrary, some authors reported that eating leftovers is not well accepted due to an aversion to reheating leftovers, as they are considered less nutritious and less fresh, and also because it was found to be boring to eat the same food twice [27,28].

The main perceived reasons for food waste were divided into two main factors, related to the lack of planning during purchase and the lack of planning during cooking, both presenting a low frequency—particularly the lack of purchase planning—which is in accordance with the buying behavior data. Eighty percent of the respondents justify the occurrence of leftovers with a consumption lower than expected, while 48% reported cooking higher quantities than necessary, with 36% of the participants indicating both reasons. It is noteworthy that less than 20% associated leftovers with the short shelf life of some items and even less with the low sensory appeal of the meals (Table 3).

Table 3. Most frequent reasons for leftovers at households.

Most Frequent Reasons	*n*	%
Consumption lower than expected *	350	79.9
Over-cooking *	212	48.4
Short shelf life	86	19.6
Meals' low acceptability	58	13.2
Other	44	10.0

* A total of 159 (36%) indicated two reasons.

According to several studies, consumers show difficulties while estimating portion sizes for the family meal, resulting in overcooking and promoting leftovers that frequently are spoiled and wasted [20]. On the other hand, families also reported a difficulty to predict children's appetite and the number of family members eating at home, resulting in leftovers of non-consumed food [21,29].

In Central Europe, the most frequently cited reasons for throwing the food away reported by Simunek et al. (2015) were similar to our findings, nevertheless in a different order. The first reason was that the food was spoilt, followed by food having been past the expiration date, an excessive amount of food having been purchased, and unpalatable food, as well as "other reasons" in the lower rank [30].

According to the research of Abeliots et al. (2015), better cooking skills are associated with better handling of leftovers and improved food waste prevention; the increased awareness of food waste is also a strong motivational factor for food waste reduction [31].

Reported waste frequency of the different food products was grouped into three dimensions: waste of vegetables, waste of protein sources, and waste of sources of carbohydrates. Waste of precooked foods emerged as an independent item, and was the individual item with the highest frequency.

Differences in frequency of food disposal as waste between food categories—namely, the lower values for the different sources of protein (Table 4)—have been attributed by others to the awareness of consumers concerning the higher environmental impact of producing meat and fish compared to other food categories, as well as to the higher cost of these. Simunek et al. (2015), in a research project developed in Central Europe [30], found that the food category most likely to be discarded was milk and dairy products, followed by fresh vegetables and mushrooms, then bread and cereals, unconsumed pre-processed foods, fruits and nuts, smoked meat, and, finally, food scraps from the plate. Contrarily, raw meat, fish, canned food, and "other" foods (such as sweets and snacks) did not appear in the food waste for any of the respondents.

Table 4. Analysis of the frequencies of types of wasted food. (n = 438).

5.	Factors related with type of food waste (explained variance, Cronbach's alpha (α)), KMO = 0.885	Mean ± SD	Loadings
	Factor 1—Waste of vegetables (var: 23.7%; α: 0.758)	**2.0 ± 0.9 [b]**	
	Vegetables	2.4 ± 1.5	0.764
	Dried and Fresh Pulses	1.7 ± 1.1	0,665
	Fruit	2.1 ± 1.1	0.640
	Soup	1.8 ± 1.0	0.601
	Factor 2—Waste of protein sources (var: 21.9%; α: 0.793)	**1.8 ± 0.8 [a]**	
	Dairy products	1.9 ± 1.1	0.751
	Fish	1.8 ± 1.1	0.714
	Eggs	1.6 ± 1.1	0.689
	Meat	1.7 ± 1.0	0.551
	Factor 3—Waste of sources of carbohydrates (var: 17.4%; α: 0.703)	**2.0 ± 0.9 [b]**	
	Potato	2.0 ± 1.1	0.810
	Bread	2.0 ±1.2	0.628
	Pasta	2.0 ± 1.2	0.510
	Pre-cooked food	**2.5 ± 1.9 [b]**	

[a, b]: homogeneous groups according to the Wilcoxon test at a 95% confidence level.

From the models in Table 5, wastage of vegetable products is mainly due to the lack of planning during purchase and buying by convenience, while it is inversely correlated to the consumption of vegetables and precooked meals.

Table 5. Multiple linear models describing the behavior of respondents concerning food waste of different groups of foods based on different determinants of food waste.

Model 1	**Waste of vegetables, food habits, and behavior $R^2_{adj} = 0.190$**	**Coefficient**	**SD**	***p*-value**
	Throwing away	0.193	0.034	0.000
	Lack of buying planning	0.084	0.027	0.002
	Consumption of vegetables	−0.124	0.038	0.001
	Purchasing by convenience	0.139	0.041	0.001
	Precooked meal	−0.038	0.017	0.023
Model 2	**Waste of protein sources, food habits, and behavior $R^2_{adj} = 0.196$**	**Coefficient**	**SD**	***p*-value**
	Throwing away	0.185	0.033	0.000
	Lack of cooking planning	0.070	0.022	0.002
	Keeping leftovers	−0.070	0.027	0.010
	Purchasing by convenience	0.097	0.038	0.011
	Animal feeding	0.038	0.019	0.042
Model 3	**Waste of carbohydrate sources, food habits, and behavior $R^2_{adj} = 0.228$**	**Coefficient**	**SD**	***p*-value**
	Throwing away	0.273	0.035	0.000
	Lack of cooking planning	0.085	0.024	0.000
	Precooked meal	−0.036	0.017	0.032
	Purchasing by convenience	0.088	0.041	0.034

Wastage of sources of protein is directly related to a lack of cooking planning, purchasing by convenience, and pet feeding, while it is inversely correlated to the use of leftovers.

Wastage of carbohydrate sources is related to a lack of cooking planning and purchasing by convenience, and it is indirectly correlated to the consumption of precooked meals.

According to the research by Visschers et al. (2016), families with children waste more food in total, namely fruits and vegetables, bakery products, and starches. Protein sources, ready-to-eat products, and dairy products seem to be the exception, probably due to consumers' appreciation of those products or to the perceived related price and the intention not to discard them [25].

Overall, these models identifying the major determinants of food waste for the different food groups are somehow low, but are significant and in line with previous studies.

There is a large number of behaviors that can have a positive impact on food waste generation and, consequently, on prevention of food waste [32]. Such behaviors include meal planning, cupboard and refrigerator checking, shopping list making, adequate storage of food items, use of food leftovers, adjust cooking amounts of food, and attention to expiration date labels [33].

It was found that respondents that do not plan purchases and that buy convenient foods waste more food. On the other hand, keeping precooked food and the consumption of vegetables are inversely associated with the amount of food waste (Table 5).

The variables that contribute the most to the waste of protein sources are lack of cooking planning, purchasing by convenience, throwing foods away, and feeding pets. The higher the amount of leftovers stored, the lower the waste of protein sources.

Carbohydrate sources are less wasted when the amount of precooked food increases. The variables of throwing away food, the lack of cooking planning, and purchasing by convenience contribute to a greater waste of carbohydrate sources.

The knowledge of both 'expiration date' or 'best used before' food labels should be improved among families. In addition, the cooking skills of consumers should also be improved, since this will enable a reduction of food waste through better handling of leftovers.

The most common strategies for reducing food waste include adjusting the quantity of food purchased to the size of household, better planning of food consumption with respect to the expiration date, and choosing high-quality foods with longer expiration dates bought from small retailers.

4. Conclusions

The sample of families studied reported a positive attitude concerning buying, consumption, and wastage, revealing a particular awareness of food waste and its social and environmental impact. These families have a privileged access to information, as they live near Lipor and have several activities in these areas, such as training courses, leaflets, cooking classes, easy access to recycling and reusing facilities, and study visits. At the same time, this is the main limitation of this study, as it impairs the generalization of the results, as there is some sample bias.

Author Contributions: Conceptualization, A.M.C.N.R. and S.F.; methodology, A.M.C.N.R. and L.M.C., formal analysis, A.M.C.N.R. and L.M.C.; investigation, T.P.; writing—original draft preparation, T.P.; writing—review and editing, A.M.C.N.R. and L.M.C.; funding, A.M.C.N.R. and L.M.C.; supervision, A.M.C.N.R. All authors have read and agreed to the published version of the manuscript.

Funding: This research was supported by national funds through FCT (Foundation for Science and Technology) within the scope of UIDB/05748/2020 and UIDP/05748/2020. Authors L.M. Cunha and A. Rocha acknowledge support from the project AgriFood XXI—NORTE-01-0145-FEDER-000041, financed through Norte2020, CCDR-N, Portugal2020, and the European Fund for Regional Development.

Conflicts of Interest: The authors declare no conflict of interest.

Appendix A

This section presents the wording of the attitudinal and behavioral questions included in the questionnaire. The original text in Portuguese is presented in italics.

"Shopping and consumption"

Frequency of shopping behaviors (7 items evaluated on an anchored scale, with answers ranging from (1) Never to (7) On every purchase)/*Com que frequência tem os seguintes comportamentos na compra de alimentos? (Sendo que "1" refere-se a "nunca" e "7" a "sempre que compra")*:

- Buying with a shopping list/Compra com lista previamente elaborada
- Purchases based on vouchers and promotions/*Compra em função dos vales de compra ou outras promoções que possui*
- Preference for a specific product or brand/*Preferência por produto especifico ou marca*
- Evaluation of contents of the refrigerator and pantry before shopping/*Antes de efetuar as compras, avalia o que existe no frigorífico e na despensa*
- Buying without a shopping list/*Compra sem preparar previamente a lista*
- Purchase of take-away meals/*Adquire refeições do take-away*
- Getting trendy foods and meals/*Adquire alimentos e refeições que estão na "moda"*

Abbreviated food frequency questionnaire (13 items answered on a fully described scale, with answers ranging from (1) Never or less than 1 time/month to (7) More than once a day)/*Com que frequência consome os seguintes alimentos? (1) Nunca ou <1 x mês; (2) 1–3 x por mês; (3) 1 x por semana; (4) 2–4 x por semana; (5) 5–6 x por semana; (6) 1 x por dia; (7) + de 1 x por dia).*

- Fruits/*Fruta*
- Seafood (fish and shellfish)/*Pescado (peixe, moluscos)*
- Vegetables and salads/*Hortícolas/saladas*
- Pulses (dried and fresh)/*Leguminosas secas/frescas*
- Eggs/*Ovos*
- Soup/*Sopa*
- Dairy products (milk, yogurt, and cheese)/*Laticínios (leite, queijo, iogurte)*
- Rice/pasta/*Massa/Arroz*
- Potatoes/*Batata*

- Bread/*Pão*
- Meat (poultry, pork, and beef)/*Carne (aves, suíno, bovino)*
- Fats (Olive oil/vegetable oils)/*Gorduras (azeite/óleo)*
- Precooked food/*Comida pré confecionada*

A food leftover frequency questionnaire (identical set of 13 items evaluated on an anchored scale, with answers ranging from (1) Never to (7) On every consumption moment)/*Indique com que frequência costuma guardar as sobras dos seguintes alimentos (Sendo que "1" refere-se a "nunca" e "7" a "sempre que consome").*

"Food and food waste habits"

Additional behaviors related to leftovers and food waste, including the usual destination of leftovers (5 items evaluated on an anchored scale, with answers ranging from (1) Never to (7) On every consumption moment)/*Indique a frequência com que realiza estes comportamentos quando tem sobras de alimentos ou refeições: (Sendo que "1" refere -se a "nunca" e "7" a "sempre que consome").*

- Preparing new culinary items/*Elabora novas preparações culinárias*
- Throwing away/*Deita para o lixo*
- Feeding animals/*Dá aos animais*
- Freezing for another occasion/*Congela para outra ocasião*
- Eating on the next day/*Consome no dia seguinte*

Most frequent reasons for leftovers in households (check all that apply)/Escolha os motivos mais frequentes para ocorrerem sobras na sua residência (Marque todas que se aplicam).

A food waste frequency questionnaire (identical set of 13 items evaluated on an anchored scale, with answers ranging from (1) Never to (7) On every consumption moment)/*Com que frequência costuma deitar fora os seguintes alimentos: (Sendo que "1"refere-se a "nunca" e "7" a "sempre que consome").*

Attitudes towards reasons for food waste (5 items evaluated on an anchored scale, with answers ranging from (1) Totally disagree to (7) Totally agree)/*Indique o seu grau de concordância com as seguintes afirmações. No meu agregado familiar ocorre desperdício alimentar quando: (Sendo que "1" refere-se a "discordo plenamente" e "7"a "concordo plenamente").*

- Excess food is made for meals/*Se confecionam alimentos em excesso para a refeição*
- No control of stored items at home/*Não se controla o que ainda existe em casa (despensa/frigorífico)*
- Foods with short shelf-life/*Se os alimentos apresentam um prazo de validade muito curto*
- No shopping list/*Ausência de lista para a realização de compras*
- Excess purchase due to promotion/*Se compra em excesso em virtude da promoção*

References

1. Bauman, Z. *A Cultura do Lixo*; Desperdiçadas, E.V., Zahar, J., Eds.; Zahar: Rio de Janeiro, Brazil, 2004; pp. 117–164.
2. Baudrillard, J. *The Consumer Society: Myths and Structures*; Sage: Thousand Oaks, CA, USA, 2016.
3. Goulart, R. Desperdício de Alimentos: Um Problema de Saúde Pública. *Revista Integração* **2008**, *54*, 285.
4. FAO. *Food Wastage Footprint: Impacts on Natural Resources*; FAO: Rome, Italy, 2013.
5. Gustavsson, J.; Cederberg, C.; Sonesson, U.; Van Otterdijk, R.; Meybeck, A. *Global Food Losses and Food Waste*; FAO: Rome, Italy, 2011.
6. Monier, V.; Mudgal, S.; Escalon, V.; O'Connor, C.; Gibon, T.; Anderson, G.; Montoux, H.; Reisinger, H.; Dolley, P.; Ogilvie, S. *Preparatory Study on Food Waste Across EU 27*; European Commission: Brussels, Belgium, 2010.
7. Agência Portuguesa do Ambiente. Redução do Desperdício Alimentar. 2014. Available online: http://www.apambiente.pt/index.php?ref=16&subref=84&sub2ref=106&sub3ref=273 (accessed on 20 August 2020).
8. European Food Information Council. *Fruit and Vegetable Consumption in Europe-Do Europeans Get Enough?* EUFIC: Brussels, Belgium, 2012.

9. Ambler-Edwards, S.; Bailey, K.S.; Kiff, A.; Lang, T.; Lee, R.; Marsden, T.K.; Simons, D.W.; Tibbs, H. *Food Futures: Rethinking UK Strategy*; A Chatham House Report; Chatham House: London, UK, 2009.
10. Farr-Wharton, G.; Fot, M.; Choi, J. Identifying factors that promote consumer behaviours causing expired domestic food waste. *J. Consum. Behav.* **2014**, *13*, 393–402. [CrossRef]
11. DECO. Preço Determina Escolhas do Consumidor. 2015. Available online: http://www.decojovem.pt/biblioteca/combateaodesperdicioalimentar/ (accessed on 30 July 2020).
12. Instituto Nacional de Estatística. *Statistical Yearbook of Portugal 2016*; Instituto Nacional de Estatística: Lisboa, Portugal, 2017.
13. Instituto Nacional de Estatística. Censos 2011. Available online: http://www.censos.ine.pt (accessed on 10 September 2020).
14. Anuário Estatístico de Portugal. 2019. Available online: https://www.ine.pt/xportal/xmain?xpid=INE&xpgid=ine_publicacoes&PUBLICACOESpub_boui=444301590&PUBLICACOEStema=55480&PUBLICACOESmodo=2 (accessed on 30 July 2020).
15. Baptista, P.; Campos, I.; Vaz, S. *Do Campo ao Garfo: Desperdício Alimentar Em Portugal*, 1st ed.; Cestras: Lisboa, Portugal, 2012.
16. Evans, D. Beyond the throwaway society: Ordinary domestic practice and a sociological approach to household food waste. *Sociology* **2012**, *46*, 41–56. [CrossRef]
17. Abeliotis, K.; Lasaridi, K.; Chroni, C.J.W.M. Research, attitudes and behaviour of Greek households regarding food waste prevention. *J. Waste Manag.* **2014**, *32*, 237–240.
18. Parizeau, K.; von Massow, M.; Martin, R.J.W.M. Household-Level dynamics of food waste production and related beliefs, attitudes, and behaviours in Guelph, Ontario. *Waste Manag.* **2015**, *35*, 207–217. [CrossRef] [PubMed]
19. Yildirim, H.; Capone, R.; Karanlik, A.; Bottalico, F.; Debs, P.; El Bilali, H. Food wastage in Turkey: An exploratory survey on household food waste. *Nutr. Res.* **2016**, *4*, 483–489.
20. Secondi, L.; Principato, L.; Laureti, T. Household food waste behaviour in EU-27 countries: A multilevel analysis. *Food Policy* **2015**, *56*, 25–40. [CrossRef]
21. Schanes, K.; Dobernig, K.; Gözet, B. Food waste matters-A systematic review of household food waste practices and their policy implications. *J. Clean. Prod.* **2018**, *182*, 978–991. [CrossRef]
22. Lanfranchi, M.; Calabrò, G.; De Pascale, A.; Fazio, A.; Giannetto, C. Household food waste and eating behavior: Empirical survey. *Br. Food J.* **2016**, *118*, 3059–3072. [CrossRef]
23. Neff, R.; Spiker, M.; Truant, P. Wasted food: US consumers' reported awareness, attitudes, and behaviors. *PLoS ONE* **2015**, *10*, e0127881. [CrossRef] [PubMed]
24. Qi, D.; Roe, B. Household food waste: Multivariate regression and principal components analyses of awareness and attitudes among US consumers. *PLoS ONE* **2016**, *11*, e0159250. [CrossRef] [PubMed]
25. Visschers, V.H.; Wickli, N.; Siegrist, M. Sorting out food waste behaviour: A survey on the motivators and barriers of self-reported amounts of food waste in households. *J. Environ. Psychol.* **2016**, *45*, 66–78. [CrossRef]
26. Direção Geral de Saúde. React Covid Inquérito Sobre Alimentação e Atividade Física em Contexto de Contenção Social. 2020. Available online: https://www.dgs.pt/programa-nacional-para-a-promocao-da-atividade-fisica/ficheiros-externos-pnpaf/rel_resultados-survey-covid-19-pdf.aspx (accessed on 30 July 2020).
27. Cappellini, B. The sacrifice of re-use: The travels of leftovers and family relations. *J. Consum. Behav.* **2009**, *8*, 365–375. [CrossRef]
28. Cappellini, B.; Parsons, E. Practising thrift at dinnertime: Mealtime leftovers, sacrifice and family membership. *Sociol. Rev.* **2012**, *60*, 121–134. [CrossRef]
29. Porpino, G.; Parente, J.; Wansink, B. Food waste paradox: Antecedents of food disposal in low income households. *Int. J. Consum. Stud.* **2015**, *39*, 619–629. [CrossRef]
30. Simunek, J.; Derflerova-Brazdova, Z.; Vitu, K. Food wasting: A study among Central European four-member families. *Int. Food Res. J.* **2015**, *22*, 2679–2683.
31. Abeliotis, K.; Lasaridi, K.; Costarelli, V.; Chroni, C. The implications of food waste generation on climate change: The case of Greece. *Sustain. Prod. Consum.* **2015**, *3*, 8–14. [CrossRef]
32. Wrap, W. *Household Food and Drink Waste in the UK*; Report Prepared by WRAP; WRAP: Banbury, UK, 2009.

33. Quested, T.; Marsh, E.; Stunell, D.; Parry, A.D. Spaghetti soup: The complex world of food waste behaviours. *Resour. Conserv. Recycl.* **2013**, *79*, 43–51. [CrossRef]

Publisher's Note: MDPI stays neutral with regard to jurisdictional claims in published maps and institutional affiliations.

Article

Driving Forces of Changing Environmental Pressures from Consumption in the European Food System

Philipp Schepelmann [1,*], An Vercalsteren [2], José Acosta-Fernandez [1], Mathieu Saurat [1], Katrien Boonen [2], Maarten Christis [2], Giovanni Marin [3], Roberto Zoboli [4] and Cathy Maguire [5]

1 Wuppertal Institut für Klima, Umwelt, Energie gGmbH, Döppersberg 19, 42103 Wuppertal, Germany; joseacosta@wupperinst.org (J.A.-F.); mathieu.saurat@wupperinst.org (M.S.)
2 Flemish Institute for Technological Research, Boeretang 200, 2400 Mol, Belgium; an.vercalsteren@vito.be (A.V.); katrien.boonen@vito.be (K.B.); Maarten.Christis@vito.be (M.C.)
3 Department of Economics, Society, Politics, Università di Urbino Carlo Bo, 61029 Urbino, Italy; giovanni.marin@uniurb.it
4 Department of International Economics, Institutions and Development, Università Cattolica del S. Cuore, 20123 Milan, Italy; roberto.zoboli@unicatt.it
5 European Environment Agency, Kongens Nytorv 6, 1050 Copenhagen, Denmark; cathy.maguire@eea.europa.eu
* Correspondence: philipp.schepelmann@wupperinst.org

Received: 31 August 2020; Accepted: 2 October 2020; Published: 8 October 2020

Abstract: The paper provides an integrated assessment of environmental and socio-economic effects arising from final consumption of food products by European households. Direct and indirect effects accumulated along the global supply chain are assessed by applying environmentally extended input–output analysis (EE-IOA). EXIOBASE 3.4 database is used as a source of detailed information on environmental pressures and world input–output transactions of intermediate and final goods and services. An original methodology to produce detailed allocation matrices to link IO data with household expenditure data is presented and applied. The results show a relative decoupling between environmental pressures and consumption over time and shows that European food consumption generates relatively less environmental pressures outside Europe (due to imports) than average European consumption. A methodological framework is defined to analyze the main driving forces by means of a structural decomposition analysis (SDA). The results of the SDA highlight that while technological developments and changes in the mix of consumed food products result in reductions in environmental pressures, this is offset by growth in consumption. The results highlight the importance of directing specific research and policy efforts towards food consumption to support the transition to a more sustainable food system in line with the objectives of the EU Farm to Fork Strategy.

Keywords: food consumption; environmentally extended input–output analysis; international trade; consumption-production perspective; structural decomposition analysis; value chain analysis; ex-post times series analysis; allocation tables

---

1. Introduction

Food is a basic human need but overconsumption, scarcity and insecurity can jeopardize health and quality of life. The food system is a complex global network of production, consumption and trade and is shaped by many factors: economic, environmental, political, technological and social, including cultural norms and lifestyles [1]. The food system inextricably links human health and social wellbeing with environmental sustainability. The EAT-Lancet Commission on Food, Planet, Health highlighted how sustainable diets have lower environmental impacts and contribute to food and nutrition security [2].

The transition to a more sustainable food system is receiving increasing policy focus as reflected in the European Union's recent Farm to Fork Strategy under the European Green Deal. Currently, the food system is responsible for a range of impacts on the environment through emissions of pollutants, depletion of resources, waste generation, loss of biodiversity and degradation of ecosystems in Europe [1,3–5]. Europe is also highly dependent on imported final and intermediate products to satisfy European domestic demand for food with trade resulting in negative impacts outside Europe [3]. Many studies have identified the important role of consumption of food and beverages in terms of generating environmental pressures. More specifically, results based on single-country environmentally extended input–output analysis (EE-IOA) for nine EU member states suggested that in 2005 final consumption in the 'food area' contributed to 21% of greenhouse gas emissions, 49% of acidifying emissions, 20% of ground ozone precursors and 40% of total material requirement [6]. Analysis using a multi-regional input–output (MRIO) model based on data from EXIOBASE 2.2 estimated that food consumption in 2007 in the EU28 was responsible for 9.5% of the carbon footprint, 51.1% of the land footprint, 26% of the material footprint and 60.7% of the water footprint [7]. Life cycle assessment (LCA) approaches used to calculate the environmental footprint of food consumption for the EU28 for 2010, with environmental impact categories with impacts broken down by food products, highlighted the large contribution of meat and dairy products to environmental pressures [8].

This paper presents an integrated assessment methodology based on EE-IOA to assess direct and indirect environmental and socio-economic effects arising from food consumption. By taking a 'consumption perspective' in an EE-IOA framework, the paper provides measures of international transfer of environmental and economic effects along the global food value chain. The assessment uses EXIOBASE 3.4 as a source of data and covers all the consumption categories within the food system. The assessment is based on the development of detailed allocation matrices, which enable the calculation in a rigorous way the environmental pressures and impacts of consumption patterns by linking the data on final consumption expenditure to EE-IO data. The procedure to develop the allocation matrices is presented in a fully transparent and detailed manner thus contributing to methodological improvements in integrated assessment. The paper also develops a decomposition analysis to measure the role of technology, consumption mix, and consumption level in driving the environmental and economic effects of food consumption. The analyses presented in the paper can support policy and decision making by providing detailed analytical knowledge on those sectors and environmental dimensions that should be targeted to reduce the environmental pressures of the food system.

2. State of the Art

A range of bottom-up and top-down approaches have been used to analyse the environmental impacts of human food consumption, including quantification of overall pressures and the impacts resulting from specific food consumption habits in households. The results of methods such as Life Cycle Assessment (LCA), input–output analysis (IOA) and other hybrid methods combining both LCA and IOA are different and not attributable solely to the characteristics of the methods. Other aspects that define the scope of the analysis, such as the geographical region, the period, the system boundaries, and the attention paid to comparing different bundles of consumed food also explain the different results.

There are a range of LCA-based studies on the environmental impacts of food consumption and nutrition [9]. The EU Joint Research Centre has taken an LCA approach to the development of Basket-of-Products (BoP) indicators, including food [6]. Although the estimation of the effects of household's consumption expenditure through IOA is based on fairly aggregated information, the availability of more detailed information on the products consumed by households enables analysis of more specific groups of products [7,10–12]. Studies have used both methodological approaches, for example, to calculate the impact of food consumption by Swiss households [13]. The results of

IOA and LCA of different European household activities including food consumption have been compared [14].

As a bottom-up approach, LCA makes it possible to quantify and evaluate in detail the different pressures and environmental impacts associated with the various processes needed to produce a given functional unit (a single product or production process). However, LCAs have strict system boundaries often neglecting a significant share of indirect impacts on the economy and the environment. As a top-down approach, MRIO analysis allows from a technical-economic point of view the estimation of the total environmental pressures and impacts induced by the consumption of a given product group along the global supply chain of all required inputs for their production. This estimation is based on information on the different monetary transactions observed in the global market between the economic activities of each country or region of the world as well as on the direct environmental pressures resulting from production activities in each of the sectors involved in the whole economic process. By using the Leontief model, moreover, there is no need to set system boundaries as all direct and indirect technical-economic requirements (and hence associated environmental effects) are considered. MRIO can also account for technological shifts in the production of intermediate inputs sourced from different countries.

In addition to these methodological differences, there are other aspects that determine the quality of the results and their subsequent interpretation. These aspects are related to the level of detail of the information available to perform both the calculations and the analysis. In the case of MRIO, one of the most important aspects is related to the allocation of the calculated effects to the different purposes, in terms of final use. The type of assessment presented in this paper requires the use of a classification of individual consumption according to purpose (COICOP) rather than a classification of products produced for consumption (CPA). This enables distribution of the environmental effects caused by the production of consumer goods among the different categories of specific consumption, as well as the analysis of the effects along the supply chain associated with changes in the consumption mode of a given product (e.g., more wood as heating fuel than wood in the form of furniture). Using a COICOP-based classification allows to account for alternative ways of satisfying the same consumer need, i.e., different mixes of (CPA or NACE) products and sectors. This allocation is typically defined in so-called allocation tables.

2.1. Allocation Tables

In general, IO tables represent the structure of production and consumption activities within the economy. Since production and consumption are determined by the use of products, the structure of the economy can be represented either according to the economic activities using and producing the products or according to the product groups produced and used in the economy. When IO analysis is used to study environmental pressures and impacts of specific consumption patterns, the data on final consumption expenditure by households presented in the IO tables (IOT) need to be linked to household expenditure data. Detailed data on goods and services purchased by households are organized according to the COICOP while the data on final consumption expenditure by households presented in the IO tables are organized according to product groups produced in the economy. The link between both is made via allocation tables or correspondence matrices, which are compiled on the basis of the product classifications underlying the IOT and COICOP.

Allocation tables allow a preliminary assignment of the several product groups represented in the IOT as used by households to the several product groups detailed in the COICOP classification. This aims to attribute the amount of private household expenditure for each of the several product groups to a certain product group consumed with a defined purpose. Depending on the group of products, the attribution of household expenditure to the several COICOP categories can be exclusive or multiple. In the case of the expenditure for product groups that are consumed for satisfying multiple purposes, additional steps are required for an accurate attribution. Due to the conceptual differences that characterize the existing data sets, additional adjustments are necessary (price transformation).

All these additional methodological steps are summarized in the "allocation tables", which detail the attribution of household consumption expenditure in terms of product groups to the COICOP categories of consumed products.

The related literature describes two procedures for the elaboration of allocation tables. In the first approach, each product group represented in the IO table is connected to a single category of consumed products (one-to-one allocation). This allows to keep the level of product groups used in the calculations. In these cases, in which a many-to-one assignment is required, several product groups are first aggregated to a single one before they are related to a particular COICOP category of products consumed. This procedure simplifies the construction of the allocation table but reduces the level of detail. In the second approach, a product group represented in the IO table is associated, if necessary, with different categories of consumed products by households. This more detailed allocation is referred to as a one-to-many allocation. By applying this procedure, a single product group can be allocated to a single category or distributed among different categories of consumed products. For example, the CPA product group 'textiles' can be allocated to the COICOP categories 'clothing' as well as 'furniture.' The CPA group 'glass products' can be associated with the COICOP categories 'construction', 'furniture' and 'packaging'.

From a theoretical point of view and because most groups of products produced in the economy are consumed for different purposes, applying the allocation approach one-to-many would allow to generate allocation tables more accurately. However, this implies the use of bottom-up data from Household Budget Enquiries (HBE), data on trade and transport margins, on value-added taxes, taxes and subsidies on products, etc. In most cases, limited data availability prevents the development of allocation tables at such detail. However, for some countries a one-to-many bottom-up approach has been used to compile a detailed allocation matrix. In the Netherlands, for example, [15] an allocation table was developed to decompose the total price paid by consumers into producer price, trade and transport margins, and value-added taxes. The update of this table [16] used detailed information from Statistics Netherlands. This enables, on the one hand, linking of the data on consumed products represented in the Budget Survey to the product groups represented in the supply table of the System of National Accounts (SNA), in which they are clustered into functional domains that differ from the COICOP categories. On the other hand, the used data enable the decomposition of the consumer price into basic price, taxes and subsidies on products, and trade and transport margins. In Germany, a similar method was applied using as classification of the consumption categories the SEA (Systematisches Verzeichnis der Einnahmen und Ausgaben der privaten Haushalte), which is the German implementation of the international COICOP standard [17]. The Federal Statistical Office has published a consumption allocation table which allows a conversion from CPA to COICOP. The Federal Statistical Office provides upon request a table with the trade margins and taxes on products for each product group in order to convert consumer prices (used to measure consumption expenditures) into producer prices (used to measure monetary flows represented in the input–output tables). Another example of such bottom-up allocation is the procedure applied for Flanders [18]. The Flemish IO-table is based on specific monetary and environmental data for Flanders and is part of an interregional IO-table, in which trade with the Brussels region and Wallonia is represented. A link to Exiobase datasets enables the quantification of import flows from outside Belgium. The household's final consumption vector in the IO table is disaggregated into different COICOP categories using a matrix in which the consumption by households is represented in purchase prices. In this way the COICOP categories are linked to the output of the sectors represented in the IO-table according to the NACE classification. The allocation table is developed bottom-up by the Federal Planning Bureau and is structured according to the different COICOP categories attributed to the NACE sectors. Bottom-up data from the Household Budget Enquiry (HBE) for Belgium [18], tax data, VAT data, trade and transport margins are used to generate the allocation tables. For example, trade margins are allocated to the trade sectors and the product value is allocated to the respective producing economic sectors.

Many examples of one-to-one or many-to-one allocation tables are available in the literature, as this approach requires a limited amount of data and is thus frequently used. For example, a many-to-one allocation table for CPA product groups to COICOP categories was developed applying a pragmatic procedure based on available data [6]. Most CPA two-digit product groups could be attributed directly to a defined COICOP category. However, missing data, e.g., from household surveys, as well as data on trade margins for a transformation of the pricing from basic to purchaser prices restricted a more detailed allocation. Thus, a few modifications in the coverage of certain COICOP categories had to be made or proxy COICOP categories were used to try to establish an imperfect but reasonable match with the given CPA product groups. For instance, in order to assign the CPA product group "food and beverages" to a COICOP category, the coverage of the original COICOP group 'Food and non-alcoholic beverages' (COICOP 01) was expanded to 'Food and beverages', i.e., including alcoholic beverages. The CPA product group 'leather and leather products' was fully assigned to COICOP category 'clothing and footwear', even though there are also other significant uses of leather in other COICOP categories. Since the 2-digit, upper level COICOP categories were the only ones that could be used, the allocation resulted in an aggregation of COICOP categories (1 digit).

The detailed allocation tables for a specific country are sometimes extrapolated to other European countries, for example, the allocation tables for the EU27 compiled based on an allocation table (in purchaser prices) for Austria [19]. For that, a correspondence matrix between the classification of supply and use tables (CPA 2002, 2-digit) and the COICOP categories was used. The authors used the RAS method for constructing the allocation matrices for the other countries based on COICOP data (in purchaser prices) from Eurostat. Additional statistical sources allowed to split the energy sector into electricity and heating. Electricity expenditure was estimated by combining IEA energy balance data (for the household sector) with IEA energy prices. Energy efficiency indices for heating and electrical appliances were taken from the ODYSSEE database, while the TREMOVE database was the primary source of the energy efficiency index for vehicles. The RAS method was also applied to construct allocation matrices for other countries, based on the 2004 German allocation matrix [20]. The author discusses the limitations of the approach, which implies that households of all countries use the same technology of converting industry products to goods. Not only may the countries be heterogeneous, but the weights may change in time. However, missing data for most countries may justify the use of the RAS method.

For some countries, allocation tables are elaborated using country-specific data on household consumption and on price components from national statistical offices. For Europe, both a simplified procedure linking one CPA to one COICOP and a more complex procedure applying RAS to extrapolate data from a specific country to other countries have been applied [6,19,20].

2.2. Limitations of Allocation Tables

Allocations of household consumption expenditure to COICOP categories are likely to change over time. The use of static allocations to calculate future scenarios can lead to potential inaccuracies [21]. Many recent studies use static coefficients [17,22–24]. However, technological change may lead to growth, reduction or substitution effects. Furthermore, the relative prices of goods may change significantly over the years. Assumptions on these dynamics are, in particular, relevant for forecasting and future scenario construction. For example, at the turn of the century, the share of air transport significantly increased at the expense of water transport. Air transport has an environmental impact substantially larger than water transport. Failing to account for this change in the transport mode would result in an underestimation of environmental impacts. There is no generally accepted approach or convention to construct allocation tables. The result is a multitude of different approaches. However, very few studies explain the underlying assumptions which were made for the construction of their allocation matrices.

As far as analyses for the food system are concerned, the literature indicates that the allocation of trade activities is a major issue. When the IO tables are in basic prices, trade margins require a correct allocation to the different COICOP categories for generating reasonable allocation tables. These trade margins include the (impact of) trade and retail (e.g., cooling, transport, warehousing), which contribute substantially to environmental pressures related to the food system. Ideally the food system is not restricted to COICOP 01 'food' but should also include COICOP 11.1 'catering' and other relevant COICOP categories. A clear definition of the food system is thus very important.

3. Research Questions

The objective of the analysis is to provide a detailed assessment of the different environmental pressures and some relevant socioeconomic effects caused by food consumption in the European Environment Agency (EEA) member countries (member countries at the time of analysis were the EU28 plus Norway, Iceland, Switzerland, Lichtenstein and Turkey. The geographical scope was selected to ensure that the analysis covered the largest share of EEA countries' economies as possible. In terms of illustrating the method and the key results and conclusions, the inclusion of non-EU countries has not significantly affected the results as they align well with other EU28 estimates). The assessment focused on the following research questions:

(i) Are European households changing their food consumption habits to goods and services that generate less environmental pressure?
(ii) What are the environmental and socio-economic effects in Europe and in the rest of the world resulting from food consumption in European households?

To carry out this analysis, COICOP was used rather than CPA to enable the distribution of the environmental effects caused by the production of consumer goods among the different categories of specific consumption, as well as analysis of the effects along the supply chain associated with changes in the consumption of a given product (e.g., more wood as heating fuel than wood in the form of furniture).

Furthermore, applying detailed environmentally and socioeconomically extended multi-regional input–output models (ESE-MRIOM), which are built from corresponding extended multi-regional input–output (MRIOT) tables, considerably increases the scope of the analysis. An ESE-MRIOM is a powerful analytical instrument for assessing different pressures and effects of production and consumption activities from a system perspective. It enables the identification of the different exporting economies in which environmental pressures and socioeconomic effects are induced by domestic consumption. Thus, the estimation of total pressures and effects occurs with greater geographical precision not only in terms of the origin, quantity and mix of products consumed, but also in terms of the technology used for their production. In this way, the identification of the hotspots of the different environmental and socioeconomic effects is characterized by a high degree of detail.

The European Topic Centre on Waste and Materials in a Green Economy has already used the ESE-MRIOM for various analytical purposes including estimating the environmental impacts of final demand in EEA member countries [1]. However, the analytical spectrum of ESE-MRIOM goes beyond this. An example of these other applications for the analysis of the food system is the structural decomposition analysis (SDA). The application of the SDA allows identification and quantification of the contribution of the main driving forces of changes in observed environmental pressures, among others, associated with food consumption. This information is important, because it indicates possible intervention points for policies aimed at reducing the environmental pressures related to food consumption.

4. Methods and Data

The method and data section first describes the EXIOBASE database used throughout this paper. Secondly, the definition and scope of the food system as applied in this paper is discussed, followed by the description of the construction of the allocation matrix. Finally, the methodologies for the input–output calculations and for the SDA are described.

4.1. *Exiobase v3.4*

EXIOBASE (v3.4) is a database containing detailed information on the world economy, the monetary flows associated with the supply and use of goods and services produced and consumed and the direct environmental and socio-economic effects of production and consumption activities [25]. The monetary flows are organized in the form of Multi-Regional Supply and Use/Input–Output Tables (MRSUT/MRIOT) with a defined industry and product group structure. The environmental and socio-economic effects are structured in the form of multiregional matrices with the same industry structure as the MRSUT/MRIOT. EXIOBASE was developed by detailing and harmonizing country-specific monetary SUT (MSUT) and data related to energy, emissions, water use, land use, resource extractions as well as employment and value added by industry. The country-specific MSUT is linked via trade and extended with environmental and socioeconomic variables (ESE-MRSUT), on which an environmentally and socioeconomically extended multi-regional input–output table (ESE-MRIOT) is built. The ESE-MRIOT can be used for an analysis along the supply chain of the environmental pressures and socio-economic effects associated with the final consumption of product groups. This version of EXIOBASE is a time series of ESE MRIOT ranging from 1995 to 2011 for 44 countries (28 EU member plus 16 major economies) and five regions of the rest of the world. The distinguishing characteristics of EXIOBASE are the high level of consistent sectoral (200 products, 163 industries for all countries and regions included) and environmental detail. For the analysis carried out here, the product-by-product tables of EXIOBASE were used.

Compared to single-country studies, the use of detailed ESE-MRIOT from EXIOBASE enables a more precise identification of the different components that contribute to (changes in) aggregate environmental or socio-economic effects. This enables the identification of policy-relevant hotspots at a detailed level.

4.2. *Definition of the Food System*

Definitions of the food system tend to be broad and can include all the elements (environment, people, inputs, processes, infrastructures, institutions, etc.) and activities that relate to the production, processing, distribution, preparation and consumption of food, and the outputs of these activities, including socio-economic and environmental outcomes [3]. However, the applied data and models require a definition of the food system in terms of COICOP categories. Table 1 presents the two-digit COICOP categories. The food system as defined in this paper includes the COICOP categories marked Italic, i.e.:

01.1 Food
01.2 Non-alcoholic beverages
02.1 Alcoholic beverages

Table 1. Definition of food system in terms of classification of individual consumption according to purpose (COICOP) categories [25].

COICOP			Description
2-digit	**3-digit**	**4-digit**	
01			Food and non-alcoholic beverages
	01.1		Food
		01.1.1	Bread and cereals
		01.1.2	Meat
		01.1.3	Fish
		01.1.4	Milk, cheese and eggs
		01.1.5	Oils and fats
		01.1.6	Fruit
		01.1.7	Vegetables
		01.1.8	Sugar, jam, honey, chocolate and confectionery
		01.1.9	Food products n.e.c.
	01.2		Non-alcoholic beverages
02			Alcoholic beverages, tobacco and narcotics
	02.1		Alcoholic beverages
03			Clothing and footwear
04			Housing, water, electricity, gas and other fuels
05			Furnishings, household equipment and routine household maintenance
06			Health
07			Transport
08			Communications
09			Recreation and culture
10			Education
11			Restaurants and hotels
12			Miscellaneous goods and services

In our allocation tables, the COICOP categories 01.2 and 02.1 are aggregated because the resolution of the product group nomenclature in EXIOBASE does not allow this distinction. The COICOP category "catering" (11.1) should ideally be included in the food system, however, in the EXIOBASE nomenclature the product group "Hotel and restaurant services", which includes catering, cannot be further disaggregated. For this reason, the analysis focusses only on the environmental pressures and socio-economic effects caused by the use and preparation of food products by private households. We acknowledge that further product groups or services should at least partly be included (e.g., energy, water, hotel and restaurant services) but cannot be allocated due to lack of detailed data. Possible over- or underestimations will be addressed in the discussion.

4.3. Construction of the Allocation Matrix

For analysis of consumption systems based on the purpose for which products are used rather than on total quantity of product groups produced for consumption in general, the establishment of the correspondence between the product classification inherent to the input–output tables used (EXIOBASE is based on CPA 2002) and the COICOP classification is required. Eurostat publishes two correspondence tables: CPA 2002 to COICOP 1999 and CPA 2008 to COICOP 1999. These correspondences help to generate the allocation tables that allow attributing the results of input–output modeling, which are structured according to the groups of products produced, to the COICOP consumption categories.

Since the research questions focused on the effects associated with the different purposes for which households consume food products, it was necessary to construct allocation tables. The procedure applied for this is summarized in the following steps:

First, we identified products exclusively for technical/industrial use or not concerning the consumption activities of households (e.g., p21.1 paper pulp) among the list of 200 goods and services of the EXIOBASE classification. This identification makes use of the Eurostat correspondence tables at the six-digit level of the CPA classification. In some cases, we assigned a product group to household consumption even though the product group should be excluded (e.g., p24.g bio gasoline). The reason is that for such product groups the household consumption column in the input–output table of EXIOBASE does contain entries. Second, we clustered the product groups that cannot be assigned to COICOP categories. Third, we had to make decisions on how to allocate the household consumption product groups to the corresponding COICOP categories. There are three cases:

1. If there is an unambiguous one-to-one correspondence between a product group of the EXIOBASE classification and a COICOP category, we assigned this product group to this one COICOP category only. For example, the product p15.b."Products of meat" is unambiguously attributed one-to-one to the COICOP category 01.1.2 "Meat".
2. If there is an unambiguous one-to-many correspondence between a product group of the EXIOBASE classification and several COICOP categories at the 4-digit level that all belong to the same category at the 3-digit level, we created a combined COICOP category at the 4-digit level and assigned the product group to that new category. For instance, the EXIOBASE products P01.d. "Vegetables, fruits, nuts" corresponds to COICOP categories 01.1.6 "Fruit" and 01.1.7 "Vegetables" and we created a combined COICOP category at the four-digit level 01.1.6_01.1.7 "Fruit and Vegetables".
3. If there is an ambiguous one-to-many correspondence between a product group of the EXIOBASE classification and several COICOP categories that cannot be reconciled as in case 2, we arbitrarily assigned the EXIOBASE product group to one COICOP category, thus turning the one-to-many into an arbitrary one-to-one attribution.

Note that case 2 and 3 may actually overlap. For instance, EXIOBASE p17 "Textiles" corresponds to the COICOP categories 03.1 "Clothing", 05.1 "furniture and furnishings, carpets and other floor coverings", 05.2 "household textiles" and 09.3 "other recreational items and equipment, gardens and pets." We arbitrarily assigned the EXIOBASE product group p17 to a new COICOP category combining 05.1 and 05.2.

Assigning an entire product group to a single COICOP category leads to over- and underestimates when the input–output analysis is carried out. On the one hand, the environmental pressure generated by the COICOP category to which all the consumption of the relevant EXIOBASE product group has been attributed will be overestimated. On the other hand, the environmental pressure of the other relevant COICOP categories from which the product group has been subtracted will be underestimated. However, country-specific additional information on the CPA 2002 classification that would allow estimating the shares of household expenditure for the several products aggregated in each EXIOBASE product group is not available. In the case of the EU Member States, such data provided by Eurostat are available only at the 2-digit level of the CPA 2008 classification and are therefore useless for a breakdown.

At the end of this process we generated an allocation table which includes:

- The unambiguous allocations of all one-to-one and one-to-many correspondences (case 1 and 2).
- The one-to-many attributions are all arbitrarily turned into one-to-one attributions, as described under case 3. For instance, the EXIOBASE product group "Products of forestry, logging and related services" (18 p02) belong to COICOP categories "Solid fuels" (04.5.4) as well as "Gardens, plants and flowers" (09.3.3). Allocation Table 1 attributes the "Products of forestry, logging and related services" fully to COICOP "Solid fuels" (04.5.4).
- The product groups that cannot be assigned to COICOP categories are clustered in a new COICOP category "Miscellaneous goods and services (not applicable for private households)" (12.) Such an EXIOBASE group is, for example, "iron ores" (33 p13.1).
- Some product groups from untypical clusters were assigned to new n.e.c. (i.e., not elsewhere classified) categories within the existing COICOP domains. This approach allows to keep them separate and to treat them analytically like category 12 describe above. For instance, EXIOBASE

"Sugar cane, sugar beet" (6 p01.f) is assigned to a new COICOP category "01 Food > 1.2 Food n.e.c. > 01.2.0 n.e.c."

- The EXIOBASE product groups "Wholesale trade and commission trade services, except of motor vehicles and motorcycles" (154 p51) and "Retail trade services, except of motor vehicles and motorcycles; repair services of personal and household goods" (155 p52) are allocated to all COICOP categories according to the share of each COICOP category in the final consumption expenditure of households. This approximation assumes that the share of the trade margins in the final consumption expenditure of households is the same (or very similar) across all COICOP categories. We use the Eurostat dataset on Final consumption expenditure of households by consumption purpose (COICOP 3 digit) [nama_10_co3_p3] for this approximation.
- The EXIOBASE product groups "Railway transportation services" (157 p60.1), "Other land transportation services" (158 p60.2), "Sea and coastal water transportation services" (160 p61.1), "Inland water transportation services" (161 p61.2), as well as "Air transport services" (162 p62) are allocated to all COICOP categories dealing with goods (i.e., services are here excluded because the term "transportation" in general is rather applicable for transport of goods and persons but not for services, note that we included repair activities of goods) according to the share of each COICOP category in the final consumption expenditure of households [26]. We used the same Eurostat dataset as above for this approximation. The allocation of the remaining items (i.e., the total shares of services in the final consumption expenditure of households) is entirely assigned to the corresponding subcategories of the COICOP category "Transport" (07). For instance, EXIOBASE "Railway transportation services" (157 p60.1) is allocated to all COICOP categories dealing with products (not services) using the Eurostat data and the rest is assigned to the COICOP category "Passenger transport by railway" (07.3.1). This relies on the assumptions that (1) the shares of the transport margins in the final consumption expenditure of households is the same (or very similar) across all COICOP categories dealing with products (services do not require transport); and (2) the private household final consumption of transport services in IOT in basic prices (such as EXIOBASE) is made of both accumulated transport margins and the direct consumption of transportation services.

4.4. Extended Input–Output Analysis

Extended input–output analysis is applied to provide the results on the ex-post time series analysis and the value chain analysis. The global pressures and effects (footprints) associated with final consumption of food products have been calculated with an extended multiregional Input-Model built from EXIOBASE data [27]. To this end, environmentally and socio-economically extended product-by-product tables were used. The core equations of the model are as follows:

$$x = A \cdot x + y \tag{1}$$

where x is the total output vector, A is the matrix of direct input coefficients (also referred to as the matrix of technological coefficients), and y is the final demand vector. Solving the equation for output transforms it into [28]:

$$x = (I - A)^{-1} \cdot y = L \cdot y \tag{2}$$

where I is the identity matrix, and L is the Leontief inverse also referred to as matrix of direct and indirect output requirements per unit produced for final demand or, more simply, multiplier matrix. The Leontief model implies the following assumptions [29]: prices are fixed in the short term, input coefficients are constant regardless of output or final demand level changes, structure of the economy is taken to be constant, at least in the reported period.

The direct environmental and socio-economic effects of national production are by definition the result of the sum of the direct effects associated with each unit produced in each industry:

$$E^T = \sum_1^n E_i = \sum_1^n e_i^{int} \cdot x_n = \left\langle e^{int} \right\rangle \cdot x \tag{3}$$

where E^T is the total environmental or socio-economic effect associated with the corresponding amounts of the final output x and e^{int} is the environmental or socio-economic effect intensity vector. Each element of e^{int} represents the amount of the effect directly caused by the production of a product group. E^T is also what we call the effect measured from the production perspective.

By substituting the vector x Equation (2) into Equation (3), an extended input–output model is created:

$$E^T = \left\langle e^{int} \right\rangle \cdot x = \left\langle e^{int} \right\rangle \cdot (I - A)^{-1} \cdot y \tag{4}$$

Applying Equation (4), the total footprint attributed to each of the different sectors of final demand is calculated. To carry out the calculation of the footprint related to each product group used to satisfy the final demand another expression of this model must be applied.

$$E^T = e^{int} \cdot x = e^{int} \cdot (I - A)^{-1} \cdot < y \geq = e^{acc} \cdot \langle y \rangle \tag{5}$$

where e^{acc} is the environmental or socio-economic effect intensity accumulated along the whole supply chain. Because all direct and indirect input requirements per unit of product group produced are represented in the Leontief inverse (L), their multiplication with the vector of the direct intensity (e^{int}) leads to the calculation of the environmental and socioeconomic accumulated intensity of each product group. The environmental or socioeconomic accumulated effect intensity (e^{acc}) is also called environmental or socioeconomic multiplier or accumulated technological effect.

The equations above are still valid in a multi-regional model such as EXIOBASE. In cases such as these, E^T simply consists of all individual country effects. We will aggregate individual country environmental or socio-economic effects into EEA countries' footprints.

The input–output calculations will generate measurements of footprints for the selected environmental and socio-economic effects associated with the final consumption of European private households. Such footprints measurements include not only production effects but also effects resulting from the ultimate use of the various products (e.g., the amounts of CO_2 resulting from gasoil production as well as from gasoil combustion). There will be a footprint for each product category available in the EXIOBASE nomenclature, distinguishing products domestically produced and imports. Each domestic and imported footprint will then be distributed to consumption purpose categories using the allocation tables described in Section 4.3. The operation is an element-wise multiplication of the environmental footprint vector (E^T) defined in Equation (5) with the "product-x-coicop category" allocation table (T^{alloc}) resulting in a (E^{alloc}) matrix as follows:

$$E_{ij}^{alloc} = E_i^T \cdot T_{ij}^{alloc} \tag{6}$$

4.5. Extrapolation

The extended input–output analysis using Exiobase v.3.4 results in ex-post time series results for pressures and effects for the period 1995–2011. A calculation of pressures and effects for the year 2017 is added via an extra extrapolation step. This extrapolation makes use of household expenditure data available from Eurostat (final consumption expenditure of households by consumption purpose (COICOP 3 digit) [nama_10_co3_p3]), which are corrected for inflation (HICP (2015 = 100)—annual data, average index and rate of change [prc_hicp_aind]), and the trend in intensities of the different pressures and effects from Exiobase. We used a linear trend based on the 1995–2011 data to estimate the 2017-intensities for household consumption of food products, total household consumption and total final demand. The linear trend is calculated through a given set of dependent y-values (i.e., intensities) and a set of independent x-values (i.e., years) and return values along the trend line, making use of the least squares method. Multiplying these extrapolated intensities with expenditures in constant prices results in total pressures and effects. Because the 2017-intensities are a result of the extrapolation, the analysis on intensities is limited to the 1995–2011 period. The 2017-footprint should be interpreted with caution, due to the uncertainty in forecasting the intensities.

4.6. Structural Decomposition Analysis

A change of a variable is a result of changes in the determinants of the variable. For instance, the production technology and the volume of final demand determine a change of the macro-economic indicator "gross production." Currently, two main methods are applied in the relevant literature to evaluate the contribution of determinants: SDA and Index Decomposition Analysis (IDA). Both methods allow to quantify the positive and negative contributions of each determinant over time. The historical variation of these determinants in turn results in the dynamics of change of macro indicators such as employment, value added, CO_2 emissions, etc. SDA uses input–output tables, IDA uses aggregate data at the sector-level. This analysis presents the results of an SDA.

The magnitude of the environmental and socio-economic effects resulting from production and consumption activities in an economy depends on:

(a) the direct effect per unit of a produced output,
(b) the technology applied in each sector, and
(c) the total volume and composition of final consumption.

Consequently, a structural decomposition of environmental and socio-economic effects of domestic demand basically is based on the quantification of three determinants of change:

- intensity effect (changes in the direct effect per unit output)
- technological effect, (changes in total requirements of intermediate per unit output)
- final demand effect (changes in total output volume and composition of final demand)

Formally, the quantification of the contributions is derived from the algebraic formula by which the environmentally extended input–output model is represented, as explained in Equation (4). However, those contributions can be reformulated when the following two aspects are taken into account. First, the final demand is determined by the absolute consumption volume as well as the structure or mix of consumed product groups. Therefore, it can be further disaggregated into both determinants. Second, both the contribution of the observed changes in the composition of the intermediate inputs required for production by each sector (Leontief-multiplier), as well as the contribution of the changes in the direct effect per unit produced in each sector (direct intensity effect) represent technological effects. Since these contributions are determined by the production characteristics prevailing in the industries of the economies from which the products originate, both "technology" contributions can be combined into a single determinant.

Rearranging the determinants this way, the quantification of the corresponding contributions that explain the change of the analyzed parameter ΔE^T (e.g., environmental pressure, valued added, employment) between two points in time (0 and t) can be expressed in its additive form, as follows:

$$\Delta E^T_{0-t} = I_{contribution} + Y^{vol}_{contribution} + Y^{str}_{contribution} \quad (7)$$

where $I_{contribution}$, $Y_{volcontribution}$ and $Y_{strcontribution}$ capture the "specific effects" in terms of accumulated technology effect (e^{acc} in Equation (5) or direct and indirect effect intensity), final demand volume effect, and final demand structure effect, respectively. For instance, if the decomposition of observed change in environmental footprint E^T caused by final demand in a defined geographical region and time period shows two specific effects with a positive value and one with a negative value, then effects with a value greater than zero have contributed to increasing the environmental or socio-economic impact, while the effect with a negative value has contributed to reducing it. In this way, the net effect is the sum of the different effects.

There are a number of algorithms available to reach the decomposition laid out in Equation (6), which can lead to significantly different results, see e.g., [30–33]. For example, when assessing the contribution resulting from changes in applied technology, we could use either beginning-of-period or end-of-period environmental multipliers for the calculation, which would lead to different results. To overcome the non-uniqueness issue, we will use the algorithm that has the particularity to be the best approximation of the average of all possible decomposition forms [34]. The mathematical formulae and their full derivations are presented in an extensive review of decomposition methods [35].

5. Results

The results section firstly describes the results of the ex-post time series analysis. It describes the contribution of food consumption to the footprints of total household consumption and total consumption in terms of two socio-economic effects, four resource use categories and three environmental impact categories. The results of other consumption domains are shown in Appendix A.

Secondly, the value chain analysis shows the share of these footprints that is located outside EEA member countries. Finally, the SDA assesses the contributions of determinants of changes in environmental or socio-economical footprints induced by private household food consumption.

5.1. Ex-Post Time Series Analysis

Ex-post time series analysis is applied to assess the contribution of food consumption to total consumption of households in EEA member countries, in terms of environmental impact, value added and jobs, and how this evolved over time. From a policy perspective, it is relevant to know whether European households are switching consumption patterns to goods and services with fewer environmental pressures, and more specifically, towards a more environmentally favorable diet.

In order to carry out the mentioned analysis, footprints for different years were calculated. For that, the ESE-MRIOM or Equation (5) was applied and the available data time series were used. Figure 1 presents a time series of the different footprints of food consumption by households, total household consumption (not only food) and total final consumption (incl. consumption by government, investments by industry, etc.) Indicators are grouped according to their focus, i.e., socio-economic parameters, resources use and impacts:

- related to food consumption by private households (COICOP 01 + 02.1);
- related to private household consumption of all other product groups in other COICOP categories;
- related to non-household final demand (e.g., investments, government expenses).

The total 2017 value is shown, which is the footprint per impact or resources use indexed to 100 on the vertical axis.

While overall final demand (in constant prices) grew by about 37% from 1995–2017, the environmental footprints of final consumption and private household consumption grew less (Global Warming Potential (GWP), energy, material and water consumption) or even decreased (acidification, eutrophication, land use). This indicates relative decoupling between GDP and the GWP, energy, material and water footprints of final consumption; with absolute decoupling occurring between GDP and acidification, eutrophication and land use footprints. The expenditure share of household consumption in total consumption was 47% in 2017 and stable compared to the 1995 and 2005 shares. This share is higher for all impacts and resource use, meaning that household consumption generates on average more impact and requires more resources compared to average investment and government expenditure.

The value added specifically for food consumption of private households also follows this overall trend, increasing by 5% between 1995 and 2017. The environmental footprints of food consumption by households increased much less, demonstrating relative decoupling with absolute decoupling for land use, acidification and eutrophication. In terms of expenditures (or gross value added) in 2017, the share of food consumption in total domestic final demand was 7% and its share in total household consumption was 14%. Compared to 1995 values, the shares slightly decreased, indicating that expenditure on food products by households has decreased in relation to total spending. For all impacts and resource use, the share caused by household food consumption in total household final demand was higher than the share of expenditures, except for energy use. The highest shares are for land use (71%), water consumption (61%), eutrophication (53%), acidification (48%) and material use (33%).

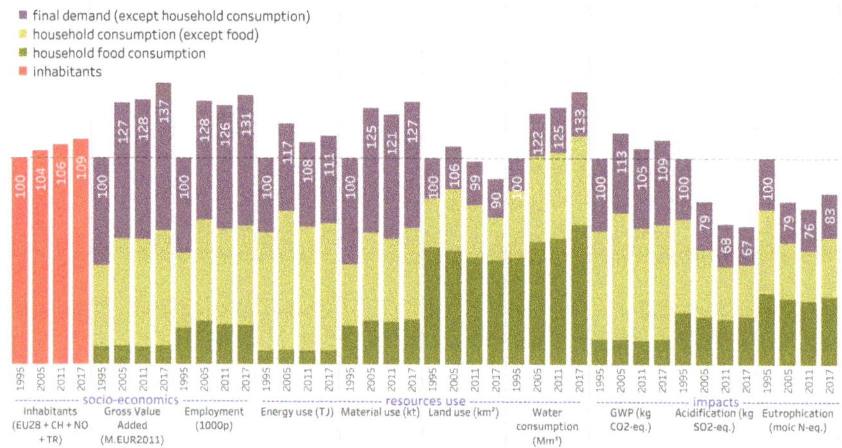

Figure 1. Impact and use of resources by consumption in EEA member countries, 1995–2005–2011–2017.

To assess if European households are switching consumption patterns to food products with fewer environmental pressures, thus towards a more environmentally favorable diet, we focus on the GWP, water and land use of household food consumption (Table 2). GWP remains more or less stable in total absolute terms although population increases, and thus shows a relative decoupling. However, absolute decoupling is close to being achieved as GWP showed a very small decrease between 1995 and 2017. The food products that contribute most to GWP are bread and cereals (01.1.1; 33–37%), meat (01.1.2; 25%) and milk, cheese and eggs (01.1.4; 17–21%), followed by fish (01.1.3; 8%) and fruit and vegetables (01.1.6 + 7; 8%). The product-specific global environmental pressures, i.e., E_{ij}^{alloc} in Equation (6), were analyzed over a shorter time period (1995–2011) (Table 2). The discussion on the trend in intensities is limited to the period 1995-2011, as the 2017 intensity is an extrapolation result of this time period (see Section 4.5). For GWP, the most important changes occurred for milk, cheese and eggs (reduction of 23%), 'other' food products (increase of 16%), oils and fats (increase of 11%), fish (reduction of 11%), bread and cereals (increase of 5%) and meat (reduction of 3%) Looking at the total water use related to household food consumption, this volume increased significantly by 30% between 1995 and 2017. The food products that contribute most to the water footprint are bread and cereals (01.1.1; 50%), fruit and vegetables (01.1.6–01.1.7; 20–25%) and meat (01.1.2; 12%). The most important relative changes in water use intensities of food products for household consumption occurred for bread and cereals (increase of 25%), fruit and vegetables (increase of 18%), fish (increase of 16%), meat (increase of 14%) and milk, cheese and eggs (reduction of 11%). The land use related to household food consumption reduced by 11% between 1995 and 2017. The food products that contributed most to land use are bread and cereals (01.1.1; 40–46%), meat (01.1.2; 21%), fruit and vegetables (01.1.6–01.1.7; 13–16%) and milk, cheese and eggs (01.1.4; 9–12%). The most important changes in land use intensities of food products for household consumption occurred for milk, cheese and eggs (reduction of 31%), fruit and vegetables (reduction of 20%) and fish (reduction of 19%). The land use related to meat was reduced slightly.

Table 2. Changes in household expenditure and intensity for different food products.

Change between 1995 and 2011	Household Expenditures	GHG Emission Intensity	Water use Intensity	Land Use Intensity
Bread and cereals (↑ 5% *)	↓ 5%	↑ 10%	↑ 32%	↑ 10%
Meat (↓ 3%)	stable	↓ 5%	↑ 12%	↓ 8%
Fish (↓ 11%)	↓ 5%	↓ 5%	↑ 25%	↓ 13%
Milk, cheese and eggs (↓ 23%)	stable	↓ 22%	↓ 10%	↓ 30%
Oils and fats (↑ 11%)	stable	↑ 9%	stable	↓ 29%
Fruit and vegetables (stable)	↓ 10%	↑ 11%	↑ 31%	↓ 12%
Sugar, jam, honey, chocolate and confectionary (↓ 6%)	↓ 6%	stable	↓ 23%	↓ 43%
Food products n.e.c. (↑ 16%)	↑ 57%	↓ 26%	↑ 12%	↓ 19%
Beverages (↑ 1%)	stable	stable	↑ 29%	↑ 8%

* Percentages relate to change in total global warming potential 'impact' between 1995 and 2011.

The changes in impact are a combined effect of the share of the type of food products in the household's diet (expenditures) and the environmental pressures (GHG emission, water and land use intensity) caused along the production chain of the respective food products. The latter can be caused by different factors not analyzed here, e.g., a changed basket of products (e.g., switch from beef to chicken meat), improved production efficiency, etc.

5.2. Value Chain Analysis

A value chain analysis is used to identify the part of the footprint, calculated by applying the ESE-MRIOM or Equation (5), that is located outside EEA member countries providing insight into which part of the world and in which sectors food consumption of households is creating environmental pressures and impacts, value added and jobs, and how this has evolved over time. From a policy perspective, it is important to know whether Europe is shifting the environmental burden to other regions by the changing food consumption patterns, and to know the related benefits i.e., employment and value added.

Figure 2 shows that a substantial and increasing (except for eutrophication) share of global environmental and socioeconomic effects caused by total final demand in EEA member countries occurs outside these countries. The contribution of resources extracted or used outside EEA member countries to the footprint of food consumption is illustrated by the estimate that more than half of total requirements for land use (57–61%) and water consumption (52–59%) occurred overseas and these are strongly correlated with agricultural production. The 'benefits' related to the reliance on imports are much less: the value added created outside EEA member countries by final demand was only 7% in 1995, increasing to 11% in 2011 (the discussion on the proportion of EEA's final demand footprint is limited to the period 1995–2011, as this proportion is derived directly from the Exiobase-model). This follows a rising trend but remains low. The same applies to the jobs created abroad due to final demand, the overseas share increased from 37 to 46% in the same period. Together, the figures on value added and employment suggest that, overall, final consumption created employment in low-value-added activities overseas.

The share of global environmental pressures and impacts generated outside EEA member countries by food consumption of households also showed an increasing trend between 1995 and 2011 (Figure 3). However, the share of resource use and environmental impacts exerted overseas from food consumption is smaller than the share generated by total final demand, with the exception of energy use where the overseas share related to food consumption is higher than that related to total final demand. The share of value added and jobs generated outside EEA member countries by households food consumption is larger than the share generated by total final demand. In 2011, 16% of the gross value added in the food production chain was generated outside EEA member countries (compared to 11% for total final demand) and 60% of the employment was located abroad. This means that food consumption in EEA member countries generates relatively less environmental impact abroad than average and creates relatively more value added there.

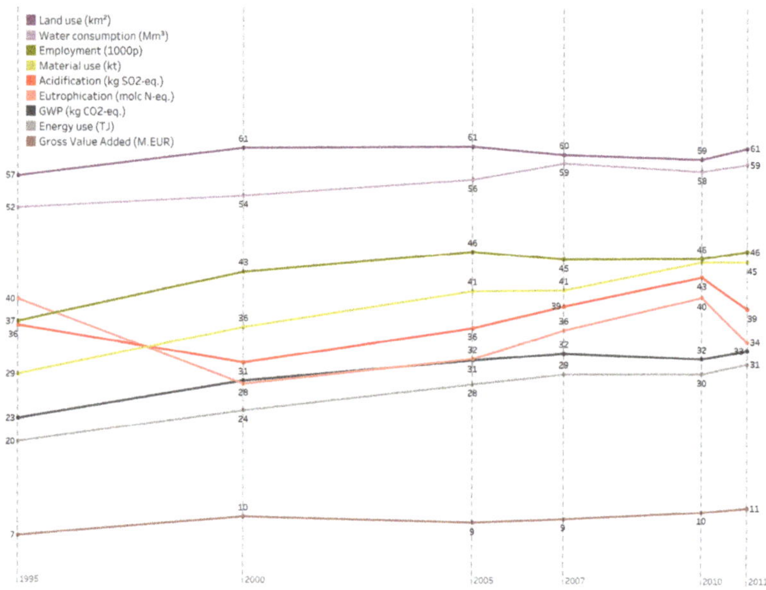

Figure 2. Proportion of EEA's final demand footprint exerted outside EEA's borders.

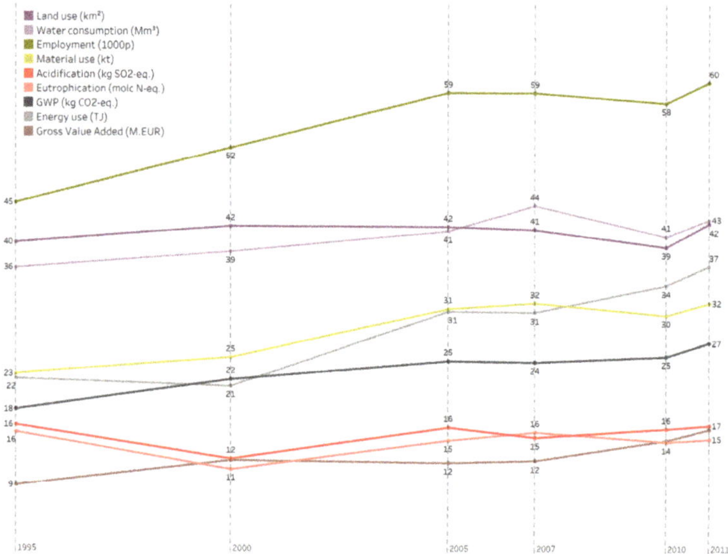

Figure 3. Proportion of EEA households' food consumption footprint exerted outside EEA's borders.

Table 3 below adds more regional detail to the share exerted outside Europe. Most of the value added by household food consumption is generated in Asia and the Pacific region (7%) along with most of the environmental impact and resource use, with the region's share increasing steadily from 1995 onwards.

Table 3. Share of impacts across world regions, generated by EEA households' food consumption. The geographical and sectoral distribution follow from the IOA, using Exiobase. Therefore, the 2011 results are presented and not the extrapolation value for 2017.

	Food Consumption by Households in EEA Countries								
Share in Geographical Regions (in 2011)	**Gross Value Added**	**Employment**	**Global Warming Potential**	**Acidification**	**Eutrophication**	**Energy Use**	**Land Use**	**Material Use**	**Water Use**
Europe	84%	40%	73%	83%	85%	63%	58%	68%	57%
North America	3%	1%	3%	1%	1%	7%	3%	2%	3%
South America	3%	5%	4%	4%	4%	2%	10%	8%	9%
Africa	2%	26%	4%	3%	2%	3%	13%	7%	11%
Asia and Pacific	7%	27%	13%	7%	6%	19%	13%	12%	16%
Middle East	2%	2%	4%	2%	1%	7%	2%	3%	3%

The value chain analysis enables a more detailed examination of the industries/sectors where food consumption causes environmental impacts and creates value added (Table 4). In 2011, 52% of the gross value added in the food production chain was linked to agriculture and food manufacturing. The other 48% was distributed across services (16%), transport (11%), trade (10%), energy (4%), plastic and chemicals (2%), metals (2%), electronics (1%) and minerals (1%). The resource use and impacts are more concentrated in the agriculture and food manufacturing, except for energy use. The relative importance of the different sectors has remained more or less stable over time.

Table 4. Share of impacts across industries/sectors worldwide, generated by EEA households' food consumption.

	Food Consumption by Households in EEA-Countries								
Share in Food Production Chain (in 2011)	**Gross Value Added**	**Employment**	**Global Warming Potential**	**Acidification**	**Eutrophication**	**Energy Use**	**Land Use**	**Material Use**	**Water Use**
Food products	52%	80%	66%	91%	94%	24%	100%	79%	100%
Textiles	0%	0%	0%	0%	0%	0%	0%	0%	0%
Paper and wood products	1%	1%	0%	1%	0%	1%	0%	1%	0%
Energy (related) products	4%	1%	16%	2%	1%	50%	0%	5%	0%
Plastics and chemicals	2%	1%	3%	1%	1%	7%	0%	3%	0%

Table 4. *Cont.*

Share in Food Production Chain (in 2011)	Food Consumption by Households in EEA-Countries								
	Gross Value Added	Employment	Global Warming Potential	Acidification	Eutrophication	Energy Use	Land Use	Material Use	Water Use
Mineral products	1%	1%	1%	0%	0%	1%	0%	11%	0%
Metal products	2%	1%	1%	1%	0%	1%	0%	2%	0%
Electronics	1%	0%	0%	0%	0%	0%	0%	0%	0%
Trade	10%	6%	2%	0%	0%	3%	0%	0%	0%
Transport	11%	4%	8%	3%	3%	9%	0%	0%	0%
Others	16%	5%	3%	0%	0%	3%	0%	0%	0%

While the value added in Europe by EEA member country households' food consumption is mainly generated in the food products industry, the value added in Asia/Pacific is generated in the paper and wood products sector as well. Employment outside Europe is located primarily in the food products industry, for which Asia/Pacific and Africa are important regions. In all regions, environmental impacts (e.g., GWP, acidification, land use and water use) are primarily caused by the food production sector, although in Asia/Pacific the share of the energy (related) products sector is equally important as the food production sector.

5.3. Structural Decomposition Analysis

The SDA enables quantification of the different contributions of three determinants to indicators of selected environmental footprints and socio-economic effects related to private household consumption in EEA member countries between 1995 and 2011. These determinants are:

- the accumulated intensity (resulting from changes in production technology applied in different industries): it reflects the total environmental or socio-economic unit per unit product used for satisfying the final demand. This accumulated intensity is the sum of the effects resulting from the use of all intermediate inputs at various stages of production along the supply chain, as well as of the effects during final use of each product group (e.g., the amount of CO_2 resulting from gasoil production, as well as from gasoil combustion by driving automobiles);
- the consumption structure (or changes in the mix of the product used by private households): it covers the overall effect of changes in the basket of consumed product groups;
- consumption volume (or changes in total volume of products used by private households): it refers to the influence of growth or reduction in total consumption expenditures of private households.

The SDA was carried out applying Equation (7) and using the version product-by-product of the multiregional input–output tables, which describes the intermediate and final use of 200 product groups in the global economy. For this assessment, an aggregated version of the EE-MRIOT was used, in which only the economies of EEA countries as whole and the rest of the world (RoW) are represented. The results of the SDA are not directly comparable with the footprint calculations based on the fully disaggregated EXIOBASE multi-regional input output table (MRIOT). To avoid confusion with absolute footprint values resulting from the fully disaggregated EXIOBASE, we normalized the SDA results. The reference (index 100) is the total footprint of private household consumption in EEA countries in 1995, calculated with the geographically aggregated (EEA+RoW) EXIOBASE.

Absolute changes in footprint levels and absolute contributions of the considered determinants are both calculated from the aggregated EXIOBASE.

5.3.1. How to Read the Charts

The solid line shows the sum of the changes caused by these factors, i.e., actual developments of the environmental or socio-economic effects compared to 1995. The bars show changes in the direct and indirect effect "intensities" (accumulated technological effect), changes in the "consumption mix" (structure of total production used by households) and changes in the "consumption volume" (total volume of household expenditure). Each factor has contributed to the change of the assessed environmental or socio-economic macro-indicator compared to 1995 levels. All results are normalized and relate to food-related consumption of private households in EEA member countries.

5.3.2. Global Warming Potential (GWP)

Compared to 1995, GHG caused by food-related consumption of households increased slightly in 2000, 2005 and 2007, and were slightly lower in 2010 and 2011 (Figure 4). In 2000, 2005 and 2007, the increase in the volume of food products for household consumption was greater than the increase in GHG emissions from food consumption. In 2010 and 2011, emissions decreased partially even though food consumption expenditure kept increasing. This, however, is not an indication of a decoupling of GWP from the volume of household food expenditure during this period. The SDA shows that the observed decline in GWP is essentially the result of technological innovations in production processes along the global supply chain that led to a reduction in cumulative emission intensities (direct and indirect emissions). Changes in the mix of products used by households also contributed, albeit to a lesser extent and with some intermittency, to the reduction of GWP. In other words, the basket of food-related products defined by private households' food consumption has partly shifted towards groups of goods and services with lower GWP compared to 1995. This shift, although significant, was not constant, with fluctuations over time.

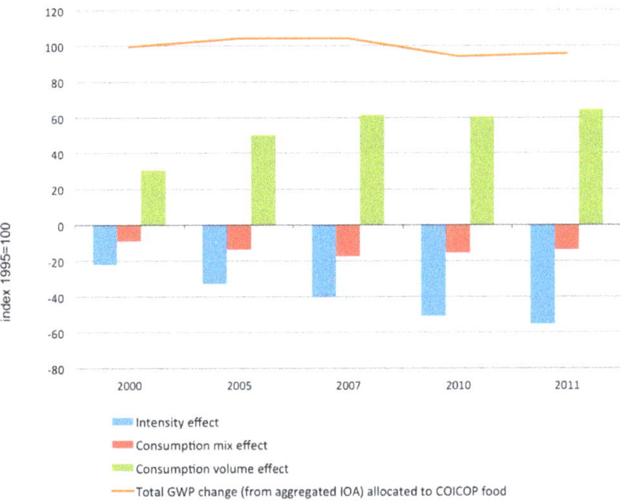

Figure 4. Normalized changes in (global) GWP footprint compared to 1995 caused by private households' food-related consumption in the EEA member countries, and decomposition into contributing factors.

In short, although the net total effect shown indicates some decoupling between GWP and food consumption, there are no clear signs of absolute decoupling due to increased total volumes of food consumed.

5.3.3. Employment

Compared to 1995, the employment generated worldwide by food consumption of private households increased in 2000 and 2005, before declining to 1995 levels and then slightly increasing between 2010 and 2011 (Figure 5). Increased productivity in global supply chains appears to significantly reduce employment induced by private household food consumption. Changes in the composition of the basket of food-related goods and services also led to declining employment levels, albeit to a lesser extent. Compared to 1995, these observed changes show that food consumption in private households has shifted slightly towards groups of goods and services produced with high labor productivity (i.e., low accumulated labor intensity). However, the increase in the volume of household food consumption expenditure pushed up employment levels, more than offsetting the decline in employment induced by the other two determinants of total change.

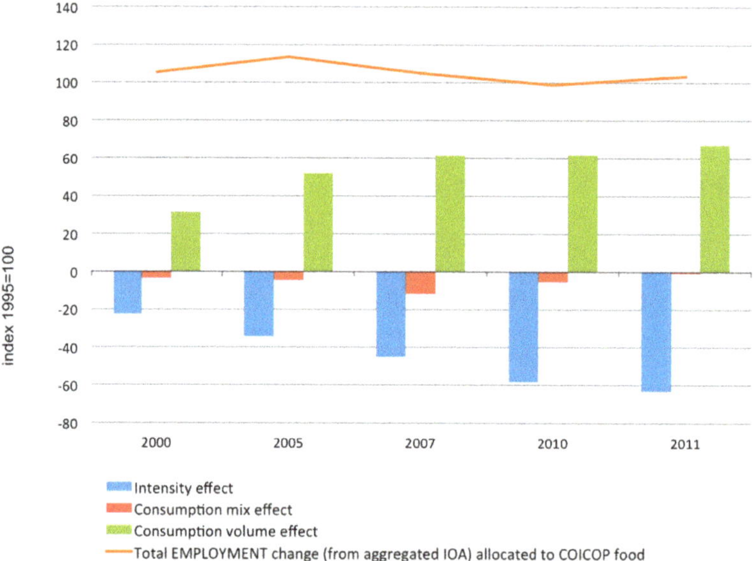

Figure 5. Normalized changes in (global) employment footprint compared to 1995 caused by private households' food-related consumption in the EEA member countries, and decomposition into contributing factors.

5.3.4. Gross Value Added (GVA)

Compared to 1995, the global GVA generated by food-related consumption of private households steadily increased in the years 2000, 2005, 2007, 2010 and 2011 (Figure 6). The trend is clearer than that for employment, culminating in a total GVA growth of more than 30% compared to 1995. The upward trend in GVA is clearly the result of the increase in the expenditure volume for food products by households which grew more than 50% in 2011 compared to 1995. In contrast, changes in the structure of the products consumed (product mix), as well as in the production technology applied along the global supply chain (cumulative technology effect), had a negative effect on the generation of value added. However, the technology-induced reduction did not go beyond 20%, resulting in a net increase of approximately 30%.

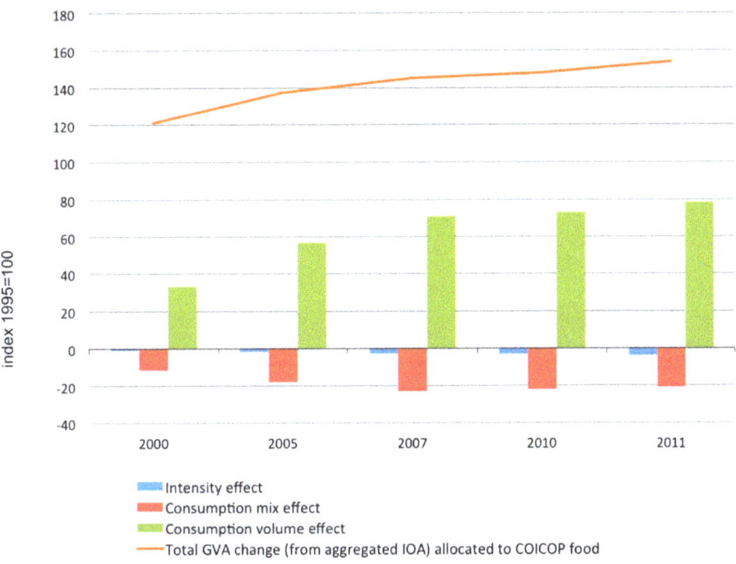

Figure 6. Normalized changes in (global) GVA footprint compared to 1995 caused by private households' food-related consumption in the EEA member countries, and decomposition into contributing factors.

In contrast to GHG emissions or employment, changes in production patterns (intensity or direct or indirect effect of the technology) hardly influence the level of GVA induced by food consumption. This might indicate a high degree of efficiency as further improvements did not have considerable economic effects. With regard to the effect induced by changes in the product mix, it can be concluded that food consumption is shifting towards a basket of products and services that has a negative impact on GVA (cheaper food products).

6. Discussion and Conclusions

This paper assessed the effects of food consumption in EEA member countries by quantifying a series of socio-economic and environmental indicators (resource use, air emissions, gross value added and employment) and analyzed the main driving forces by means of SDA. The results are based on detailed input–output data from Exiobase, having attributed products to different consumption purposes by means of an ad-hoc allocation. Most studies only briefly discuss the conversion of household consumption expenditure for products supplied by sectors to the COICOP categories. Articles provide neither detailed allocation tables nor supplementary material on the allocation process which limits the reproducibility of results. Since different allocations lead to different results, this can lead to erroneous conclusions about the reliability of assessments and give the impression of a lack of transparency or even arbitrariness of allocations. The missing additional information, the rather sparse discussion of analytical consequences and the missing standardization is a considerable gap in the literature. Therefore, this paper has carefully described the methodology with emphasis on the allocation. As indicated in Section 4.3, assumptions had to be made during construction of the allocation matrix, which, in combination with a lack of detailed data, resulted in some limitations of the analysis. For example, there was a focus on private households rather than a wider range of product groups or services. However, the procedure to develop the allocation matrices is presented in a fully transparent and detailed manner, thus contributing to methodological improvements in integrated assessment, and the application of the different analytical methods answers the main research questions.

6.1. Ex-Post Time Series Analysis

The results of the ex-post time series analysis are in line with other studies based on similar types of assessments. These also demonstrate that food consumption is a predominant driver of environmental impacts in categories such as acidification, eutrophication and land use, which are typically consequences of agricultural activities [7,36]. Figure 1 summarizes the assessment of the main socio-economic and environmental indicators of consumption. Animal-based products such as meat, dairy and eggs are the food products which have been identified as responsible for a major part of the impacts (more than 50%) [35]. The analysis in this study identifies the same type of food products as the highest contributors, although not so distinctly and not for all impact categories (Table 2).

With regard to the evolution over time, the growth in consumption in combination with an increasing population have been confirmed as drivers of increased environmental impact in most categories [35]. However, the increase in environmental pressures is overall lower than the increase in consumption which indicates a relative decoupling. Our findings also suggest an absolute decoupling in some impact categories such as land use. A relative decoupling is observed in relation to GWP, which remains more or less stable in absolute terms even though population has increased. Between 1995 and 2017 our assessment suggests even slight tendencies of absolute decoupling. The dynamics of the GWP are mainly influenced by the share of specific types of food products in household expenditures and the GHG emission intensity of the production chain of the respective food products.

6.2. Value Chain Analysis

The value chain analysis emphasizes the importance of imports for European household consumption, resulting in environmental pressures being exerted outside of Europe. Our analysis shows, however, that food consumption in EEA member countries generates relatively less environmental pressure abroad than on average, and creates relatively more value added. From the perspective of trade, imports of agricultural and food products in Europe are important contributors to eutrophication, water and land use induced by imports, and these impacts are increasing over time, however less than their value added [35]. There has been growth of emissions in imports between 1995 and 2015 along with growing importance of imports from Asia and related pressures in that region [36]. The latter study observes that 37% of agricultural emissions caused by European consumption patterns occur outside of the EU, which confirms our findings.

6.3. Structural Decomposition Analysis

The results of the SDA highlight the environmental and socioeconomic effects resulting from technological development and changing mix of consumed products, and the counteracting increases resulting from expanded consumption.

6.3.1. Technology and Mix of Consumed Products (Cumulative Intensity and Effects of the Consumption Mix)

Technological innovation along the global supply chain, as well as changes in the mix of consumed products, has substantially reduced direct and indirect greenhouse gas emissions. However, the effect of the changing consumption mix was intermittent and much less prominent than the cumulative intensity effect resulting from technological innovation in production processes. In other words, since 1995, the basket of food-related products consumed by private households has partly shifted towards goods and services with lower GWP. However, technological progress has been the most important determinant in reducing greenhouse gas emissions. With regard to employment, the results show a pattern similar to the changes observed in terms of GWP. Technological innovation in production processes significantly reduces employment along the supply chain. This effect was supported by changes in the mix of consumed products, however, with a much lower impact. The SDA results show that, in contrast to GHG or employment, the mix of products consumed by households has a much greater impact on the gross value added induced by the consumption of food products than technological developments. The changes observed for the period between 2000 and 2007 associated with the product mix indicated that household food consumption changed so it is made up of products that induce directly and indirectly a lower value added than in previous periods; however, this trend has reversed since 2007. On the other hand, productivity improvements in production systems have avoided a substantial increase in the costs of direct and indirect intermediate products and services associated with the production of food consumed by households. These trends could be further improved through policy instruments, such as circular economy initiatives which would make it possible to reduce production costs without affecting value added and with a product mix that includes labor-intensive foods with a higher value added (for example, due to the consumption of organic products).

Comparing the SDA results on employment, GVA and GWP highlight that technological progress has resulted in a productivity increase. In combination with changes in consumption patterns, this has increased GVA and reduced GHG emissions as well as employment. In other words, food consumption by households is associated with a production of goods and services with lower production costs, which are significantly cleaner, but at the same time induce less employment.

6.3.2. Growth of Consumption

Regardless of the variable, the results of the footprint analysis clearly show that the net effect of food consumption by households is essentially determined by the total volume of consumption of the related product groups. Although there are considerable uncertainties related to the presented SDA, it confirms the role of growth in consumption volumes offsetting reductions in environmental pressures from improved productivity and changes in the product mix. Even though products have become relatively cheaper and less labor-intensive, the increase in consumption to a large extent compensates for the productivity increases, resulting in a more or less constant employment and an absolute growth in gross value added.

6.3.3. Conclusions

The analyses presented here contribute to the knowledge base demonstrating that food consumption is an important driver of environmental pressures and impacts. This is the case particularly for impact categories such as acidification, eutrophication and land use, which are related to agricultural activities, with animal-based products like meat, dairy and eggs responsible for a major part of these impacts. However, there has been relative decoupling between environmental pressures and consumption over time, and European food consumption generates relatively less environmental pressures outside of Europe (due to imports) than the average European consumption.

The results of the footprint analysis clearly show that the net effect of food consumption by households is essentially determined by the total volume of consumption of the related product groups. The SDA confirms the role of growth in consumption volumes offsetting reductions in environmental pressures from improved productivity and changes in the product mix.

These analyses also highlight trade-offs rather than synergies between employment, environment and economic growth, especially between economic growth and the environment in a "full world" [37,38]. This is in line with recent findings of the International Resource Panel who, in the latest resource outlook, analyzed drivers of the material footprint in seven world regions and globally [39]. For two periods (1990–2000 and 2000–2016), economic growth ("affluence") was by far the most prominent driver of domestic extraction of natural resources in Europe, offsetting considerable technological gains.

Even though we acknowledge that our findings are rather pessimistic in relation to mainstream discussions about the potentials of a European green economy, the empirical pattern seems to be clear and in line with similar assessments, e.g., [36,39,40]. This highlights the need to increase research on IO data and models and highly detailed analytical approaches that enable analysis of direct and indirect impacts of and interlinkages within the food system. These approaches and models can have an important role within the new EU research and innovation agenda associated with the EU Farm to Fork Strategy and Horizon Europe 2021–2027 (Cluster 6: Food, bio-economy, natural resources, agriculture and environment). The Farm to Fork Strategy (COM(2020)381) adopts an objective of achieving a neutral or positive environmental impact food chain, and Horizon Europe's Mission Board for 'Soil health and food' indicates a "20–40% reduced global footprint of EU's food and timber imports on land degradation" as a 'mission' for EU research and innovation [5]. Such tools used in combination with behavioral and social research can provide an advanced knowledge base to support these ambitious policy and research objectives.

Author Contributions: All authors contributed equally to the conceptualization, methodology, formal analysis, data curation, visualization, draft preparation, review and editing. All authors have read and agreed to the published version of the manuscript.

Funding: This research received no external funding and was undertaken as part of the joint work programme of the European Environment Agency and European Topic Centre on Waste and Materials in a Green Economy.

Acknowledgments: The authors would like thank colleagues at the European Environment Agency, Joint Research Centre and Eurostat for useful discussions. Usual disclaimer apply.

Conflicts of Interest: The authors declare no conflict of interest.

Appendix A. Direct and Indirect Pressures per Euro Expenditure within Different Household Consumption Categories

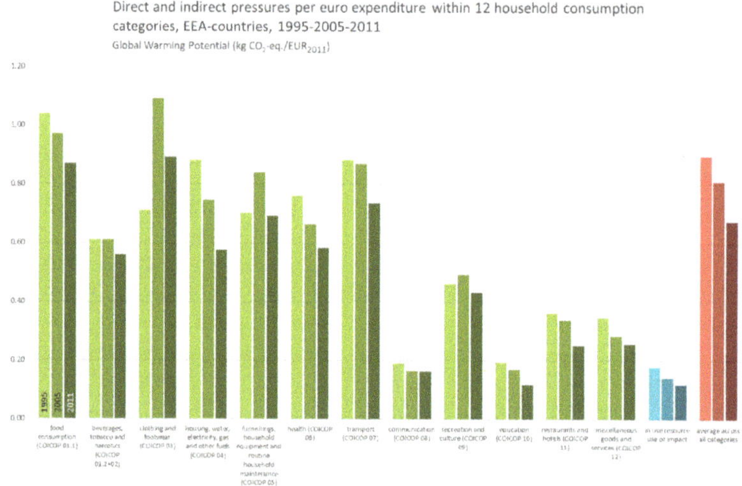

Figure A1. Direct and indirect pressures (global warming potential) per euro expenditure within 12 household consumption categories, EEA countries, 1995–2005–2011.

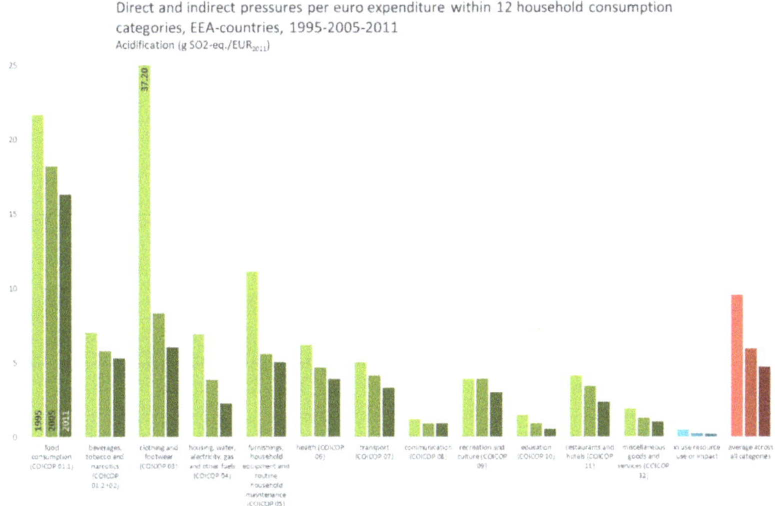

Figure A2. Direct and indirect pressures (acidification) per euro expenditure within 12 household consumption categories, EEA countries, 1995–2005–2011.

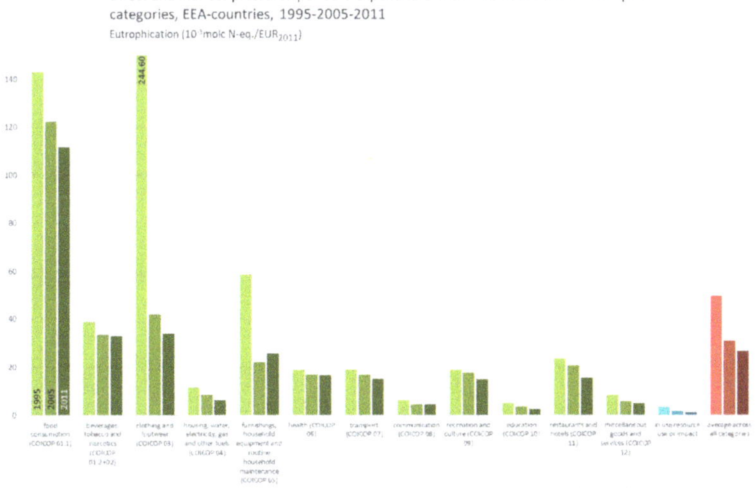

Figure A3. Direct and indirect pressures (eutrophication) per euro expenditure within 12 household consumption categories, EEA countries, 1995–2005–2011.

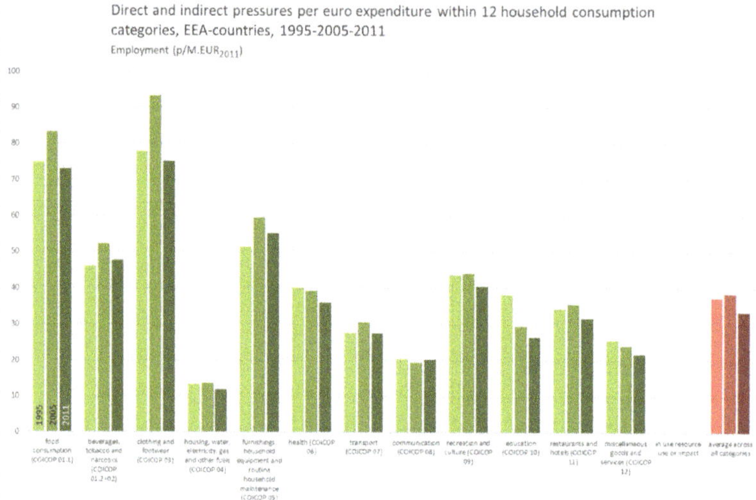

Figure A4. Direct and indirect pressures (employment) per euro expenditure within 12 household consumption categories, EEA countries, 1995–2005–2011.

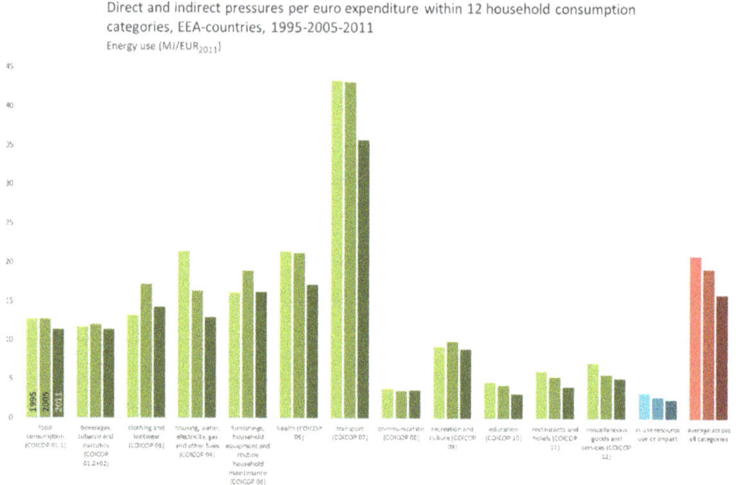

Figure A5. Direct and indirect pressures (energy use) per euro expenditure within 12 household consumption categories, EEA countries, 1995–2005–2011.

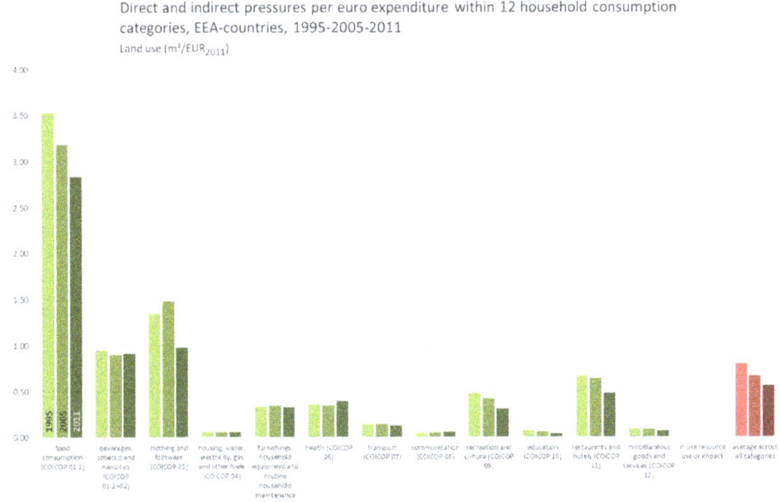

Figure A6. Direct and indirect pressures (land use) per euro expenditure within 12 household consumption categories, EEA countries, 1995–2005–2011.

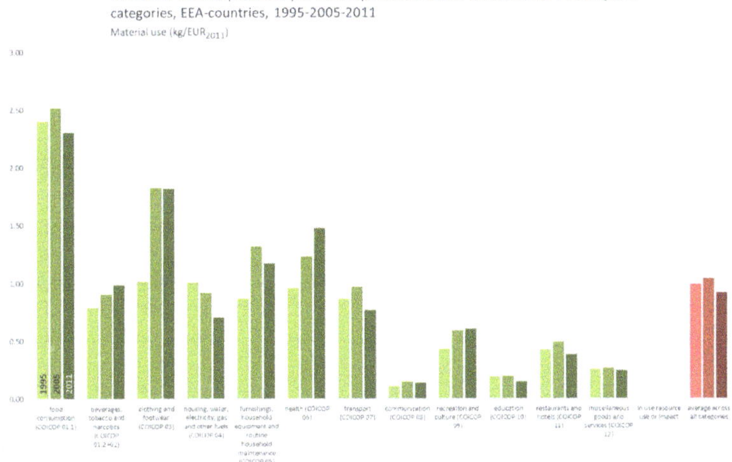

Figure A7. Direct and indirect pressures (material use) per euro expenditure within 12 household consumption categories, EEA countries, 1995–2005–2011.

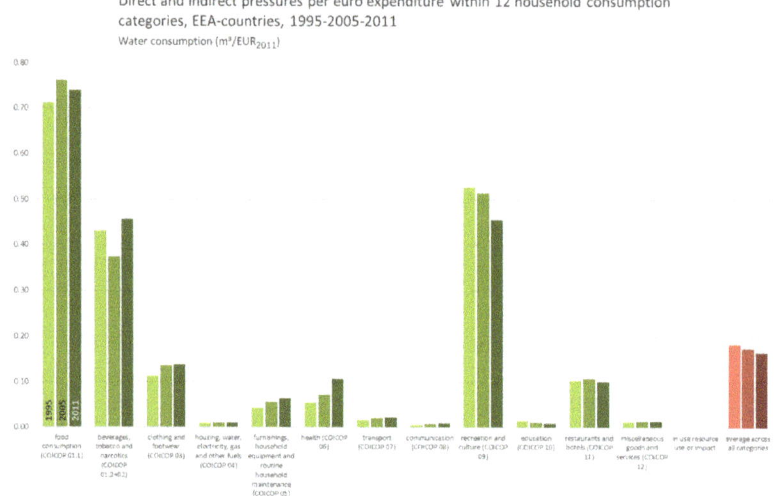

Figure A8. Direct and indirect pressures (water consumption) per euro expenditure within 12 household consumption categories, EEA countries, 1995–2005–2011.

References

1. EEA. *The European Environment–State and Outlook 2020. Knowledge for Transition to a Sustainable Europe*; European Environment Agency: Copenhagen, Denmark, 2019.
2. Willett, W.; Rockström, J.; Loken, B.; Springmann, M.; Lang, T.; Vermeulen, S.; Garnett, T.; Tilman, D.; DeClerck, F.; Wood, A.; et al. Food in the Anthropocene: The EAT–Lancet Commission on healthy diets from sustainable food systems. *Lancet* **2019**, *393*, 447–492. [CrossRef]
3. EEA. Food in a green light. A system approach to sustainable food. In *EEA Report 16/2017*; European Environment Agency: Copenhagen, Denmark, 2017.
4. IPES FOOD. *Towards a Common Food Policy for the EU-The Policy Reform and Realignment that is Required to Build Sustainable Food Systems in Europe*; International Panel of Experts on Sustainable Food Systems: Brussels, Belgium, 2019.
5. European Commission. Caring for Soil is Caring for Life Ensure 75% of Soils Are Healthy by 2030 for Healthy Food, People, Nature and Climate. Interim Report of the Mission Board for Soil Health and Food, Independent Expert Report. 2020. Available online: https://op.europa.eu/en/web/eu-law-and-publications/publication-detail/-/publication/32d5d312-b689-11ea-bb7a-01aa75ed71a1 (accessed on 5 October 2020).
6. EEA. Environmental pressures from European consumption and production—A study in integrated environmental and economic analysis. In *EEA Technical Report No 2/2013*; European Environment Agency: Copenhagen, Denmark, 2013; ISSN 1725-2237.
7. Ivanova, D.; Stadler, K.; Steen-Olsen, K.; Wood, R.; Vita, G.; Tukker, A.; Hertwich, E.G. Environmental Impact Assessment of Household Consumption. *J. Ind. Ecol.* **2016**, *23*, 526–536. [CrossRef]
8. Notarnicola, B.; Tassielli, G.; Renzulli, P.A.; Castellani, V.; Sala, S. Environmental impacts of food consumption in Europe. *J. Clean. Prod.* **2017**, *140*, 735–765. [CrossRef]
9. Nemecek, T.; Jungbluth, N.; i Canals, L.M.; Schenck, R. Environmental impacts of food consumption and nutrition: Where are we and what is next? *Int. J. Life Cycle Assess.* **2016**, *21*, 607–620. [CrossRef]
10. Tukker, A.; Baush-Goldbohm, S.; Verheijden, M.; Koning, A.; Kleijn, R.; Wolf, O.; Dominguez, L. *Environmental impacts of diet changes in the EU*; EUR 23783 EN; European Commission, Joint Research Center, Institute for Prospective Technological Studies: Seville, Spain, 2009.
11. Rahmani, R.; Bakhshoodeh, M.; Zibaei, M.; Heijman, W.; Eftekhari, M.H. Economic and Environmental Impacts of Dietary Changes in Iran: An Input-Output Analysis. *Int. J. Food Syst. Dyn.* **2011**, *2*, 447–463.

12. Reynolds, C.J.; Piantadosi, J.; Buckley, J.; Weinstein, J.D.; Boland, P. Evaluation of the environmental impact of weekly food consumption in different socio-economic households in Australia using environmentally extended input–output analysis. *Ecol. Econ.* **2015**, *111*, 58–64. [CrossRef]
13. Jungbluth, N.; Itten, R.; Schori, S. Environmental impacts of food consumption and its reduction potentials. In Proceedings of the 8th International Conference on LCA in the Agri-Food Sector, Rennes, France, 2–4 October 2012.
14. Huysman, S.; Schaubroeck, T.; Goralczyk, M.; Schmidt, J.; Dewulf, J. Quantifying the environmental impacts of a European citizen through a macro-economic approach, a focus on climate change and resource consumption. *J. Clean. Prod.* **2016**, *124*, 217–225. [CrossRef]
15. Nijdam, D.; Wilting, H.; Goedkoop, M.; Madsen, J. Environmental load from Dutch private consumption. How much damage takes place abroad? *J. Ind. Ecol.* **2005**, *9*, 147–168. [CrossRef]
16. Vringer, K.; Benders, R.; Wilting, H.; Brink, C.; Drissen, E.; Nijdam, D.; Hoogervorst, N. A hybrid multi-region method (HMR) for assessing the environmental impact of private consumption. *Ecol. Econ.* **2010**, *69*, 2510–2516. [CrossRef]
17. Kronenberg, T. The impact of demographic change on energy use and greenhouse gas emissions in Germany. *Ecol. Econ.* **2009**, *68*, 2637–2645. [CrossRef]
18. Bilsen, V.; Vincent, C.; Vercalsteren, A.; Van der Linden, A.; Geerken, T.; Vandille, G.; Avonds, L. *Het Vlaams Uitgebreid Milieu Input-Output Model, Idea Consult*; Vito en Federaal Planbureau: Mol, Belgium, 2010.
19. Kratena, K.; Streicher, G.; Temurshoev, U.; Amores, A.F.; Arto, I.; Mongelli, I.; Neuwahl, F.; Rueda-Cantuche, J.M.; Andreoni, V. *FIDELIO 1–Fully Interregional Dynamic Econometric Long-term Input-Output Model for the EU27, JRC Scientific and Policy Reports, JRC81864*; Joint Research Center/Institute for Prospective Technology Studies, European Commission: Seville, Spain, 2013.
20. Jokubauskaite, S. The integration of (QU)AIDS and input-output analysis in a panel data setting. Master's Thesis, Universität Wien, Vienna, Austria, 2015.
21. Kronenberg, T. *On the Intertemporal Stability of Bridge Matrix Coefficients, STE Preprint 17/2011*; Forschungszentrum Jülich: Jülich, Germany, 2011.
22. Washizu, A.; Nakano, S. On the environmental impact of consumer lifestyles–using a Japanese environmental Input–Output table and the linear expenditure system demand function. *Econ. Syst. Res.* **2010**, *22*, 181–192. [CrossRef]
23. Druckman, A.; Jackson, T. The bare necessities: How much household carbon do we really need? *Ecol. Econ.* **2010**, *69*, 1794–1804. [CrossRef]
24. Mongelli, I.; Neuwahl, F.; Rueda-Cantuche, J.M. Integrating a household demand system in the input–output framework. Methodological aspects and modelling implications. *Econ. Syst. Res.* **2010**, *22*, 201–222. [CrossRef]
25. Stadler, K.; Wood, R.; Bulavskaya, T.; Södersten, C.-J.; Simas, M.; Schmidt, S.; Usubiaga, A.; Acosta-Fernández, J.; Kuenen, J.; Bruckner, M.; et al. EXIOBASE 3: Developing a Time Series of Detailed Environmentally Extended Multi-Regional Input-Output Tables. *J. Ind. Ecol.* **2018**, *22*, 502–515. [CrossRef]
26. Eurostat. *Eurostat Manual of Supply, Use and Input-Output Tables*; Statistical Office of the European Union: Luxembourg, 2008; ISBN 978-92-79-04735-0.
27. Tukker, A.; de Koning, A.; Wood, R.; Hawkins, T.; Lutter, S.; Acosta, J.; Rueda Cantuche, J.M.; Bouwmeester, M.; Oosterhaven, J.; Drosdowski, T.; et al. EXIOPOL–Development and illustrative analyses of a detailed global MR EE SUT/IOT. *Econ. Syst. Res.* **2013**, *25*, 50–70. [CrossRef]
28. Miller, R.; Blair, P.R. *Input–Output Analysis: Foundations and Extensions*, 2nd ed.; Cambridge University Press: Cambridge, UK, 2009.
29. Oosterhaven, J. Leontief versus Ghoshian price and quantity models. *South. Econ. J.* **1996**, *62*, 750–759. [CrossRef]
30. Dietzenbacher, E.; Los, B. Structural Decomposition Techniques: Sense and sensitivity. *Econ. Syst. Res.* **1998**, *10*, 307–323. [CrossRef]
31. De Haan, M. A structural decomposition analysis of pollution in the Netherlands. *Econ. Syst. Res.* **2001**, *13*, 181–196. [CrossRef]
32. De Boer, P. Additive structural decomposition analysis and index number theory: An empirical application of the Montgomery decomposition. *Econ. Syst. Res.* **2008**, *20*, 97–109. [CrossRef]

33. Ang, B.W.; Huang, H.C.; Mu, A.R. Properties and linkages of some index decomposition analysis methods. *Energy Policy* **2009**, *37*, 4624–4632. [CrossRef]
34. Hoekstra, R.; van der Bergh, J.C.J.M. Comparing structural and index decomposition analysis. *Energy Econ.* **2003**, *25*, 39–64. [CrossRef]
35. Sala, S.; Beylot, A.; Corrado, S.; Crenna, E.; Sanyé-Mengual, E.; Secchi, M. *Indicators and Assessment of the Environmental Impact of EU Consumption*; Consumption and Consumer Footprint for Assessing and Monitoring EU Policies with Life Cycle Assessment; European Commission: Luxembourg, 2019; ISBN 978-92-79-99672-6. [CrossRef]
36. Wood, R.; Neuhoff, K.; Moran, D.; Simas, M.; Grubb, M.; Stadler, K. The Structure, Drivers and Policy Implications of the European Carbon Footprint. *Climate Policy* **2019**, *20*, S39–S57. [CrossRef]
37. Daly, H. A new economics for our full world. In *Handbook on Growth and Sustainability*; Edward Elgar Publishing: Cheltenham, UK, 2017.
38. Jackson, T. *Prosperity without Growth: Economics for a Finite Planet*; Routledge: Abingdon-on-Thames, UK, 2009.
39. Oberle, B.; Bringezu, S.; Hatfield-Dodds, S.; Hellweg, S.; Schandl, H.; Clement, J.; Ekins, P. *Global Resources Outlook 2019: Natural Resources for the Future We Want*; United Nation Environment Programme: Nairobi, Kenya, 2019.
40. Vivanco, D.F.; Kemp, R.; van der Voet, E. How to deal with the rebound effect? A policy-oriented approach. *Energy Policy* **2016**, *94*, 114–125. [CrossRef]

Article

A Worldwide Hotspot Analysis on Food Loss and Waste, Associated Greenhouse Gas Emissions, and Protein Losses.

Xuezhen Guo *, Jan Broeze, Jim J. Groot, Heike Axmann and Martijntje Vollebregt

Fresh Food & Chains Department, Wageningen Food & Biobased Research, PO Box 17, 6700 AA Wageningen, The Netherlands; jan.broeze@wur.nl (J.B.); jim.groot@wur.nl (J.J.G.); heike.axmann@wur.nl (H.A.); martijntje.vollebregt@wur.nl (M.V.)

* Correspondence: Xuezhen.guo@wur.nl

Received: 21 August 2020; Accepted: 7 September 2020; Published: 11 September 2020

Abstract: Reducing food loss and waste (FLW) is prioritized in UN sustainable development goals (SDG) target 12.3 to contribute to "ensure sustainable consumption and production patterns". It is expected to significantly improve global food security and mitigate greenhouse gas (GHG) emissions. Identifying "hotspots" from different perspectives of sustainability helps to prioritize the food items for which interventions can lead to the largest reduction of FLW-related impacts. Existing studies in this field have limitations, such as having incomplete geographical and food commodity coverage, using outdated data, and focusing on the mass of FLW instead of its nutrient values. To provide renewed and more informative insights, we conducted a global hotspot analysis concerning FLW with its associated GHG emissions and protein losses using the most recent data (the new FAO Food Balance Sheets updated in 2020). The findings of this research are that there were 1.9 Gt of FLW, 2.5 Gt of associated GHG emissions, and 0.1 Gt of associated protein losses globally in 2017. The results of the FLW amounts, GHG emissions, and protein losses per chain link are given on the scale of the entire world and continental regions. Next to this, food items with relatively high FLW, GHG emissions, and protein losses are highlighted to provide the implications to policymakers for better decision making. For example, fruits and vegetables contribute the most to global FLW volumes, but the product with the highest FLW-associated GHG emissions is bovine meat. For bovine meat, FLW-associated GHG emissions are highest at the consumer stage of North America and Oceania. Oil crops are the major source of protein losses in the global food chain. Another important finding with policy implications is that priorities for FLW reduction vary, dependent on prioritized sustainability criteria (e.g., GHG emissions versus protein losses).

Keywords: food loss and waste; GHG emissions; protein losses; global food supply chains; hotspots

---

1. Introduction

According to World Hunger Statistics, about 1 in 9 people globally do not have enough food to lead a healthy active life. The situation can be even worse in the future with the fast-growing world population. The 2050 world population is estimated at 9 billion and it requires a 70% increase in food production to meet people's demands for food [1]. Despite the precious value of food, it is estimated that a quarter to one-third of food produced for human consumption is lost or wasted. This makes food loss and waste (FLW) a hot research topic, especially during the last five years (e.g., Fiore et al. [2], Fiore et al. [3], Spada et al. [4], Adamashvili et al. [5], Pellegrini et al. [6], Vittuari et al. [7], Vittuari et al. [8], Pagani et al. [9], Ishangulyyev et al. [10]). FLW does not only escalate the issue of food security but also significantly contributes to anthropogenic greenhouse gas (GHG) emissions; many, including Springmann et al. [11], estimated that "halving food loss and waste would reduce environmental pressures by 6–16% compared with the baseline projection". Based on expected production volumes and global dietary changes, Hiç et al. [12] estimated that in 2050 the global FLW-associated GHG emissions in the production phase only will equal the total GHG emissions of the US in 2011. Therefore, FLW is also a serious problem related to climate change.

Given the importance of the topic, efforts have been made for quantifying FLW and associated GHG emissions to provide data-based decision support. However, much of the existing literature just focuses on a limited set of chain stages and specific countries with limited detailing of food commodities [13]. For example, the studies by Quested et al. [14], Wenlock and Buss [15], and Wenlock et al. [16] limited the scope to the household level. Buzby and Hyman [17] and Kantor et al. [18] studied FLW in the US downstream supply chain stages.

Xue et al. [19] conducted a literature review on FLW-related studies. They find that existing studies' spatial coverage is quite limited, with a strong focus on the developed world (e.g., Kling [20], Harrington et al. [21], Engström and Carlsson-Kanyama [22], Williams et al. [23], Leal Filho and Kovaleva [24]). Moreover, existing studies emphasize the retail and consumer stages but overlook the upstream food supply chain (e.g., Hodges et al. [25], Fehr and Romão [26], Jones [27], Loke and Leung [28], Stenmarck et al. [29]). Those studies generate useful information on FLW in the specific country and food commodity combinations but fail to show a bigger picture on the super-nation scales with a holistic commodity and chain coverage. Other relevant studies to address the FLW quantification issues in a partial way include Åhnberg and Strid [30], Beretta et al. [31], Caldeira et al. [32], and Lanfranchi et al. [33], etc.

In addition to the aforementioned country and chain stage-specific studies, limited research has been conducted at continental and global levels with a comprehensive geographical and commodity coverage (hereafter called comprehensive studies). Figure 1 lists the relevant comprehensive studies (including this paper) in chronological order.

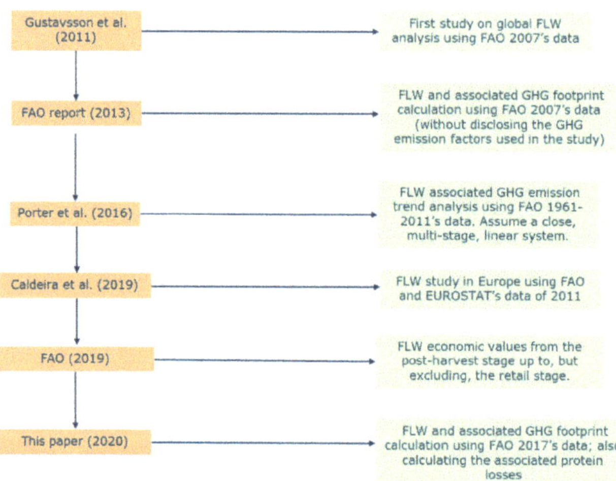

Figure 1. Comparing the relevant comprehensive studies on food loss and waste (FLW) and associated greenhouse gas (GHG) emissions.

Gustavsson et al. [34] is recognized as the first integrated study on global FLW. They specified FLW percentages to individual stages of the food chain based on FAO Food Balance Sheets (FBS) data from 2007. The FLW percentages used in that study were partly derived from the literature; when literature data were not available, the authors used assumptions and estimations to fill the gap. As a follow-up, two years later, FAO issued a technical report that investigated a food wastage footprint on natural resources [35]. It used the FAO FBS's data of 2007 with the same FLW percentages in

Gustavsson et al. [34] as the basis to calculate GHG emissions associated with FLW. However, this study does not disclose the used emission factors to the readers. Following the two FAO studies, Porter, Reay, Higgins and Bomberg [13] made important contributions to this research line by conducting a more comprehensive literature review on FLW percentages and primary-production-phase GHG emission factors along all supply chain stages, across different supranational regions. They used the FAO FBS's data from 1961 to 2011 to investigate the trend development of FLW and associated GHG emissions. A big advantage of Porter's study is that the used FLW percentages and emission factors were derived from more recent literature; these are transparently listed in the article. Despite its significant scientific contribution, there is one point to be improved for Porter's study. Porter, Reay, Higgins and Bomberg [13] assumed a closed, multi-stage, linear system (i.e., a linear chain) in which the input food mass to each chain stage equals the output food mass (after losses) of its previous stage. This implies that all the raw products need to go through the five supply chain stages including processing. However, many raw products can skip the processing stage and directly go to the food distribution stage as fresh produce. Another limitation of Porter's work is that it only presents the GHG emissions from all FLW but not the chain-wise overview.

After Porter's work, Caldeira, De Laurentiis, Corrado, van Holsteijn and Sala [32] performed a mass flow analysis to quantify FLW in the European Union (EU) and concluded that the stage contributing the most to FLW varies amongst product groups (e.g., for many food groups, the highest share of food waste is at consumption stage, but this is not the case for fish or oil crops). Also in 2019, FAO issued another report to introduce their Global Food Loss Index [36,37]. Instead of looking into the FLW mass, it calculated the economic value of FLW. The limitation of this research is that it was based on the survey data from only 23 countries for 10 key commodities and all other data were estimated. Moreover, this analysis only includes part of the food chain (from the post-harvest stage up to, but excluding, the retail stage).

Through the literature review, we observed that existing studies have limitations of lacking comprehensive coverage (e.g., geography, chain stages, food commodities), and using outdated data (e.g., the food volume data before 2011) and literature. Moreover, the existing comprehensive studies hardly addressed the protein loss issue associated with the FLW (except for Alexander et al. [38]), which is very relevant to global food security.

To fill the knowledge gap, we carried out a comprehensive global hotspot analysis on FLW, as well as the associated GHG emissions and protein losses. We used the most up-to-date data (i.e., the new Food Balance Sheets published in 2020) and literature as the basis to provide the policymakers with an updated overview. This study enriches the research line of the comprehensive FLW studies, as shown in Figure 1, along the entire food chain at both regional and global levels.

2. Materials and Methods

Similar to the majority of the relevant studies in this field, the primary source to obtain data in this research was the FAO FBS. However, all the previous studies used the old version of FBS with the most recent data year 2011. This research is the first to use the new FBS (updated in 2020) that covers 2017's data. Different from Porter's study, which only considered the primary-production-phase GHG emissions, we also incorporated the emissions due to international transportation in global food trades. The Detailed Trade Matrix (DTM) from FAODATA was used to map the global food trade flows. Because other post-harvest operations (like processing, packaging, refrigerated storage, etc.) largely differ between and even within product categories, we did not include those effects. However, such emissions may be significant (see e.g., Scherhaufer et al. [39]).

A schematic structure of the methodology employed in this research is presented in Figure 2.

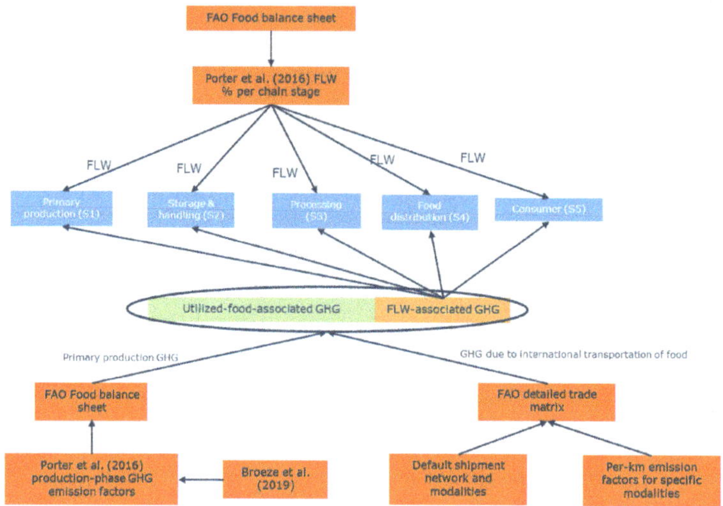

Figure 2. The schematic structure of the methodology applied in this paper.

From the FAO FBS, the most important categories of data include primary production, import and export quantities, processing, food, and feed. The data were retrieved for each country and commodity combination for 2017. Porter, Reay, Higgins and Bomberg [13] is the source from which we derived the FLW percentages. The primary-production-phase GHG emission factors that were applied to the corresponding food chain stages were also obtained from Porter, Reay, Higgins and Bomberg [13]. GHG emissions are expressed as CO_2 equivalents. The transportation-related GHG emission factors were from CO2emissiefactoren.nl. The DTM was used to derive import profile (percentages) for each food commodity and each reporter country (i.e., mapping the sourcing flows in percentages from all trade partner countries). The reason why we did not directly use the import quantity numbers registered in the DTM is that those numbers differ from the ones registered in FAO FBS. Since FBS is the leading data source in our study, we decided to use the DTM in an indirect way (i.e., generating the import profile) to keep the data consistent with the FBS data.

The analysis starts with the agricultural production data which we derived from the FBS. Since the production volumes registered in FBS also include the produce for feed, seeds, and other uses, we need to subtract them from the total primary production. To realize that, we applied the same multiplier as defined in Porter's work (see Equation (1)).

$$\text{FOF} = 1 - \left(\frac{\text{Feed} + \text{Seed} + \text{OtherUses}}{\text{Domestic Supply}}\right) \tag{1}$$

where FOF indicates the fraction of total primary production that is for food use.

Our analysis is targeted at a food supply chain with 5 stages: primary production (S1), storage and handling (S2), processing (S3), food distribution (S4), and consumer (S5). Here, two features of the FBS are essential: (1) there are no data records available for S2 and S5 and (2) the quantity registered for S1, S3, and S4 is the "net quantity" after subtracting the losses in the referred stage. These features determine that different formulas need to be applied to calculate the FLW for the two types of stages. For S2 and S5, their input food mass is just equal to the registered quantity in S1 and S4 because this registered quantity is the "net quantity" after losses. Multiplying the input food mass of S2 and S5 by the associated FLW percentage, we can obtain the FLW in S2 and S5 using Formula (2a):

$$\text{FLW} = \text{FM} \times \text{FLWP} \tag{2a}$$

where FLWP is the FLW percentage and FM is the input food mass of S2 or S5. An FLWP is associated with a specific food item and a specific supply chain stage varied by the different regions.

Differently, the FLW losses for S1, S3, and S4 are calculated by solving Equation (2b), also because the registered quantity for those stages is the "net quantity" after losses:

$$FLWP = \frac{FLW}{NFM + FLW} \tag{2b}$$

where NFM denotes the net food mass after subtracting FLW of the current stage (S1, S3, S4).

Regarding the GHG, only the emissions related to primary production and trade transportation were taken into account due to limited data availability. For the total primary-production-phase GHG emissions, we multiplied the GHG emission factor with the total food mass of the primary production (see Equation (3)).

$$PGHG = PFM \times GHGFP \tag{3}$$

where PGHG is the primary-production-phase GHG emissions and GHGFP denotes the GHG emission factor for primary production. PFM is the total food mass in the primary production stage. Here, it is necessary to note that since GHGFP does not include the emissions in the post-harvest chain stages, PGHG only accounts for GHG emissions due to the on-farm activities.

Except for the GHG emissions from the on-farm activities, the emissions from international trade transportation are also considered. To calculate the transportation-related GHG emissions, Equation (4) is used:

$$TGHG = TFM \times D \times GHGFT \tag{4}$$

where TGHG is the transportation-related GHG emissions, TFM is the transported food mass, D is the transportation distance, and GHGFT is the modality-dependent GHG emission factor for food transportation (per km per ton). The transport distances and modality choices were determined by reasonable assumptions for all trade streams listed in the DTM. For countries located on the same continent (or generally reachable by truck), truck transport was presumed. In case a flow goes from one continent to another, the first part of the transport was assumed by truck from the capital city of the origin country to its closest seaport, then by sea freight shipping to the closest seaport of the capital city of the destination country, followed by truck transport from the port to the destination city. For the sea freight, the following two assumptions were made: ambient bulk transport for robust products (e.g., cereals) and reefer transport for perishable products (e.g., meat, dairy, fruit, and vegetables). In the latter case, additional fuel use for reefer cooling was taken into account, i.e., assuming 20% more GHG emissions than the ambient sea container.

The total GHG emissions were allocated to different supply chain stages according to the amounts of FLW yielded in those stages. The allocated GHG emissions are therefore called FLW-associated GHG emissions, which are the GHG footprints of FLW. For S1 and S2, the corresponding fraction of the primary-production-phase GHG emissions was allocated based on the quantity of FLW generated at that stage. For the remaining stages, the corresponding fraction of both primary production and transportation-related GHG emissions were allocated in accordance with the quantity of FLW per stage.

To calculate the associated protein losses of the FLW, the "FAO Food Composition Table" was used. In the "FAO Food Composition Table", the protein content per food item is listed. We have already obtained the mass of FLW of the food items in different supply chain stages. Therefore, we multiplied the mass of FLW to the protein content of the corresponding food items to derive the protein losses associated with those FLW.

3. Results

3.1. Food Loss and Waste and the Associated GHG Emissions

The general results of this research show that the total global FLW in 2017 was about 1.9 Gt of food, which accounted for 29% of the total primary food production. The GHG emissions associated with these FLW in 2017 are estimated at 2.5 Gt. Note, however, that since GHG emissions due to other post-harvest operations are left out of these analyses, the actual FLW-associated GHG emissions are expected to be somewhat higher than this value.

The international transport-related GHG emissions were very small and only equaled 3% of the GHG emissions of the total primary production. This shows that the primary agricultural production plays a much more important role in GHG emissions than international food transportation.

Figure 3 presents the overview of major contributors, the hotspots, to global FLW and associated GHG emissions in 2017. Vegetables and fruits contributed the most to FLW, accounting for almost half of the total FLW. Their contributions to the FLW-associated GHG emissions were relatively small, about 16.8%. Bovine meat was not a hotspot at all with respect to FLW (only for 0.7% of the total FLW), but it was the largest hotspot for the FLW-associated GHG emissions and contributed as much as vegetables and fruits combined (16.3%). Dairy accounted for 6.8% of the total FLW and 10.2% of the associated GHG emissions. Roots and tubers were a hotspot for FLW (12.1%) but not for the associated GHG emissions (2.7%). The FLW-associated GHG emissions for fish and seafood and rice were 11% and 10%, respectively, of the total. Wheat accounted for 6.6% of the total FLW and poultry accounted for 6% of the total FLW-associated GHG emissions. Thus, these food categories are also important food items to be considered for FLW reduction.

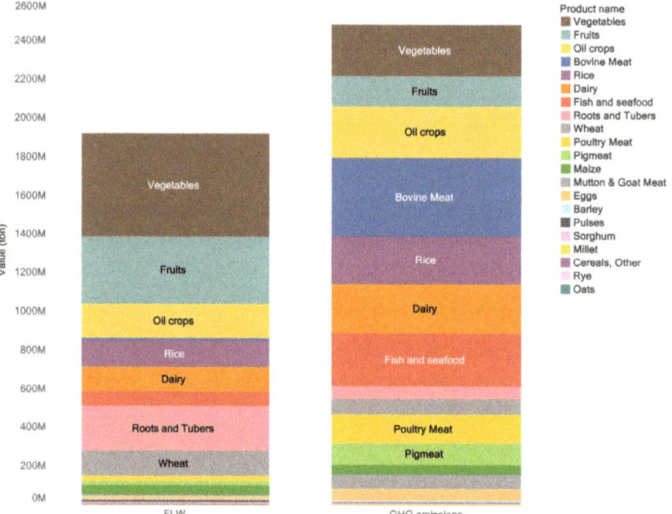

Figure 3. Global hotspots for FLW and associated GHG emissions in 2017.

Figure 4 demonstrates the global hotspots overview by the chain stage. In general, primary production and consumer stages yielded slightly higher FLW and associated GHG emissions than the storage and handling and food distribution stages. The processing stage generated much lower FLW and associated GHG emissions compared to other stages. Specific to food items, it shows that bovine meat was responsible for the highest FLW-associated GHG emissions in the consumer stage. FLW for vegetables and fruits was prominent for all chain stages except for processing. FLW for roots and tubers mainly occurred in the primary production and storage and handling stage. On the other hand, dairy

FLW happened more in the food distribution and consumer stages. Oil crops (including soybeans) are the major source of FLW and associated GHG emissions in the processing stage. Mutton and goat meat pops up as a high emission item associated with FLW in the primary production stage but remains insignificant in other stages.

Figure 4. Global hotspots for FLW and associated GHG emissions by chain stage in 2017.

FAO has grouped the countries of the world into seven region categories based on their relative development phases. The countries within a region category are expected to have a comparable stage of development, natural conditions, and climate; therefore, it is expected that they feature comparable FLW percentages and primary-production-phase GHG emission factors. The seven regions are: Europe; Industrialized Asia; Latin America; North Africa, West and Central Asia; North America and Oceania; South and South-East Asia; Sub-Saharan Africa. The previous comprehensive studies in this field used those regions as references as well (e.g., Gustavsson, Cederberg, Sonesson, van Otterdijk and Meybeck [34], Porter, Reay, Higgins and Bomberg [13]).

Figure 5 indicates that Industrialized Asia generated the largest volumes of FLW and South and South-East Asia generated the highest FLW-associated GHG emissions. Through zooming into different supply chain stages, as shown in Figure 6, we can see that Industrialized Asia and South and South-East Asia were dominating FLW and associated GHG emissions at all but the consumer stage. In the consumer stage, FLW of Europe was larger than that of South and South-East Asia, and the FLW-associated GHG emissions of North America and Oceania were the largest among the seven regions. The top 10 countries are presented in Figure 7, which accounted for approximately 60% of the global FLW and associated GHG emissions. China was the one with the largest FLW and associated GHG emissions globally. It was followed by India and the United States.

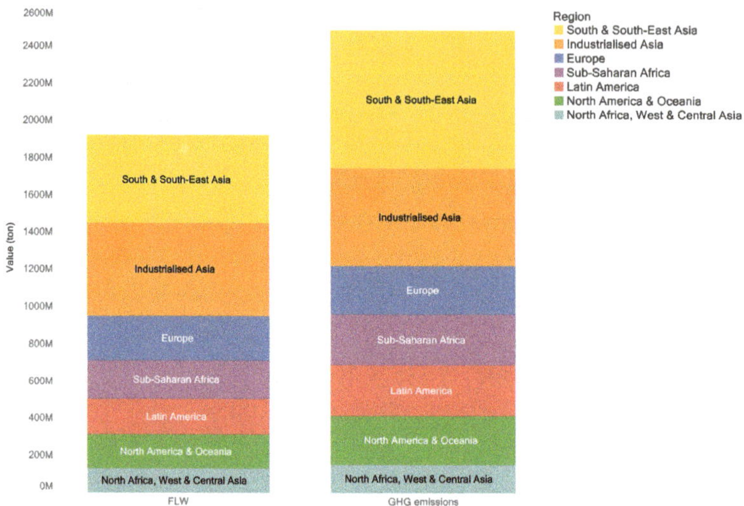

Figure 5. The overview of FLW and associated GHG emissions for FAO-defined regions in 2017.

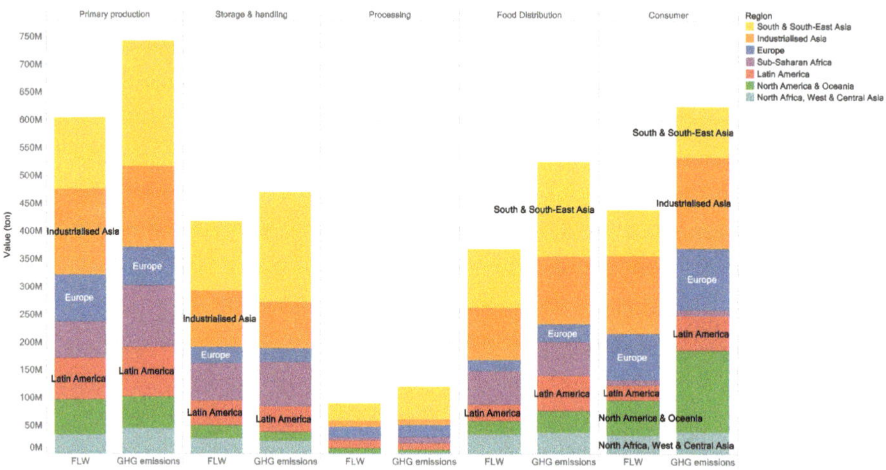

Figure 6. The overview of FLW and associated GHG emissions for FAO-defined regions by chain stage in 2017.

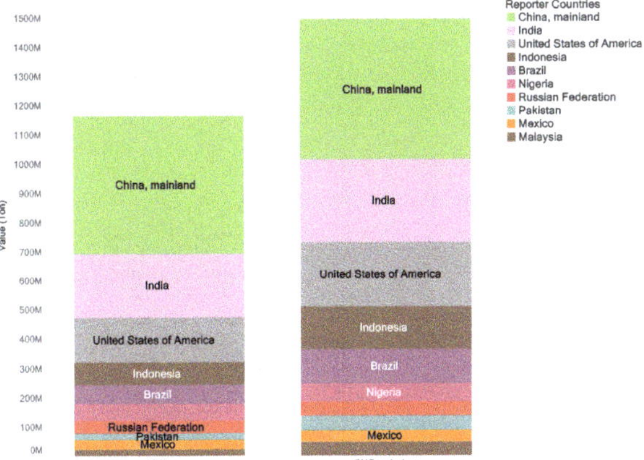

Figure 7. The top 10 countries in the world that generated the most FLW and associated GHG emissions in 2017.

Figures 8–14 present region-based overviews for the seven FAO-defined regions, specified to the distinguished chain stages. The values shown in the graphs are the per-capita values (kg/person) to eliminate the effect of population size and make the figures more comparable.

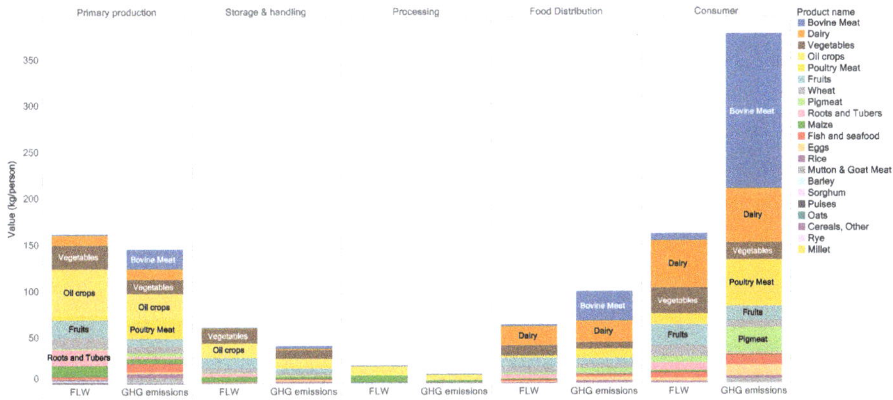

Figure 8. FLW and associated GHG emissions for North America and Oceania by chain stage in 2017.

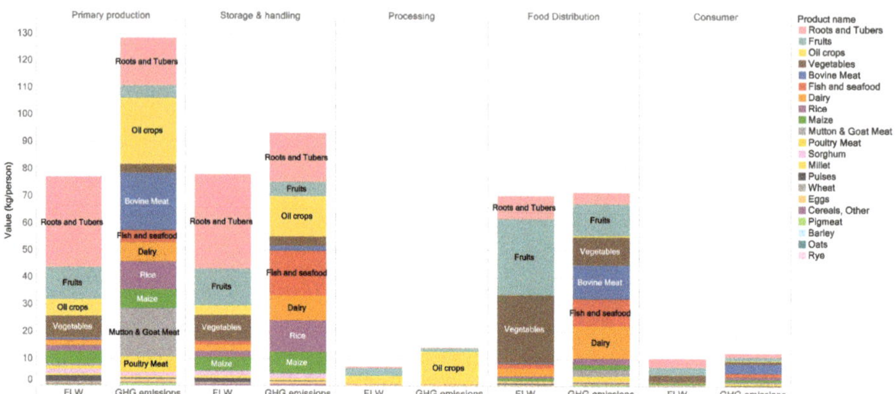

Figure 9. FLW and associated GHG emissions for Sub-Saharan Africa by chain stage in 2017.

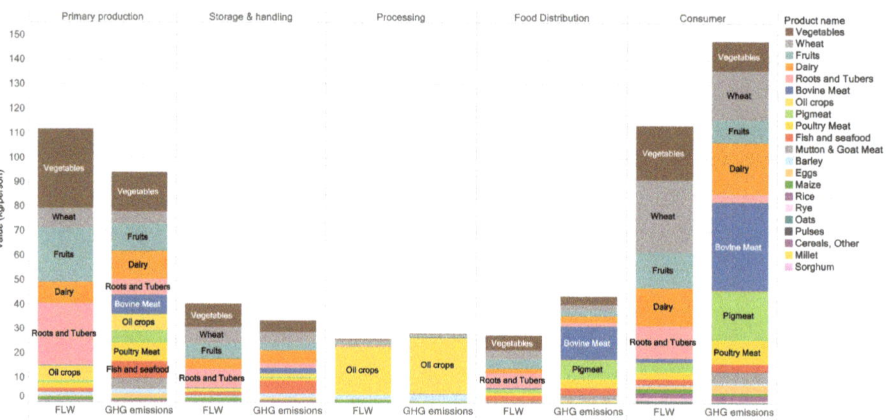

Figure 10. FLW and associated GHG emissions for Europe by chain stage in 2017.

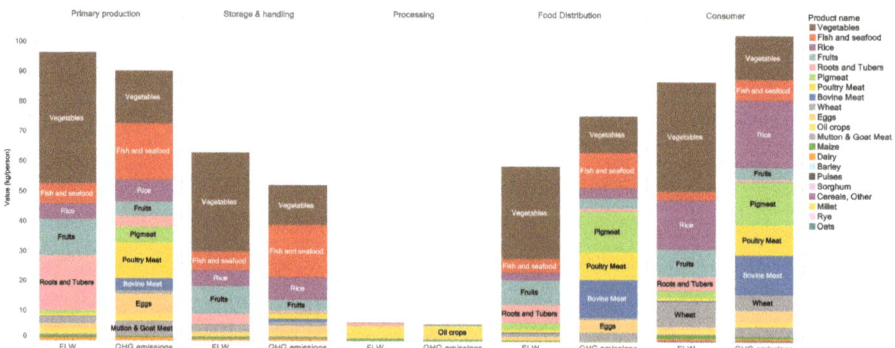

Figure 11. FLW and associated GHG emissions for Industrialized Asia by chain stage in 2017.

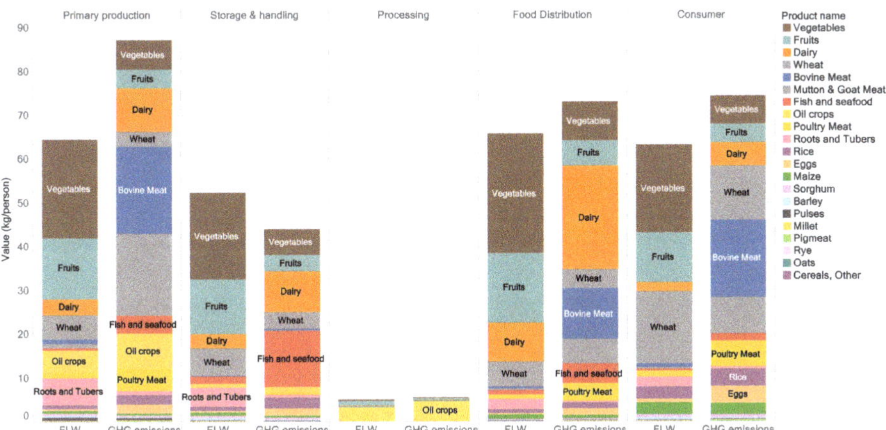

Figure 12. FLW and associated GHG emissions for North Africa, West and Central Asia by chain stage in 2017.

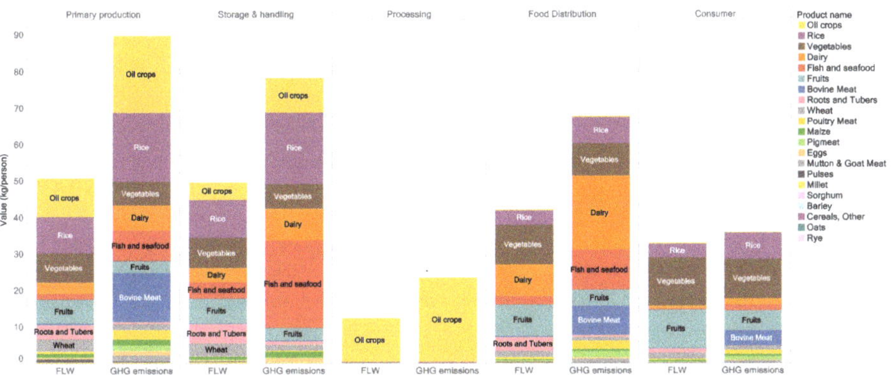

Figure 13. FLW and associated GHG emissions for South and South-East Asia by chain stage in 2017.

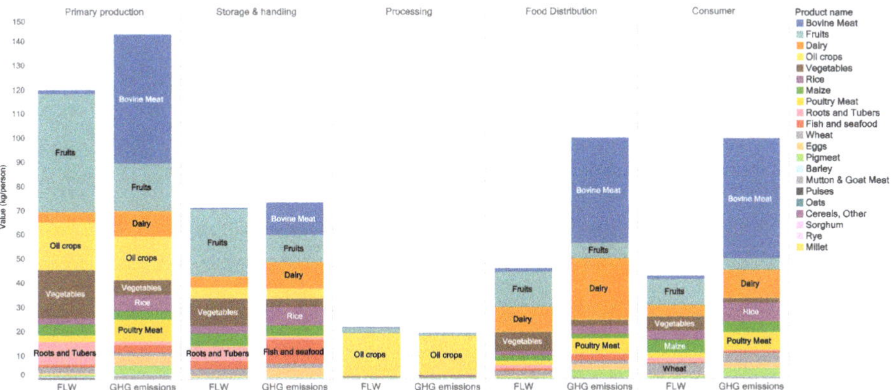

Figure 14. FLW and associated GHG emissions for Latin America by chain stage in 2017.

Figures 8 and 9 show that North America and Oceania generated much more per-capita FLW in the primary production and consumer stages than Sub-Saharan Africa. For other stages including storage and handling, processing, and food distribution, Sub-Saharan Africa had more FLW than North America and Oceania. Another prominent observation in Figure 8 is that bovine meat accounted for 44% of the FLW-associated GHG emissions in the consumer stage of North America and Oceania, and combined with dairy, the ruminants-related number becomes 60%. FLW of roots and tubers for Sub-Saharan Africa was significant in the primary production and storage and handling stages, while for fruits and vegetables, it was prominent in the food distribution stage.

Although both are developed areas, as shown in Figure 10, Europe generated less FLW than North America and Oceania in almost all supply chain stages. Moreover, wheat in the consumer stage of Europe popped up as a high-FLW item, reflecting the large wastage of bread in European households. Furthermore, the same pattern of stage-specific FLW in North America and Oceania is observed.

Significant amounts of vegetables were wasted in Industrialized Asia, followed by fish and seafood, rice, and fruits, as shown in Figure 11. Vegetables, dairy, and fruit are also the major contributors to FLW and the associated GHG emissions in North Africa, West and Central Asia, as shown in Figure 12. For Latin America, fruits become the major source of FLW and the importance of vegetables drops notably, as shown in Figure 14. Bovine meat plays a dominating role in generating FLW-associated GHG emissions in Latin America. Oil crops and rice are the top two contributors in South and South-East Asia, as shown in Figure 13.

3.2. Protein Losses

Protein losses associated with food losses are an important issue related to food security because protein is the essential nutrient to build up human bodies and assure a healthy life. In terms of the FLW-associated protein losses at the global scale, oil crops (including soybeans) and wheat are the top two items, followed by vegetables, rice, and fish and seafood, as shown in Figure 15. Here, it looks a bit surprising that not all the protein crops (such as pulses) are the hotspot items. The reason is that the total protein losses for a specific item are determined by both the content of protein and the mass of FLW. Since the FLW for pulses is very small, as shown in Figure 3, even though they contain high protein content, the total protein losses are still relatively low. Compared to Figure 3, the importance of fruits and vegetables becomes lower due to the low protein content.

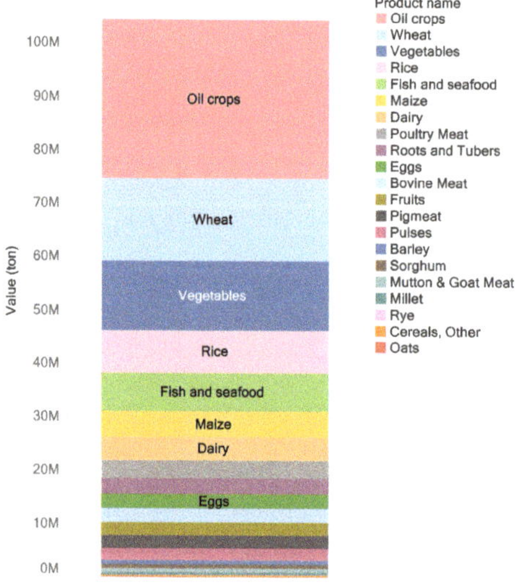

Figure 15. Global protein losses associated with the FLW in 2017.

Looking into the different stages of the global food chains, as shown in Figure 16, the protein losses associated with oil crops mainly occurred in the primary production and processing stages. The wheat-related protein losses happened mostly in the consumer stage. This is contrary to the overview in Figure 4, where the FLW for the top two items (i.e., fruits and vegetables) are quite evenly distributed in all chain stages except for processing. Oil crops are the hotspots in the processing stage in terms of FLW, the associated GHG emissions, and protein losses.

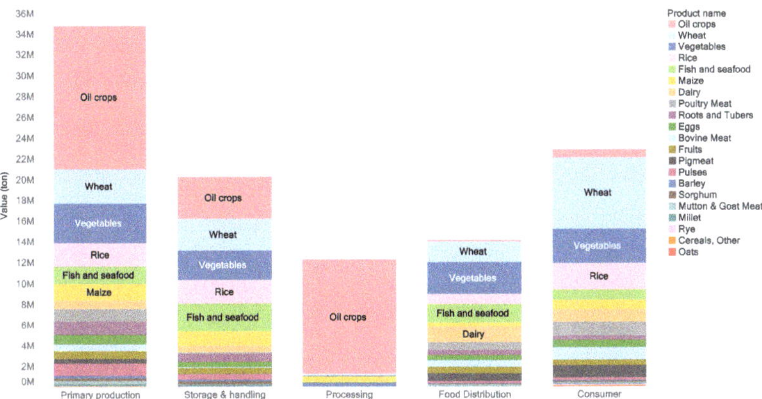

Figure 16. Global protein losses associated with the FLW by chain stage in 2017.

If zooming into the seven regions, we can see that the largest protein losses in 2017 for North America and Oceania and Latin America were from oil crops and concentrated in the primary production, storage and handling, and processing stages, as shown in Figures 17 and 18. Bovine meat, the dominating item for GHG emissions, as shown in Figure 8, becomes a minor source for protein losses. It is even less important than poultry meat. The same chain-wise distribution pattern for the oil crops in North America and Oceania and Latin America was observed in Sub-Saharan Africa but with smaller magnitudes, as shown in Figure 19. Maize was the second-largest source for protein losses in Sub-Saharan Africa, while it contributes a very small amount to the FLW and associated GHG emissions, as shown in Figure 9. For Europe, wheat takes over the position of the oil crops and becomes the largest source, especially in the consumer stage, as shown in Figure 20. This is in line with its importance in terms of FLW and associated GHG emissions, ranking second on the list, as shown in Figure 10. The chain-wise distribution of wheat-related protein losses in North Africa, West and Central Asia shows a similar pattern as in Europe, as shown in Figure 21. For Industrialized Asia, oil crops' position is replaced by vegetables as the number one item with the highest protein losses, as shown in Figure 22. This is because vegetables are the dominating source for FLW in Industrialized Asia, as shown in Figure 11. Rice and wheat are the top two sources for protein losses in South and South-East Asia, as shown in Figure 23, while the top two items for FLW and associated GHG emissions are oil crops and rice, as shown in Figure 13.

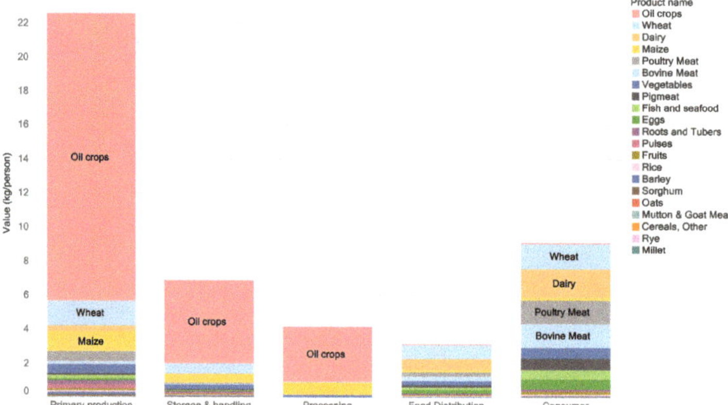

Figure 17. The protein losses associated with the FLW by chain stage in North America and Oceania in 2017.

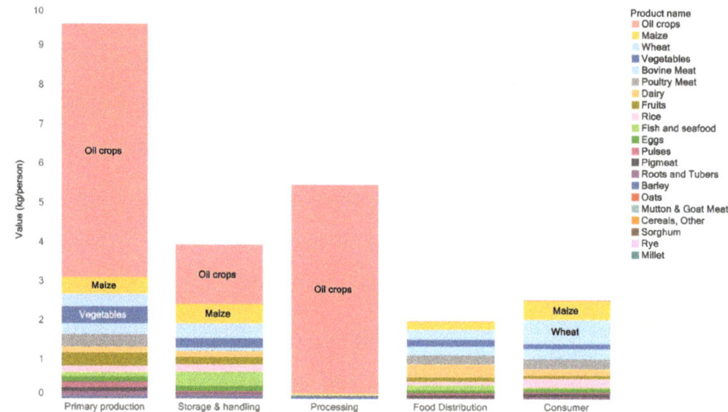

Figure 18. The protein losses associated with the FLW by chain stage in Latin America in 2017.

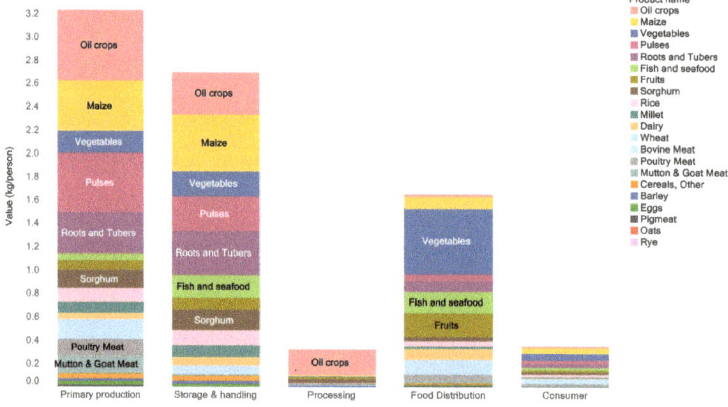

Figure 19. The protein losses associated with the FLW by chain stage in Sub-Saharan Africa in 2017.

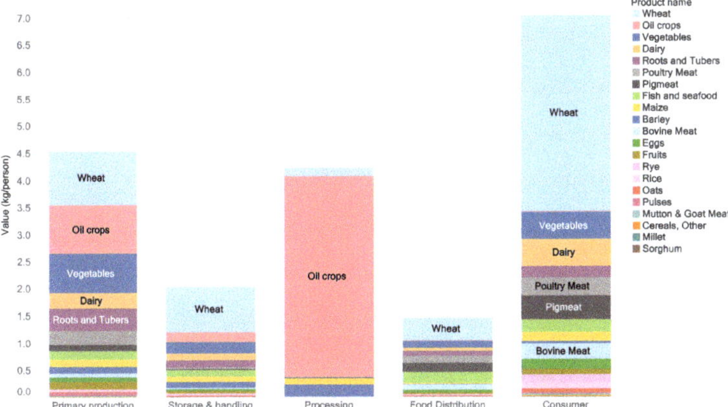

Figure 20. The protein losses associated with the FLW by chain stage in Europe in 2017.

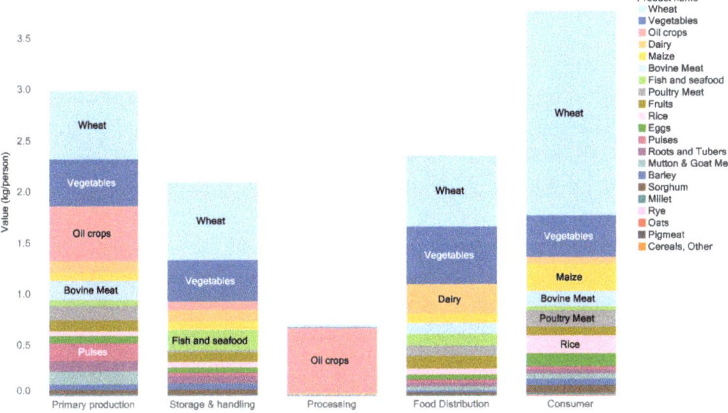

Figure 21. The protein losses associated with the FLW by chain stage in North Africa, West and Central Asia in 2017.

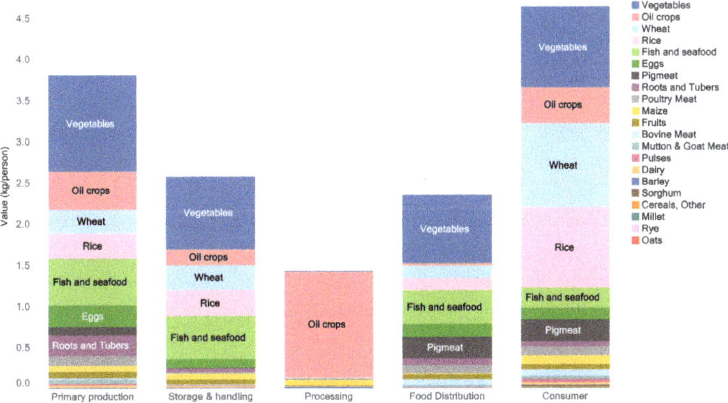

Figure 22. The protein losses associated with the FLW by chain stage in Industrialized Asia in 2017.

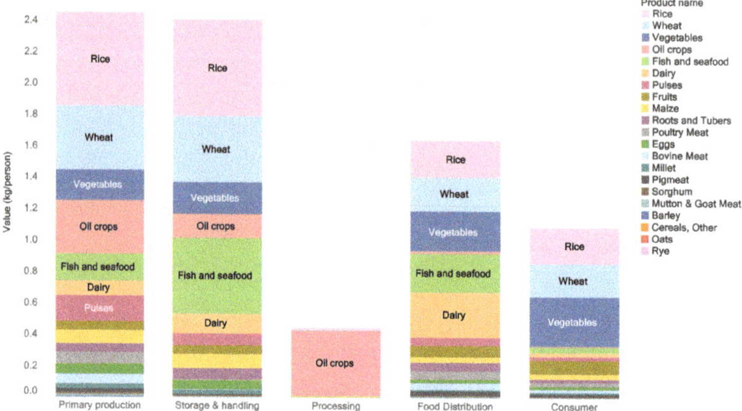

Figure 23. The protein losses associated with the FLW by chain stage in South and South-East Asia in 2017.

4. Discussion

In this research, we conducted a global hotspot analysis by quantifying the FLW as well as the associated GHG emissions and protein losses. Different from the traditional climate-smart agriculture viewpoint, we adopted the prospect of climate-smart food chains. This research is of high significance because of the fast-growing world population with more middle classes favoring higher-income dietary patterns. It requires more fresh and nutritional food supplies (e.g., fresh fruit and vegetables, milk, meat, oil crops). Those are responsible for large amounts of FLW associated with huge GHG emissions and protein losses, as well as other nutrients. The hotspot analysis can help the policymakers identify the crops and animal products that have the highest potentials on FLW. It therefore contributes to food and nutrition security, as well as climate change mitigation.

4.1. Conclusions and Implications

The total values on FLW and associated GHG emissions are quite comparable to other studies. The computed global FLW in this research is equal to 1.9 Gt for 2017. This complies with the 1.6 Gt for 2011 calculated by Porter, Reay, Higgins and Bomberg [13], taking the different years in use into account. The calculated FLW accounts for 29% of the total primary food production. This is in line with the widely accepted concept that roughly a quarter to one-third of the food produced in the world is lost or wasted based on the research of Gustavsson, Cederberg, Sonesson, van Otterdijk and Meybeck [34] and others. The total calculated FLW-associated GHG emissions in 2017 equals 2.5 Gt, which also makes sense compared to 2.2 Gt in 2011 estimated by Porter, Reay, Higgins and Bomberg [13].

Specific to the food items, fruits and vegetables are the major contributors to global FLW. Bovine meat leads to significant FLW-associated GHG emissions, especially in the consumer stage of North America and Oceania (i.e., GHG emission footprint of the FLW at the consumer level considering the emissions from the primary production activities and international food transportation). Moreover, oil crops are the dominating source of protein losses. The different hotspots identified by different criteria underpin the importance of addressing the FLW issues from various aspects of sustainability. Namely, from the point of view of reducing GHG emissions, policymakers should focus on designing intervention strategies to reduce beef consumption; however, from a nutrient perspective, it is more sensible to put more resources in reducing the losses for oil crops. The policymakers therefore should make a balanced plan considering the trade-offs between different intervention points given the intervention resources are limited.

The total GHG emissions due to international food transportation are marginal compared to the primary production-related GHG emissions. This confirms the alleged dominating role for primary production in GHG emissions of the global food system. For example, Hamerschlag and Venkat [40] indicated that the primary-production-phase GHG emissions for meat products including beef, lamb, pork, and poultry account for 50% to 90% of the full-life GHG emissions. For the non-meat products, including milled cereals, dairy, and various types of fruits and vegetables, the percentages are between 60% to 85% [41–43]. Scherhaufer, Lebersorger, Pertl, Obersteiner, Schneider, Falasconi, De Menna, Vittuari, Hartikainen and Katajajuuri [39] demonstrated that for the EU countries, the primary-production-phase GHG emissions are as much as 93% of the full-life GHG emissions for beef. As to other animal products, the percentages are 78%, 75%, 64%, and 63% for milk, pork, fish, and chicken, respectively. The number is 86% for the greenhouse-grown tomatoes and 60% for bread.

In the aforementioned primarily animal-related data, the high emission factor of the agricultural production is dominant. As the opposite, in plant-based products, with much lower crop emission factors, post-harvest operations have relatively large contributions and therefore deserve close attention. However, owing to the diverse characteristics of post-harvest chains and the lack of relevant data, we have not yet included them into our calculations.

4.2. Limitations and Future Research

There are also limitations to this study. Firstly, the data used for the hotspot analysis are secondary data. They may have relatively low accuracy compared to the primary data from direct measurement. The used information from FAO Food Balance Sheets is not amazingly accurate, particularly for developing countries where data along the supply chain have limitations. Secondly, the used FLW percentages are based on FAO-defined regions because those are by far the most complete data available to conduct global and super-national studies. However, within these regions, there could be still some variations between different countries. For example, for Industrialized Asia, the same FLW percentages were applied to Japan, South Korea, and China, which should actually be deviated in reality. For example, the fruit and vegetable post-harvest losses for China (widely believed between 20% and 30%) should be much higher than those in Japan and South Korea due to the much poorer cold chain infrastructure in China. The same issue also applies to the region-based GHG emission factors. Thirdly, the FLW percentages are aggregated for some food product groups in Porter, Reay, Higgins and Bomberg [13]. For example, all the fruits and vegetables share one set of FLW percentages. This could cause some levels of inaccuracy when there are big variations between different fruit and vegetable products (e.g., the on-farm losses for apples could be much lower than that for strawberries).

Future research should focus on generating more national-level data on FLW percentages and GHG emission factors, preferably through more frequent field surveys within individual countries. This can enable the country-specific analysis of the identification of hotspots. When the primary data are not available, smart use of the secondary data with careful validation is required. It would be desirable to have the FLW percentages and GHG emission factors at the specific product level instead of the aggregated product group level. Moreover, to have the GHG emission factors for the full life cycles of products will make the calculation on FLW-associated GHG emissions more complete. It should, therefore, be a future research direction to go as well.

Author Contributions: Conceptualization, X.G., J.B., M.V. and H.A.; methodology, X.G., J.B. and J.J.G.; software, X.G. and J.J.G.; formal analysis, X.G.; data curation, X.G., J.J.G. and J.B.; writing—original draft preparation, X.G.; writing—review and editing, J.B., M.V., H.A. and J.J.G.; visualization, X.G.; project administration, H.A. All authors have read and agreed to the published version of the manuscript.

Funding: CGIAR Research Program on Climate Change, Agriculture and Food Security (CCAFS).

Acknowledgments: This work was implemented as part of the CGIAR Research Program on Climate Change, Agriculture and Food Security (CCAFS), which is carried out with support from CGIAR Fund Donors and through bilateral funding agreements. For details please visit https://ccafs.cgiar.org/donors. The views expressed in this document cannot be taken to reflect the official opinions of these organizations.

Conflicts of Interest: The authors declare no conflict of interest.

References

1. FAO. *Global Agriculture towards 2050*; FAO: Rome, Italy, 2009.
2. Fiore, M.; Contò, F.; Pellegrini, G. Reducing food losses: A (dis)-opportunity cost model. *Riv. Di Studi Sulla Sostenibilita* **2015**, *1*. [CrossRef]
3. Fiore, M.; Pellegrini, G.; Sala, P.L.; Conte, A.; Liu, B. Attitude toward food waste reduction: The case of Italian consumers. *Int. J. Glob. Small Bus.* **2017**, *9*, 185–201. [CrossRef]
4. Spada, A.; Conte, A.; Del Nobile, M.A. The influence of shelf life on food waste: A model-based approach by empirical market evidence. *J. Clean. Prod.* **2018**, *172*, 3410–3414. [CrossRef]
5. Adamashvili, N.; Chiara, F.; Fiore, M. Food Loss and Waste, a global responsibility?! *Econ. Agro Aliment.* **2019**, 825–846. [CrossRef]
6. Pellegrini, G.; Sillani, S.; Gregori, M.; Spada, A. Household food waste reduction: Italian consumers' analysis for improving food management. *Br. Food J.* **2019**, *121*, 1382–1397. [CrossRef]
7. Vittuari, M.; De Menna, F.; García-Herrero, L.; Pagani, M.; Brenes-Peralta, L.; Segrè, A. Chapter 17—Food systems sustainability: The complex challenge of food loss and waste. In *Sustainable Food Supply Chains*; Accorsi, R., Manzini, R., Eds.; Academic Press: Cambridge, MA, USA, 2019; pp. 249–260.
8. Vittuari, M.; Pagani, M.; Johnson, T.G.; De Menna, F. Impacts and costs of embodied and nutritional energy of food waste in the US food system: Distribution and consumption (Part B). *J. Clean. Prod.* **2020**, *252*, 119857. [CrossRef]
9. Pagani, M.; De Menna, F.; Johnson, T.G.; Vittuari, M. Impacts and costs of embodied and nutritional energy of food losses in the US food system: Farming and processing (Part A). *J. Clean. Prod.* **2020**, *244*, 118730. [CrossRef]
10. Ishangulyyev, R.; Kim, S.; Lee, S.H. Understanding food loss and waste—Why are we losing and wasting food? *Foods* **2019**, *8*, 297. [CrossRef]
11. Springmann, M.; Clark, M.; Mason-D'Croz, D.; Wiebe, K.; Bodirsky, B.L.; Lassaletta, L.; De Vries, W.; Vermeulen, S.J.; Herrero, M.; Carlson, K.M. Options for keeping the food system within environmental limits. *Nature* **2018**, *562*, 519–525. [CrossRef]
12. Hiç, C.; Pradhan, P.; Rybski, D.; Kropp, J.P. Food surplus and its climate burdens. *Environ. Sci. Technol.* **2016**, *50*, 4269–4277. [CrossRef]
13. Porter, S.D.; Reay, D.S.; Higgins, P.; Bomberg, E. A half-century of production-phase greenhouse gas emissions from food loss & waste in the global food supply chain. *Sci. Total Environ.* **2016**, *571*, 721–729. [PubMed]
14. Quested, T.; Ingle, R.; Parry, A. *Household Food and Drink Waste in the United Kingdom 2012*; WRAP: London, UK, 2013.
15. Wenlock, R.; Buss, D. Wastage of edible food in the home: A preliminary study. *J. Hum. Nutr.* **1977**, *31*, 405–411. [PubMed]
16. Wenlock, R.; Buss, D.; Derry, B.; Dixon, E. Household food wastage in Britain. *Br. J. Nutr.* **1980**, *43*, 53–70. [CrossRef] [PubMed]
17. Buzby, J.C.; Hyman, J. Total and per capita value of food loss in the United States. *Food Policy* **2012**, *37*, 561–570. [CrossRef]
18. Kantor, L.S.; Lipton, K.; Manchester, A.; Oliveira, V. Estimating and addressing America's food losses. *Food Rev. Natl. Food Rev.* **1997**, *20*, 2–12.
19. Xue, L.; Liu, G.; Parfitt, J.; Liu, X.; Van Herpen, E.; Stenmarck, Å.; O'Connor, C.; Östergren, K.; Cheng, S. Missing food, missing data? A critical review of global food losses and food waste data. *Environ. Sci. Technol.* **2017**, *51*, 6618–6633. [CrossRef]
20. Kling, W. Food waste in distribution and use. *J. Farm Econ.* **1943**, *25*, 848–859. [CrossRef]
21. Harrington, J.M.; Myers, R.A.; Rosenberg, A.A. Wasted fishery resources: Discarded by-catch in the USA. *Fish Fish.* **2005**, *6*, 350–361. [CrossRef]
22. Engström, R.; Carlsson-Kanyama, A. Food losses in food service institutions Examples from Sweden. *Food Policy* **2004**, *29*, 203–213. [CrossRef]
23. Williams, H.; Wikström, F.; Otterbring, T.; Löfgren, M.; Gustafsson, A. Reasons for household food waste with special attention to packaging. *J. Clean. Prod.* **2012**, *24*, 141–148. [CrossRef]

24. Leal Filho, W.; Kovaleva, M. *Food Waste and Sustainable Food Waste Management in the Baltic Sea Region*; Springer: Berlin/Heidelberg, Germany, 2015.
25. Hodges, R.J.; Buzby, J.C.; Bennett, B. Postharvest losses and waste in developed and less developed countries: Opportunities to improve resource use. *J. Agric. Sci.* **2011**, *149* (Suppl. S1), 37–45. [CrossRef]
26. Fehr, M.; Romão, D. Measurement of fruit and vegetable losses in Brazil: A case study. *Environ. Dev. Sustain.* **2001**, *3*, 253–263. [CrossRef]
27. Jones, T.W. Using Contemporary Archaeology and Applied Anthropology to Understand Food Loss in the American Food System. Ph.D. Thesis, Bureau of Applied Research in Anthropology, University of Arizona, Tucson, AZ, USA, 2004.
28. Loke, M.K.; Leung, P. Quantifying food waste in Hawaii's food supply chain. *Waste Manag. Res.* **2015**, *33*, 1076–1083. [CrossRef] [PubMed]
29. Stenmarck, Å.; Jörgen Hanssen, O.; Silvennoinen, K.; Katajajuuri, J.-M.; Werge, M. *Initiatives on Prevention of Food Waste in the Retail and Wholesale Trades*; Nordic Council of Ministers: Copenhagen, Denmark, 2011.
30. Åhnberg, A.; Strid, I. *When Food Turns into Waste—A Study on Practices and Handling of Losses of Fruit and Vegetables and Meat in Willys Södertälje Weda*; Swedish University of Agricultural Sciences: Uppsala, Sweden, 2010.
31. Beretta, C.; Stoessel, F.; Baier, U.; Hellweg, S. Quantifying food losses and the potential for reduction in Switzerland. *Waste Manag.* **2013**, *33*, 764–773. [CrossRef]
32. Caldeira, C.; De Laurentiis, V.; Corrado, S.; van Holsteijn, F.; Sala, S. Quantification of food waste per product group along the food supply chain in the European Union: A mass flow analysis. *Resour. Conserv. Recycl.* **2019**, *149*, 479–488. [CrossRef] [PubMed]
33. Lanfranchi, M.; Giannetto, C.; De Pascale, A. Analysis and models for the reduction of food waste in organized large-scale retail distribution in eastern Sicily. *Am. J. Appl. Sci.* **2014**, *11*, 1860. [CrossRef]
34. Gustavsson, J.; Cederberg, C.; Sonesson, U.; van Otterdijk, R.; Meybeck, A. Global food losses and food waste: Extent, causes and prevention. In *SAVE FOOD: An Initiative on Food Loss and Waste Reduction*; FAO: Rome, Italy, 2011.
35. FAO. *Food Wastage Footprint: Impacts on Natural Resources*; FAO: Rome, Italy, 2013.
36. FAO. *The State of Food and Agriculture: Moving Forward on Food Loss and Waste Reduction*; FAO: Rome, Italy, 2019.
37. FAO. Food Loss Index. Online Statistical Working System for Loss Calculations. 2019. Available online: http://www.fao.org/food-loss-and-food-waste/flw-data) (accessed on 19 July 2020).
38. Alexander, P.; Brown, C.; Arneth, A.; Finnigan, J.; Moran, D.; Rounsevell, M.D. Losses, inefficiencies and waste in the global food system. *Agric. Syst.* **2017**, *153*, 190–200. [CrossRef]
39. Scherhaufer, S.; Lebersorger, S.; Pertl, A.; Obersteiner, G.; Schneider, F.; Falasconi, L.; De Menna, F.; Vittuari, M.; Hartikainen, H.; Katajajuuri, J.-M. *Criteria for and Baseline Assessment of Environmental and Socio-Economic Impacts of Food Waste*; BOKU University of Natural Resources and Life Sciences, Institute of Waste Management: Vienna, Austria, 2015.
40. Hamerschlag, K.; Venkat, K. *Meat Eater's Guide to Climate Change+ Health: Lifecycle Assessments: Methodology and Results 2011*; Environmental Working Group: Washington, DC, USA, 2011.
41. Cellura, M.; Longo, S.; Mistretta, M. Life Cycle Assessment (LCA) of protected crops: An Italian case study. *J. Clean. Prod.* **2012**, *28*, 56–62. [CrossRef]
42. Sheane, R.; Lewis, K.; Hall, P.; Holmes-Ling, P.; Kerr, A.; Stewart, K.; Webb, D. *Identifying Opportunities to Reduce the Carbon Footprint Associated with the Scottish Dairy Supply Chain—Main Report*; Scottish Government: Edinburgh, UK, 2011.
43. Shi, C.W.P.; Rugrungruang, F.; Yeo, Z.; Song, B. Carbon footprint analysis for energy improvement in flour milling production. In *Glocalized Solutions for Sustainability in Manufacturing*; Springer: Berlin/Heidelberg, Germany, 2011; pp. 246–251.

Article

The Effects of Fast Food Restaurant Attributes on Customer Satisfaction, Revisit Intention, and Recommendation Using DINESERV Scale

Se-Hak Chun * and Ariunzaya Nyam-Ochir

Department of Business Administration, Seoul National University of Science and Technology, Seoul 01811, Korea; n.ariuna9211@gmail.com
* Correspondence: shchun@seoultech.ac.kr; Tel.: +82-2-970-6487; Fax: +82-2-973-1349

Received: 19 August 2020; Accepted: 5 September 2020; Published: 10 September 2020

Abstract: The fast food restaurant business is one of the fastest-growing industries in the world. International and local restaurant chains are trying to satisfy the demands of customers for a variety of products and services. Along with changing market trends, customers are now becoming more sophisticated and demanding. Customer satisfaction is an essential business issue, as entrepreneurs have realized that favorable customer feedback is key for a long-term sustainable operation. Customers who have an excellent experience at a restaurant may recommend the restaurant to others, spread positive information, or become a loyal customer. The fast food industry has only recently developed in Mongolia and an increasing number of global fast food chains are now entering the market every year. The purpose of this paper is to examine and evaluate the factors affecting customer satisfaction, revisit intention, and likelihood of recommendation for Mongolian fast food restaurants, as well as a global fast food restaurant in Mongolia using the DINESERV scale. This study focuses on comparing directly competing food chains; only two brands were studied because of the limited fast food presence in Ulaanbaatar. Then, it aims to analyze how satisfaction levels influence a customer's revisit intention and likelihood of recommending a restaurant. Furthermore, an in-depth analysis of the difference between local and global fast food brands is a key element that this paper analyzes. Moreover, this paper investigates how results can be different according to whether the respondent resides in Mongolia or Korea and discusses business implications. The results of this paper show that four factors (food quality, service quality, price, and atmosphere of a restaurant) positively influence customer satisfaction, revisit intention, and likelihood of recommendation for Mongolian and global fast food restaurants, and customer satisfaction has a positive influence on customer revisit intention and likelihood of recommendation for both types of restaurants. However, depending on whether it is a Mongolian fast food restaurant or a global fast food restaurant, the factors affecting customer satisfaction, revisit intention, and recommendation are different.

Keywords: fast food restaurant; customer satisfaction; revisit intention and recommendation; regression analysis; factor analysis

1. Introduction

The fast food restaurant business is one of the fastest-growing industries in the world and the global fast food market is expected to grow at a compound annual growth rate (CAGR) of 5.1% from 2020 to 2027 [1]. International and local restaurant chains are trying to satisfy the demand of customers for a variety of products and services. People prefer to use fast food restaurants for their convenience and to save time, and there have been certain changes in consumer trends that have increased the popularity of eating out; therefore, the fast food restaurant industry is growing rapidly. Like other industries, customer satisfaction is an essential business issue for restaurant businesses.

In the competitive hospitality sector, customer satisfaction has become a key element of business strategy. Restaurants wish to maximize the positive experience of a consumer in order to increase their intentions of revisiting [2,3]. Entrepreneurs have realized that favorable customer feedback is key for developing a long-term sustainable operation. A thorough understanding and knowledge of the factors that influence customer satisfaction is useful to allow restaurant owners and managers to design and deliver the right products to customers. Thus, customer satisfaction plays an important role in every business organization, whether it is providing a service or a product. The obvious reason for satisfying the firm's customers is to allow the business to expand and gain a higher market share, leading to improved profitability.

In recent years, domestic and foreign direct investment has drastically increased, becoming an important driver of economic growth [4]. Moreover, the GDP growth rate in Mongolia averaged 5.45 percent from 1991 until 2018, reaching an all-time high of 17.50 percent in the fourth quarter of 2011 and a record low of −30 percent in the fourth quarter of 1992 [5].

Although the franchise industry in Mongolia is in its infancy, the fast food industry has grown in Mongolia in the last few years. KFC was the first Western food chain to open a store in Mongolia in 2013. In 2015, Burger King opened its first store in Mongolia, following the opening of KFC and Pizza Hut [6]. For Burger King, there were a total of nine outlets as of December, 2018 [7]. In 2018, the South Korean Lotte's fast food chain Lotteria opened its first Mongolian store in the country's capital city of Ulaanbaatar [8]. Furthermore, Mongolia's food and beverage business is one of the most advanced local industries in terms of technology, equipment, and know-how and seeks to capitalize on the nation's abundant agricultural resources to not only meet the local market demand, but also export to neighboring countries. Many leading companies of the Food & Beverages sector, such as APU JSC, SUU JSC, are listed on the Mongolian Stock Exchange [4]. Mongolian brands include BD's Mongolian Barbeque and Berlin Burger, the latter being Mongolia's very first fast food restaurant, which opened in 1992. As of 2020, there are a total of nine Berlin Burger stores in Ulaanbaatar. Multinational chains, such as Burger King, and local chains, such as Berlin Burger, are growing in Mongolia because of their product development, quality standards, and effective localization [9]. With an increasing number of people eating out, the industry offers a major opportunity to capture a larger consumer base. As a result of the trend, international food chains are investing huge amounts of money to grab a share of this highly lucrative market. They are spending all their resources and efforts to understand their customers better and give them the best possible services.

Currently, research in the Mongolian fast food industry is lacking, largely due to the early stage the industry is in. Thus, in this paper, we analyze various factors for the success of the fast food market in Mongolia. We examine and evaluate the factors affecting customer satisfaction, revisit intention, and likelihood of recommendation for Mongolian fast food restaurants and global fast food restaurants in Mongolia and compare the satisfaction levels of customers with reference to restaurants in Mongolia. We investigate which factors among food quality, service quality, atmosphere and interior, and price and value affect customer satisfaction using a factor analysis and regression analysis. We also analyze how customer satisfaction is related to customer revisit intention and likelihood of recommendation. For this, we aggregate the data through Google's survey form and email methods.

The results of this paper show that four factors (food quality, service quality, price, and atmosphere of a restaurant) positively influence customer satisfaction, revisit intention, and likelihood of recommendation for Mongolian and global fast food restaurants, and customer satisfaction has a positive influence on customer revisit intention and likelihood of recommendation for both types of restaurants. Moreover, the results show that preference for global food chains over local ones owes greatly to customer experiences regarding the atmosphere of a restaurant.

The remainder of the paper is organized as follows. Section 2 reviews previous literature and presents the theoretical background. Section 3 presents the research methodology, and Section 4 discusses the results and implications. Section 5 concludes the study and discusses future research directions.

2. Research Background

The DINESERV is well known to be a reliable, relatively simple measurement tool for determining how consumers view a restaurant's quality [10]. Similar to SERVQUAL [11], DINESERV was developed by Stevens et al. [12] to assess customers' perceptions of restaurant service quality. In many previous studies, the DINESERV instrument has been used to investigate ways to improve a restaurant's quality and increase customer satisfaction, which, in turn, determines revisit intentions and likelihood of recommendation [10,12–14].

2.1. Service Quality

Service quality is a measure of how well a service conforms to the customer's expectations and the success factor of a fast food restaurant [11]. Service quality is the main component of a fast food restaurant that can be measured and improved continuously. When there is a close interaction between a service employee and a customer, the perception of what is being delivered is as important as what is actually delivered [15]. In other words, it is the result of the comparison between expectations about service and perceptions of the way the service has been performed that customers make. Therefore, the employees' behaviors and attitudes can influence a customer's perception of quality for the service offered [16].

2.2. Food Quality

Food quality is an important component and it has been constantly shown to be a core value that a customer considers in deciding which fast food restaurant to eat at [17]. Many studies have investigated food quality characteristics, such as the freshness of food, food presentation, food taste, a variety of menu options, and food temperature. Food quality is considered to be a key foundation for customer satisfaction and customers' revisit intention [18–20]. According to Peri [21], food quality is an absolute requirement to satisfy the needs and expectations of restaurant customers. Youth-aged customers who mostly prefer eating delicious food require good quality food and beverages to satisfy them. Analysts have stated that the quality of menu items affects customers' revisit intention [22].

2.3. Atmosphere

Nowadays, people prefer to eat a lot more often. Customers are more aware of the atmosphere in which they are dining in than they were before. This requires restaurant owners to put more effort into designing and providing more comfortable surroundings for their customers. The atmosphere of a restaurant can be as important as the food itself [23]. The restaurant atmosphere is influenced by several factors, such as the interior design, temperature, cleanliness, music, and table arrangement.

2.4. Price

The price of a product or service can affect the level of satisfaction among customers because it has an associated sense of fairness. A customer's perceptions of unfair pricing lead to negative outcomes, such as a lower level of revisit intention, dissatisfaction, and negative word of mouth. The pricing of restaurant items also varies according to the type of restaurant. If the price is high, customers are likely to expect high quality, otherwise, it can induce a sense of being "ripped off." Likewise, if the price is low, customers may question the ability of the restaurant to deliver product and service quality. Moreover, due to the competitiveness of the restaurant industry, customers are able to establish internal reference prices. When establishing prices for a restaurant, an internal reference price is defined as a price in a buyer's memory that serves as a basis for judging or comparing actual prices [24].

2.5. Customer Satisfaction

Customer satisfaction is becoming a common goal for businesses. Customer satisfaction, as defined by Oliver [25], is the after-purchase judgment or evaluation of a product or service. It is also frequently described as the extent to which the chosen product meets or exceeds consumer expectations. It is, hence, a comprehensive domain that is the result of several inter-related variables impacting each other on an ongoing basis, rather than a single variable [23]. Customer satisfaction is an overall evaluation that compares post-purchase perceived performance with purchase expectations [26].

Taking the past as an example, when consumers decide to have a meal in a specific restaurant, they will have an expectation of how they will be served. After the meal, they will compare the serving experience with their level of anticipation. If the service quality the restaurant offered is equal to or higher than expected, they will be satisfied with this restaurant and likely come to the same restaurant again [27]. Based on this theory, customer satisfaction is the measure of the gap between a customer's expectations and perceived performance. Therefore, to enlarge the market segments in the restaurant industry, customer satisfaction is a powerful predictor of customer intent to repurchase [28].

In Qu's study, by analyzing data from Chinese restaurants in Indiana, it was found that the higher a customer's satisfaction with food and environment, service and courtesy, price and value, location, and advertising and promotion, the greater the likelihood of the customer returning [13]. Different from Qu's conclusion, Weiss et al. [29] found that customer revisit intention is only influenced by satisfaction with the restaurant food quality and atmosphere. Although dimensions used to estimate customer satisfaction in different studies have not been identical, the use of satisfaction as a determinant factor of customer revisit intention has been consistent across different studies. Many studies have identified factors that influence customer satisfaction, including service quality, variety of the menu, price, food quality, food presentation, ambience, and convenience.

2.6. Revisit Intention and Recommendation

When a company offers a product or service, it is possible that there are many similar products or services on the market provided by their competitors. Customers usually have many alternative choices. Therefore, it is important for companies to improve the value experienced by existing customers and take effective steps to encourage their repurchase behaviors, as well as attract new customers. Repeat customers are more profitable than new customers. Chen and Hu [30] described customer revisit intention as a customer's intention to revisit the same restaurant and recommend it to members of their circle. Customers that have an excellent experience at the restaurant will recommend the restaurant to others, spread positive information, or become a loyal customer. Customer revisit intention has been studied in many domains, such as tourism services, catering services, hospital services, retail business, bank services, and telecom businesses. A number of models of factors driving revisit intention have been constructed by means of structural equation modeling or logistic regression. The factors considered in these models include satisfaction, number of previous visits, cost, and customer value. Among the factors influencing revisits is customer satisfaction [27].

3. Research Model and Methodology

Like Kim and Choi [31], we investigated which factors—such as food quality, service quality, atmosphere and interior, and price and value—affect customer satisfaction, which, in turn, determines revisit intentions and likelihood of recommendation, using a factor analysis and a regression analysis.

There is a variety of measurement tools and techniques for assessing service quality. One of the most popular and widely used is the SERVQUAL (service quality) instrument [32]. In restaurant settings, service quality is usually measured with an adapted version of SERVQUAL, called DINESERV [12]. A modified version of DINESERV is applied in this study as well. The DINESERV is considered a reliable and relatively simple tool for determining how customers view a restaurant's quality [33] and has been used in many restaurant settings, such as fine dining [34], casual dining [35], fast food [36,37], food courts [38], and chain restaurants [39]. During the last two decades, SERVQUAL and DINESERV have been widely used to measure service quality in the hospitality industry [40]. The DINESERV tool was applied to assess managers' perceptions of service quality, and firms' financial reports were used to analyze operational efficiency and profitability [41].

3.1. Research Model

The DINESERV, constructed by Stevens et al. [12], consists of service-quality standards that fall into DINESERV factors: assurance, empathy, reliability, responsiveness, and tangibles. Kim et al. [42] found that the five restaurant factors—food quality, service quality, price and value, atmosphere, and convenience—had a significant impact on the customer's satisfaction with dining facilities [43]. Kim and Choi [31] used four factors—food quality, service quality, interior, and price and value—to investigate the perception gap of service attributes between operators and customers. Like Kim and Choi [31], we investigated how four factors—food quality, service quality, atmosphere and interior, and price and value—affect customer satisfaction, which, in turn, determines revisit intentions and likelihood of recommendation, using a factor analysis and a regression analysis. Table 1 shows the four service factors of restaurants used in this study and related results in previous studies.

Table 1. Factors, items, and questionnaire sources.

Factors	Items	Sources
Food quality	Taste of food	[43] [44] [42] [45]
	Freshness of food	
	Menu variety	
	Good portions	
Service quality	Kindness	[15] [46] [19]
	Good attitude	
	Quick service response	
	Well trained	
	Chef's knowledge	
Atmosphere	Good interior and decoration	[43] [47] [48]
	Clean dining areas and restroom	
	Comfortable seats	
	Comfortable temperature	
	Music and pleasant feeling	
Price	Valuable price	[46] [42,44]
	Discount	
Satisfaction	Overall satisfaction	[10]
Revisit	Revisit	[27,29]
Recommendation	Recommendation	[30]

Figure 1 shows the conceptual research model proposed to investigate how four service factors of restaurants, such as food quality, service quality, atmosphere, and price, affect the customer's satisfaction, revisit intention, and likelihood of recommendation.

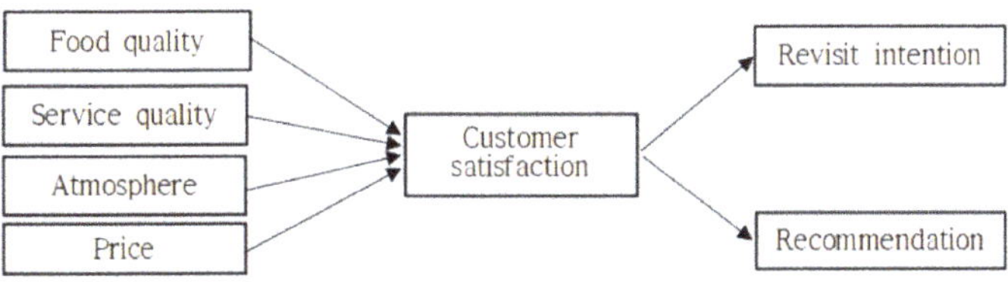

Figure 1. Research model.

Based on the above review of literature regarding DINESERV factors, the following research questions were derived:

Research Question 1: Which institutional DINESERV factors will have greater impacts on customer satisfaction?
Research Question 2: Which institutional DINESERV factors will have greater impacts on revisit intention?
Research Question 3: Which institutional DINESERV factors will have greater impacts on recommendations?
Research Question 4: Do DINESERV factors and the customers' overall satisfaction levels differ with respect to different restaurants (between global restaurants and local restaurants)?
Research Question 5: Are there any significant relationships among customer satisfaction, revisit intention, and likelihood of recommendation?

3.2. Data Collection and Methods

Two of the most popular fast food restaurants in Ulaanbaatar were selected for the study: namely, Burger King and Berlin Burger. This study focused on comparing directly competing food chains; only two brands were studied. The limited fast food presence in Ulaanbaatar inevitably led to this limitation, but further research may take a more comprehensive approach of studying the entire fast food environment in Mongolia. Data were collected over a two-week period in October, 2018. The data analysis was based on 151 valid questionnaire responses collected through Google surveys. The period of data collection was short, and for this reason, only 151 questionnaires were returned.

Data were analyzed using the statistical package SPSS. A structured questionnaire using a five-point Likert scale was used to collect the data. The content of the questionnaire was divided into four sections. The first section, related to the respondent's demographic profile, included their age, gender, marital status, occupation, and income. The second section focused on how often the respondent eats out and what influences him/her to visit that restaurant. The third section measured the respondent's perceptions of the independent variables in the fast food restaurant: atmosphere, service quality, price, and food quality. A five-point Likert scale ranging from 1 (strongly disagree) to 5 (strongly agree) was used to measure the customer's perceptions of the factors. This was intended to help respondents make their choice for each question. The fourth section measured the respondent's willingness to revisit and recommend the restaurant. The questionnaire consisted of 19 questions.

4. Results of the Study and Implications

4.1. Demographic Profile of Respondents

A descriptive statistical analysis was run on respondents' demographic profiles. To gain a better understanding of the customer's level of satisfaction from fast food restaurants, the demographic characteristics of the respondents were analyzed. The results are shown in Table 2. The demographic data of respondents, including gender, age, marital status, occupation, and income, are shown in Table 2. The sample contained more females (53.6%) than males (46.4%). Almost 52 percent of them were aged between 26 and 40 years old. Ninety of the respondents (59.6%) lived in Mongolia and 61 of them (40.4%) lived in South Korea. In terms of occupation, 40.4% of them were students and 40.4% of respondents worked in the private sector. A total of 38.4% of the respondents had an annual income of more than 1,500,000 tugruk, and 27.2% of them had an income between 1,000,000 and 1,500,000 tugruk.

Table 2. Demographic characteristics of the respondents (n = 151).

Items		Frequency	Percent
Country of residence	Mongolia	90	59.6%
	South Korea	61	40.4%
Gender	Male	70	46.4%
	Female	81	53.6%
Age	18–25	64	42.4%
	26–40	78	51.7%
	41–60	9	6%
	61 and above	0	0%
Marital status	Married	61	40.4%
	Single	90	59.6%
Occupation	Student	61	40.4%
	Civil servant	12	7.9%
	Private sector	61	40.4%
	Others	17	11.3%
Income (tugruk)	<500,000	12	7.9%
	500,000–1,000,000	40	26.5%
	1,000,000–1,500,000	41	27.2%
	>1,500,000	58	38.4%

4.1.1. Frequency of Visits of Respondents to Fast Food Restaurants

Table 3 shows the responses to how often the participants reported eating at the fast food restaurant. A total of 22.5% of Burger King customers reported eating out once a week, and 13.2% of them reported eating out once every two weeks. A total of 37.1% reported eating out once a month. The results also show that 7.3% of Mongolian fast food restaurant respondents reported eating out once a week, 10.6% reported eating out once every two weeks, and 30.5% reported eating out once a month.

Table 3. Demographic profile of respondents.

Frequency of Visit	Burger King		Berlin Burger	
	No. of Respondents	Percent	No. of Respondents	Percent
Once a week	27	22.5	46	7.3
Once every two weeks	32	13.2	16	10.6
Once a month	66	37.1	11	30.5
Total	125	72.8	73	48.3
System (missing)	26	27.2	78	51.7
Total	151	100	151	100

4.1.2. Comparison of Mean for Each Item (t-Test)

The results of the t-tests on food quality, service quality, atmosphere, customer satisfaction, revisit intention, and likelihood of recommendation for Mongolian and global fast food restaurants are shown in Table 4.

Table 4. *t*-test of means between Burger King vs. Berlin Burger.

Factors	Items	Burger King vs. Berlin Burger		
		Residence (Mongolia) $n = 90$	Residence (Korea) $n = 61$	Total $n = 151$
Food quality	Taste of food	3.4889 vs. 2.6556 *	3.6721 vs. 3.3443 **	3.5629 vs. 2.9338 *
	Freshness	3.4111 vs. 2.6111 *	3.5410 vs. 3.2787 ***	3.4636 vs. 2.8808 *
	Menu variety	3.2360 vs. 2.8000 *	3.6885 vs. 3.3770 **	3.4200 vs. 3.0331 *
	Good Portions	3.1910 vs. 2.8000 **	3.6885 vs. 3.3607 **	3.3933 vs. 3.0265 *
Service quality	Kindness	3.4889 vs. 2.6222 *	3.6230 vs. 2.9836 *	3.5430 vs. 2.7682 *
	Good attitude	3.4778 vs. 2.6333 *	3.6393 vs. 3.0000 *	3.5430 vs.2.7815 *
	Quick service	3.3778 vs. 2.6111 *	3.5902 vs. 3.0656 *	3.4636 vs. 2.7947 *
	Well trained	3.4000 vs. 2.6333 *	3.6066 vs. 2.9508 *	3.4834 vs. 2.7616 *
	Chef's knowledge	3.3778 vs. 2.6556 *	3.5738 vs. 3.0000 *	3.4570 vs. 2.7947 *
Atmosphere	Interior/decoration	3.4000 vs. 2.5111 *	3.6885 vs. 3.0000 *	3.5166 vs. 2.7086 *
	Clean dining areas	3.3556 vs. 2.5333 *	3.6230 vs. 2.9016 *	3.4636 vs. 2.6821 *
	Comfortable seats	3.4444 vs. 2.5333 *	3.6885 vs. 3.0328 *	3.5430 vs. 2.7351 *
	Temperature	3.4889 vs. 2.7111 *	3.7213 vs. 3.0164 *	3.5828 vs. 2.8344 *
	Music and feeling	3.2333 vs. 2.4000 *	3.6721 vs. 2.9016 *	3.4106 vs. 2.6026 *
Price	Valuable price	3.0222 vs. 2.7333 **	3.6230 vs. 3.1311 *	3.2649 vs. 2.8940 *
	Discount	2.7333 vs. 2.3333 *	3.5246 vs. 2.7705 *	3.0530 vs. 2.5099 *
Satisfaction		3.2111 vs. 2.5111 *	3.5410 vs. 3.0000 *	3.3444 vs. 2.7086 *
Revisit intention		3.1444 vs. 2.4333 *	3.6557 vs. 3.1148 *	3.3510 vs. 2.7086 *
Recommendation		3.0333 vs. 2.3000 *	3.5574 vs. 3.0000 *	3.2450 vs. 2.5828 *

* denotes $p < 0.01$, ** denotes $p < 0.05$, *** denotes $p < 0.1$.

Table 4 shows that the mean scores were higher for Burger King than Berlin Burger in all groups—that is, whether the respondent resided in Mongolia or Korea, they preferred Burger King, a global fast food chain. This trend shows consistency over all items. The restaurant temperature (3.5828) and food taste (3.5629) had the highest mean scores for the global fast food restaurant, while the variety of menu options (3.0331) and portions (3.0265) had the highest mean scores for the Mongolian fast food restaurant. Discount had the lowest mean score for both restaurants: 3.0530 for the global fast food restaurant and 2.5099 for the Mongolian fast food restaurant. Burger King outperformed Berlin Burger most in the "atmosphere" factor. Moreover, for the "good portions" and "valuable price" items, the score between the two fast food chains was smallest. Berlin Burger seemed to have an advantage in price relative to other items; although the difference was statistically significant, it was also the smallest.

In addition, Table 4 exhibits a difference in perception scores depending on whether the respondent was living in Korea or Mongolia. Overall, respondents residing in Korea gave higher scores for all items. Moreover, they regarded Burger King's relative superiority over Berlin Burger less significant for most items; the score difference for the two fast food chains were smaller for this group. Such perception difference depending on the respondent's residence may be due to several factors, including experience of other global fast food chains and long term perception. The score difference between Burger King and Berlin Burger was relatively smaller for Korea-residing respondents, especially for freshness and taste of food, and less so for items such as music and feeling, temperature, clean dining areas, and well-trained employees. These findings suggest that the positive reputation of Burger King in atmosphere-related areas had a lasting effect in comparison to food-related items.

Furthermore, one can also infer that such differences may be due to the fact that Mongolians residing in Korea have greater accessibility to fast food and are more familiar with it; therefore, these respondents may generally have more lenient standards for judging fast food options. Such findings imply the following: first, the preference for Burger King over Berlin Burger owes greatly to perception differences for the "atmosphere" factor. This suggests that Berlin Burger should focus on enhancing customer experiences regarding service and ambience. This is especially important as these items seemed to have a greater lasting effect on customer perception. Furthermore, as Burger King's superiority in food quality was least significant, especially in the long-term perspective, it would be prudent for the brand to invest more resources in related areas.

4.2. Factor Analysis

The principal component analysis started with 16 items. After performing the principal component analysis with varimax rotation, the results revealed that the 16-item scale fell into four factors. All items had loadings of greater than 0.5, and there were no items that needed to be removed. Tables 5 and 6 show the results of the four factors for Burger King and Berlin Burger.

Table 5. Rotated Component Matrix (Burger King).

Items	Factors				Cronbach's Alpha
	Service Quality (SQ)	**Atmosphere (A)**	**Food Quality (FQ)**	**Price (P)**	
Well trained (SQ4)	0.823	0.379	0.282	0.210	0.977
Attitude (SQ2)	0.808	0.396	0.245	0.257	
Availability (SQ3)	0.805	0.375	0.311	0.216	
Staff knowledge (SQ5)	0.795	0.312	0.338	0.227	
Kindness (SQ1)	0.776	0.367	0.304	0.250	
Temperature (A4)	0.331	0.796	0.392	0.159	0.969
Comfortable seats (A3)	0.383	0.783	0.351	0.218	
Interior (A1)	0.428	0.771	0.224	0.252	
Cleanliness (A2)	0.397	0.752	0.263	0.316	
Music (A5)	0.403	0.721	0.235	0.361	
Fresh (FQ2)	0.301	0.406	0.767	0.175	0.931
Tasty (FQ1)	0.441	0.346	0.736	0.196	
Variety of menus (FQ3)	0.283	0.277	0.704	0.429	
Good portions (FQ4)	0.298	0.199	0.693	0.457	
Discount (P2)	0.246	0.287	0.264	0.839	0.897
Value (P1)	0.295	0.302	0.362	0.757	
KMO					0.942
Bartlett's Test of Sphericity				Chi-square	3368.902
				df (sig.)	120 (0.000)

Table 6. Rotated Component Matrix (Berlin Burger).

Items	Component				Cronbach's Alpha
	Service Quality (SQ)	Atmosphere (A)	Food Quality (FQ)	Price (P)	
Staff knowledge (SQ5)	0.779	0.358	0.318	0.329	0.983
Availability (SQ3)	0.776	0.351	0.402	0.240	
Attitude (SQ2)	0.774	0.388	0.373	0.202	
Well trained (SQ4)	0.755	0.374	0.427	0.242	
Kindness (SQ1)	0.733	0.389	0.450	0.202	
Variety of menu (FQ3)	0.362	0.842	0.266	0.180	0.968
Tasty (FQ1)	0.334	0.798	0.340	0.240	
Good portions (FQ4)	0.332	0.782	0.329	0.243	
Fresh (FQ2)	0.340	0.765	0.346	0.288	
Music (A5)	0.381	0.211	0.779	0.302	0.967
Comfortable seats (A3)	0.339	0.438	0.769	0.208	
Cleanliness (A2)	0.418	0.335	0.742	0.250	
Interior (A1)	0.406	0.426	0.725	0.232	
Temperature (A4)	0.425	0.389	0.621	0.304	
Discount (P2)	0.271	0.259	0.428	0.767	0.871
Value (P1)	0.366	0.489	0.244	0.687	
KMO					0.941
Bartlett's Test of Sphericity				Chi-square	3964.552
				df (sig.)	120(0.000)

A KMO test is a measure of how suited the data are for factor analysis. The test measures the sampling adequacy for each variable in the model and for the complete model. If the value for the KMO test is greater than 0.50, then a factor analysis can be done for the same data set. It should also be significant at the 5% level and the p-value should be less than 0.05. Based on Table 5, it can be observed that the KMO measure was 0.942, which meant that the variables were suitable for factor analysis.

Cronbach's alpha was used to examine the internal reliability of the 16 items used to measure the four factors. The Cronbach's alpha varies from 0 to 1, and a value of 0.6 or less indicates unsatisfactory reliability. Table 5 shows that all factors had values exceeding 0.6. The service quality was measured by five items and had the highest alpha coefficient of 0.977. Atmosphere was measured by five items and had an alpha coefficient of 0.969. Price was measured by two items and showed the lowest alpha coefficient of 0.897.

Table 6 shows that the KMO measure for Berlin Burger was 0.941, which meant that the variables used were suitable for factor analysis. It was also significant at the 5% level because the p-value was 0.000, which is less than 0.05. Table 6 also shows that the Cronbach's alpha exceeded 0.6 for all factors. Service quality was measured by five items and showed the highest alpha coefficient of 0.983. Price was measured by two items and showed the lowest alpha coefficient of 0.871.

4.3. Factors Affecting Customer Satisfaction, Revisit Intention, and Recommendation

4.3.1. Customer Satisfaction

A regression analysis was conducted to investigate the influence of the institutional DINESERV factors on customer satisfaction. Table 7 shows the results of the regression analysis with the four factors as independent variables and customer satisfaction as the dependent variable.

Table 7. Regression result for customer satisfaction (Burger King).

Factors	Customer Satisfaction				
	Unstandardized Coefficient		Standardized Coefficients	*t*-Value	Sig.
	B	Std. Error	Beta		
Constant	3.347	0.041		82.217 *	0.000
Food quality	0.543	0.041	0.457	13.295 *	0.000
Service quality	0.394	0.041	0.332	9.646 *	0.000
Atmosphere	0.653	0.041	0.550	15.989 *	0.000
Price	0.539	0.041	0.454	13.205 *	0.000

* $p < 0.01$; $R^2 = 0.828$, Adjusted $R^2 = 0.824$.

Table 7 shows the regression results indicating that four factors had significant and positive effects on customer satisfaction. The adjusted R square value of this model, which was a more conservative estimate of the variance by considering error variance, was found to be 0.824, indicating that 82.4% of customer satisfaction could be explained by the four independent factors. Thus, the explanatory power of the model was satisfactory. Moreover, the coefficients of the four factors were significant at the 1% level, suggesting a positive relationship between customer satisfaction and food quality, service quality, atmosphere, and price. In other words, all factors (food quality, service quality, price and value, and atmosphere) were found to be significant predictors affecting customer satisfaction. Meanwhile, the "atmosphere" variable was shown to have the highest standardized coefficient ($\beta = 0.550$, $p < 0.01$) with regards to customer satisfaction for the global fast food restaurant, which implied that atmosphere was the most influential factor for predicting customer satisfaction in the case of Burger King. Next was food quality ($\beta = 0.457$, $p < 0.01$), followed by atmosphere ($\beta = 0.454$, $p < 0.01$) and service quality ($\beta = 0.332$, $p < 0.01$).

Table 8 indicates that the coefficients of the four factors were significant at the 1% level, suggesting a positive relationship between customer satisfaction and food quality, service quality, atmosphere, and price. The adjusted R square value of 0.768 indicated that 76.8% of the customer satisfaction could be explained by the four independent variables. The results showed that service quality was the highest standardized coefficient ($\beta = 0.502$, $p < 0.01$) of customer satisfaction for the Mongolian fast food restaurant, which implied that service quality was the most influential factor for Berlin Burger, followed by food quality ($\beta = 0.441$, $p < 0.01$), atmosphere ($\beta = 0.415$, $p < 0.01$), and service quality ($\beta = 0.395$, $p < 0.01$). All factors (food quality, service quality, price and value, and atmosphere) were found to be significant predictors affecting customer satisfaction for Berlin Burger.

Table 8. Regression result for customer satisfaction (Berlin Burger).

Factors	Customer Satisfaction				
	Unstandardized Coefficient		Standardized Coefficients	*t*-Value	Sig.
	B	Std. Error	Beta		
Constant	2.709	0.044		61.204 *	0.000
Food quality	0.498	0.044	0.441	11.217 *	0.000
Service quality	0.566	0.044	0.502	12.752 *	0.000
Atmosphere	0.468	0.044	0.415	10.544 *	0.000
Price	0.446	0.044	0.395	10.042 *	0.000

* $p < 0.01$; $R^2 = 0.774$, adjusted $R^2 = 0.768$.

4.3.2. Revisit Intention

Table 9 shows that the coefficients of the four factors were significant at the 1% level, suggesting a positive relationship between customer revisit intention and food quality, service quality, atmosphere, and price. The results indicated that price attained the highest beta coefficient, which implied that price was the most influential factor ($\beta = 0.528$, $p < 0.01$) in predicting customer revisit intention for the global fast food restaurant. The second highest was atmosphere ($\beta = 0.475$, $p < 0.01$), followed by food quality ($\beta = 0.407$, $p < 0.01$) and service quality ($\beta = 0.275$, $p < 0.01$). The adjusted R square value of 0.739 indicated that 73.9% of customer revisit intention was explained by the four factors. The results indicated that these factors had a positive impact on customer revisit intention.

Table 9. Regression result for customer revisit intention (Burger King).

Factors	Revisit Intention				
	Unstandardized Coefficient		Standardized Coefficients	*t*-Value	Sig.
	B	Std. Error	Beta		
Constant	3.353	0.051		65.243 *	0.000
Food quality	0.502	0.052	0.407	9.730 *	0.000
Service quality	0.339	0.052	0.275	6.567 *	0.000
Atmosphere	0.585	0.052	0.475	11.352 *	0.000
Price	0.651	0.052	0.528	12.624 *	0.000

* $p < 0.01$; $R^2 = 0.746$, adjusted $R^2 = 0.739$.

Table 10 shows that the coefficients of the four factors for Berlin Burger were significant at the 1% level, suggesting a positive relationship between customer revisit intention and food quality, service quality, atmosphere, and price. The results indicate that service quality attained the highest standardized coefficient, which implies that service quality was the most influential factor ($\beta = 0.517$, $p < 0.01$) of customer revisit intention for the Mongolian fast food restaurant. The next was food quality ($\beta = 0.416$, $p < 0.01$), followed by price ($\beta = 0.385$, $p < 0.01$) and atmosphere ($\beta = 0.377$, $p < 0.01$). The adjusted R square value of 0.723 indicated that 72.3% of the customer revisit intention could be explained by the four independent variables. Research Question 2 asked whether food quality, service quality, atmosphere, and price have positive influences on the customer revisit intention. The results indicated that these factors had positive impacts on customer revisit intentions, supporting a positive answer to Research Question 2.

Table 10. Regression result for customer revisit intention (Berlin Burger).

Factors	Revisit Intention				
	Unstandardized Coefficient		Standardized Coefficients	*t*-Value	Sig.
	B	Std. Error	Beta		
Constant	2.709	0.050		54.321 *	0.000
Food quality	0.484	0.050	0.416	9.674 *	0.000
Service quality	0.602	0.050	0.517	12.024 *	0.000
Atmosphere	0.439	0.050	0.377	8.767 *	0.000
Price	0.448	0.050	0.385	8.953 *	0.000

* $p < 0.01$; $R^2 = 0.730$, adjusted $R^2 = 0.723$.

4.3.3. Recommendation

Table 11 shows that the coefficients of the four factors were significant at the 1% level, suggesting a positive relationship between customer likelihood of recommendation and food quality, service quality, atmosphere, and price. The regression results in Table 11 indicate that price attained the highest standardized coefficient, which implies that price was the most influential factor ($\beta = 0.552$, $p < 0.01$) affecting the likelihood of a customer recommending the global fast food restaurant. The next highest was atmosphere ($\beta = 0.464$, $p < 0.01$), followed by food quality ($\beta = 0.406$, $p < 0.01$) and service quality ($\beta = 0.302$, $p < 0.01$). The adjusted R square value of 0.770 indicates that 77% of the customer likelihood of recommendation could be explained by the four factors.

Table 11. Regression result for customer recommendation (Burger King).

Factors	Customer Recommendation				
	Unstandardized Coefficient		Standardized Coefficients	*t*-Value	Sig.
	B	Std. Error	Beta		
Constant	3.247	0.048		67.590 *	0.000
Food quality	0.498	0.048	0.406	10.324 *	0.000
Service quality	0.371	0.048	0.302	7.689 *	0.000
Atmosphere	0.569	0.048	0.464	11.804 *	0.000
Price	0.676	0.048	0.552	14.028 *	0.000

* $p < 0.01$; $R^2 = 0.776$, adjusted $R^2 = 0.770$.

Table 12 indicates that the coefficients of the four factors were significant at the 1% level, suggesting a positive relationship between customer revisit intention and food quality, service quality, atmosphere, and price. The regression results indicate that service quality attained the highest standardized coefficient, which implies that service quality was the most influential factor ($\beta = 0.464$, $p < 0.01$) affecting the likelihood of a customer recommending the Mongolian fast food restaurant. The next highest was the price ($\beta = 0.427$, $p < 0.01$), followed by food quality ($\beta = 0.396$, $p< 0.01$) and atmosphere ($\beta = 0.356$, $p < 0.01$).

Table 12. Regression result for customer recommendation (Berlin Burger).

Factors	Customer Recommendation				
	Unstandardized Coefficient		Standardized Coefficients	*t*-Value	Sig.
	B	Std. Error	Beta		
Constant	2.583	0.054		47.645 *	0.000
Food quality	0.460	0.054	0.396	8.455 *	0.000
Service quality	0.539	0.054	0.464	9.907 *	0.000
Atmosphere	0.414	0.054	0.356	7.617 *	0.000
Price	0.496	0.054	0.427	9.120 *	0.000

* $p < 0.01$; $R^2 = 0.680$, adjusted $R^2 = 0.672$.

4.3.4. Comparison of Two Restaurants

Table 13 indicates that all four factors (food quality, service quality, atmosphere, and price) had significant effects on customer satisfaction, revisit intention, and customer likelihood of recommendation. However, differences between the global fast food restaurant and Mongolian fast food restaurant were found in terms of t-values. The t-values of the food quality, atmosphere, and price for the global fast food restaurant were higher than those of the Mongolian fast food restaurant, and the t-value of service quality was lower than the t-value of the Mongolian fast food restaurant. The results answered research question 4, showing that the DINESERV factors and the customers' overall satisfaction level differed between the global and Mongolian restaurants.

Table 13. Comparison of regression results for two restaurants.

Model	Burger King (p-Value)			Berlin Burger (p-Value)		
	Customer Satisfaction	Revisit Intention	Customer Recommendation	Customer Satisfaction	Revisit Intention	Customer Recommendation
Constant	3.347	3.353	3.247	2.709	2.709	2.583
Food Quality	13.295 * (0.000)	9.730 * (0.000)	10.324 * (0.000)	11.217 * (0.000)	9.674 * (0.000)	8.455 * (0.000)
Service Quality	9.646 * (0.000)	6.567 * (0.000)	7.689 * (0.000)	12.752 * (0.000)	12.024 * (0.000)	9.907 * (0.000)
Atmosphere	15.989 * (0.000)	11.352 * (0.000)	11.804 * (0.000)	10.544 * (0.000)	8.767 * (0.000)	7.617 * (0.000)
Price	13.205 * (0.000)	12.624 * (0.000)	14.028 * (0.000)	10.042 * (0.000)	8.953 * (0.000)	9.120 * (0.000)

* $p < 0.01$.

4.4. The Relationship between Customer Satisfaction, Revisit Intention, and Recommendation

Research Question 5 asked whether there are any significant relationships among customer satisfaction, revisit intention, and likelihood of recommendation. The question can be interpreted through the following hypotheses.

Hypothesis 1. *Customer satisfaction will have a positive influence on customer revisit intention and customer likelihood of recommendation.*

Hypothesis 2. *Customer revisit intention will have a positive influence on customer likelihood of recommendation.*

In order to answer the research question about the association between customer satisfaction and revisit intention and likelihood of recommendation, the results of Pearson correlations for three variables are presented in Table 14.

Table 14. Pearson correlations for pairs.

Pairs	Burger King (p-Value)	Berlin Burger (p-Value)
Customer satisfaction vs. revisit intention (r1)	0.907 *(0.000)	0.919 *(0.000)
Customer satisfaction vs. recommendation (r2)	0.909 *(0.000)	0.913 *(0.000)
Recommendation vs. revisit intention (r3)	0.884 *(0.000)	0.915 *(0.000)

$n = 151$, * $p < 0.01$.

Customer satisfaction was found to have a strong, positive relationship (r1 = 0.907 for Burger King and 0.919 for Berlin Burger, $p < 0.01$) with revisit intention. This result indicates a strong relationship between the satisfaction level of the respondents with their revisit intention. Customer satisfaction was also strongly positively correlated (r1 = 0.909 for Burger King and 0.913 for Berlin Burger, $p < 0.01$) with likelihood of recommendation. This high correlation shows that high customer satisfaction leads to a high likelihood of recommendation. In addition, likelihood of recommendation was strongly correlated (r1 = 0.884 for Burger King and 0.915 for Berlin Burger, $p < 0.01$) with revisit intention.

The finding implies that likelihood of recommendation helped to enhance the revisit intention of fast food restaurant customers. Thus, Hypothesis 1, customer satisfaction will have a positive influence on revisit intention, was supported ($p < 0.000$). The results show that customer satisfaction had a significant effect on customer revisit intention for both restaurants. Furthermore, the hypothesis that customer satisfaction will have a positive influence on customer likelihood of recommendation was supported ($p < 0.000$) for both restaurants. Table 14 shows that the level of satisfaction had an effect on the global fast food restaurant and the Mongolian fast food restaurant. However, the t-values of revisit intention and likelihood of recommendation for the Mongolian fast food restaurant were higher than those for the global fast food restaurants. This means that customer satisfaction influenced customer revisit intention and likelihood of recommendation for the Mongolian fast food restaurant to a greater extent. Moreover, Hypothesis 2, customer revisit intention will have a positive influence on customer likelihood of recommendation, was supported ($p < 0.000$) for both restaurants.

4.5. Discussion and Implications

This research delves into the competitive advantages of global and local food chains for the sustainability of the Mongolian fast food industry, including suggestions regarding which factors are relatively more essential from the management perspective. Both the t-test and regression analysis show that the "atmosphere" factor, which includes music, comfortable seats, cleanliness, interior, and temperature, is a key asset of global franchises. Although brand perceptions of the global food chain were more positive than the local brand for all categories, both score differences and regression coefficients imply that atmosphere is the global food chain's prime advantage over local ones. Furthermore, the results of this study show that this factor is especially important, since it seems to have an enduring effect on positive customer perceptions. These findings relay significant business implications; in order for local food chains to thrive, they must invest more resources into enhancing customer experiences related to the "atmosphere" category, and should continue to achieve comparative advantages in "service quality," which was found to be the key element in predicting customer satisfaction for the local food chain.

This study also provides evidence regarding perception differences between customer groups that have greater accessibility to various franchises and those that have not. Compared to South Korea, residents of Mongolia have less experience in global food chains as the introduction of such franchises is very recent, and the simple presence of stores in Seoul and Ulaanbaatar is incomparable. Such environmental differences lead to different customer perceptions; customers residing in Korea appear to harbor more positive attitudes towards fast food in general, as scores for both franchises were higher than for respondents residing in Mongolia. Furthermore, the results show a difference in perception scores depending on whether the respondent was living in Korea or Mongolia. The score difference between Burger King and Berlin Burger was relatively smaller for Korea-residing respondents, especially for freshness and taste of food, and less so for items such as music and feeling, temperature, clean dining areas, and well-trained employees. These findings suggest that the positive reputation of Burger King in atmosphere-related areas has a lasting effect in comparison to food-related items. This suggests that Berlin Burger should focus on enhancing customer experiences regarding service and ambience. Furthermore, as Burger King's superiority in food quality is least significant, especially in the long-term perspective, it would be prudent for the brand to invest more resources in related areas.

5. Conclusions

This study was conducted to gain a better understanding of customer satisfaction in restaurants by studying the factors of food quality, service quality, atmospherics, and price. It fulfilled its aims of identifying the relationships among the four variables with customer satisfaction, which leads to revisit intention and likelihood of recommendation. The results show that all four factors (food quality, service quality, price, and atmosphere of a restaurant) positively affect customer satisfaction, revisit intention and likelihood of recommendation for the global fast food and Mongolian restaurants. Moreover, the results show that customer satisfaction will have a positive influence on revisit intention and likelihood of customer recommendation for both restaurants. However, depending on whether it is a Mongolian fast food restaurant or a global fast food restaurant, the level of factors affecting customer satisfaction were different. For a global fast food chain, the restaurant atmosphere was considered the most important factor influencing customer satisfaction. Price was also considered the most important factor for customer revisit intention and likelihood of recommendation. This shows that customers of global fast food restaurants in Mongolia consider atmosphere to be the most important factor for customer satisfaction, revisit intention, and likelihood of customer recommendation. Thus, managers of global fast food restaurants in Mongolia should pay attention to the polite behavior of staff and whether there is a comfortable atmosphere, which, in turn, enhances customer satisfaction. For the Mongolian fast food restaurant, service quality was considered the most important factor influencing customer satisfaction, revisit intention, and likelihood of customer recommendation. Food quality was the second most important factor affecting customer satisfaction and revisit intention, and price was the second most important factor affecting customer recommendation. This means that customers of the Mongolian fast food restaurant put more focus on service quality and food quality. Thus, restaurant owners need to make a constant effort to improve service quality and offer delicious meals at valuable prices to their customers.

This study also showed a difference in perception scores depending on whether the respondent was living in Korea or Mongolia. Overall, respondents residing in Korea gave higher scores for all items. They regarded Burger King's relative superiority over Berlin Burger less significant for most items. The score difference between Burger King and Berlin Burger was relatively smaller for Korea-residing respondents, especially for freshness and taste of food, and less so for items such as music and feeling, temperature, clean dining areas, and well-trained employees. These findings suggest that the positive reputation of Burger King in atmosphere-related areas has a lasting effect in comparison to food-related items. One can also infer that such differences may be due to the fact that Mongolians residing in Korea have greater accessibility to fast food and are more familiar with it; therefore, these respondents may generally have more lenient standards for judging fast food options.

This study aimed to investigate factors affecting customer satisfaction, revisit intention, and recommendations for fast food restaurants in Mongolia. As the industry has only recently developed, research is currently lacking in the field. This paper provides a starting point in research in the Mongolian fast food market, as well as customer perceptions. Furthermore, this study conducts an in-depth analysis of the difference between local and global fast food chains, especially depending on whether the customer resides in a fast food-friendly environment or not. The findings in this study also lead to important managerial implications regarding competitive advantages.

There are some limitations associated with this study. First, the sample size could have been larger. The period of data collection was short, and for this reason, only 151 questionnaires were returned. This suggests that the research should take place over a couple of months. Secondly, as our study focused on comparing directly competing food chains, only two brands were studied. The limited fast food presence in Ulaanbaatar inevitably led to this limitation, but further research may take a more comprehensive approach of studying the entire fast food environment in Mongolia. Further studies may also examine the restaurant service quality for a particular type of restaurant using a larger sample size or using different sets of factors for each type of restaurant. An extension of the study range and inclusion of other study methodologies to analyze the restaurant service quality may also develop the implications of this study.

Author Contributions: Conceptualization, A.N.-O. and S.-H.C.; formal analysis, A.N.-O. and S.-H.C.; methodology, S.-H.C.; software, A.N.-O. and S.-H.C.; visualization, A.N.-O. All authors have read and agreed to the published version of the manuscript.

Funding: This work was supported by the Ministry of Education of the Republic of Korea and the National Research Foundation of Korea (NRF-2019S1A5A2A01046398).

Conflicts of Interest: The authors declare no conflict of interest.

References

1. Grand View Research. Fast Food & Quick Service Restaurant Market Size, Share & Trends Analysis Report by Type (Chain, Independent), by Cuisine (American, Turkish & Lebanese), by Region, and Segment Forecasts, 2020–2027. Available online: https://www.grandviewresearch.com/industry-analysis/fast-food-quick-service-restaurants-market (accessed on 6 August 2020).
2. Abdelkafi, N.; Täuscher, K. Business Models for Sustainability from a System Dynamics Perspective. *Organ. Environ.* **2015**, *29*, 74–96. [CrossRef]
3. Gupta, S.; McLaughlin, E.; Gomez, M. Guest Satisfaction and Restaurant Performance. In *The Next Frontier of Restaurant Management: Harnessing Data to Improve Guest Service and Enhance the Employee Experience*; Cornell University Press: Ithaca, NY, USA, 2019; pp. 33–53.
4. Mongolian Business Sectors. Available online: http://www.mse.mn/uploads/files/mongolian%20business%20sectors%201_compressed.pdf (accessed on 6 August 2020).
5. Mongolia GDP Growth Rate YoY. Available online: https://tradingeconomics.com/mongolia/gdp-growth (accessed on 6 August 2020).
6. Zhu, Z. Burger King Opens in Mongolia, Bringing Burgers to Their Birthplace. Available online: https://asiamattersforamerica.org/articles/burger-king-opens-in-mongolia-bringing-burgers-to-their-birthplace (accessed on 6 August 2020).
7. Great Food Comes First. Available online: http://www.burgerking.mn (accessed on 6 August 2020).
8. Yonhap, S. Korean Fast-Food Chain Lotteria Opens in Mongolia. Available online: http://www.koreaherald.com/view.php?ud=20180625000515 (accessed on 6 August 2020).
9. Gikonyo, L.; Berndt, A.; Wadawi, J. Critical Success Factors for Franchised Restaurants Entering the Kenyan Market. *SAGE Open* **2015**, *5*, 1–8. [CrossRef]
10. Markovic, S.; Raspor, S.; Sergaric, K. Does restaurant performance meet customers' expectations? An assessment of restaurant service quality using a modified DINESERV approach. *Tour. Hosp. Manag.* **2010**, *16*, 81–195.
11. Parasuraman, A.; Zeithaml, V.A.; Berry, L.L. SERVQUAL: A Multiple-Item Scale for Measuring Consumer Perceptions of Service Quality. *J. Retail.* **1988**, *64*, 12–40.
12. Stevens, P.; Knutson, B.; Patton, M. Dineserv: A Tool for Measuring Service Quality in Restaurants. *Cornell Hotel. Restaur. Adm. Q.* **1995**, *36*, 56–60. [CrossRef]
13. Qu, H. Determinant Factors and Choice Intention for Chinese Restaurant Dining. *J. Restaur. Foodserv. Mark.* **1997**, *2*, 35–49. [CrossRef]
14. Pettijohn, L.S.; Pettijohn, C.E.; Luke, R.H. An Evaluation of Fast Food Restaurant Satisfaction. *J. Restaur. Foodserv. Mark.* **1997**, *2*, 3–20. [CrossRef]
15. Rashid, I.; Abdullah, M. The impact of service quality and customer satisfaction on customer's loyalty: Evidence from Fast Food restaurant of Malaysia. *Int. J. Inf. Bus. Manag.* **2015**, *7*, 784–793.
16. Brady, M.K.; Cronin, J.J. Customer Orientation. *J. Serv. Res.* **2001**, *3*, 241–251. [CrossRef]
17. Namin, A. Revisiting customers' perception of service quality in fast food restaurants. *J. Retail. Consum. Serv.* **2017**, *34*, 70–81. [CrossRef]
18. Namkung, Y.; Jang, S. Does Food Quality Really Matter in Restaurants? Its Impact on Customer Satisfaction and Behavioral Intentions. *J. Hosp. Tour. Res.* **2007**, *31*, 387–409. [CrossRef]
19. Ha, J.; Jang, S. The effects of dining atmospherics on behavioral intentions through quality perception. *J. Serv. Mark.* **2012**, *26*, 204–215. [CrossRef]
20. Ryu, K.; Han, H. Influence of the Quality of Food, Service, and Physical Environment on Customer Satisfaction and Behavioral Intention in Quick-Casual Restaurants: Moderating Role of Perceived Price. *J. Hosp. Tour. Res.* **2009**, *34*, 310–329. [CrossRef]
21. Peri, C. The universe of food quality. *Food Qual. Prefer.* **2006**, *17*, 3–8. [CrossRef]
22. Kumar, I.; Garg, R.; Rahman, Z. Influence of retail atmospherics on customer value in an emerging market. *Great Lakes Her.* **2010**, *4*, 1–13.

23. Banerjee, S.; Singhania, S. Determinants of Customer Satisfaction, Revisit Intentions and Word of Mouth in the Restaurant Industry–Study Conducted In Selective Outlets of South Kolkata. *Int. J. Bus. Manag. Invent.* **2018**, *6*, 63–72.
24. Andaleeb, S.S.; Conway, C. Customer satisfaction in the restaurant industry: An examination of the transaction-specific model. *J. Serv. Mark.* **2006**, *20*, 3–11. [CrossRef]
25. Oliver, R.L. A Cognitive Model of the Ancedents and Consequences of Satisfaction Decisions. *J. Mark. Res.* **1980**, *17*, 460–469. [CrossRef]
26. Fornell, C. A national customer satisfaction barometer: The Swedish experience. *J. Mark.* **1992**, *56*, 6–21. [CrossRef]
27. Yan, X.; Wang, J.; Chau, M. Customer revisit intention to restaurants: Evidence from online reviews. *Inf. Syst. Front.* **2013**, *17*, 645–657. [CrossRef]
28. Oh, H. Diners' Perceptions of Quality, Value, and Satisfaction. *Cornell Hotel. Restaur. Adm. Q.* **2000**, *41*, 58–66. [CrossRef]
29. Weiss, R.; Feinstein, A.H.; Dalbor, M. Customer Satisfaction of Theme Restaurant Attributes and Their Influence on Return Intent. *J. Foodserv. Bus. Res.* **2004**, *7*, 23–41. [CrossRef]
30. Chen, P.-T.; Hu, H.-H. How determinant attributes of service quality influence customer-perceived value: An empirical investigation of the Australian coffee outlet industry. *Int. J. Contemp. Hosp. Manag.* **2010**, *22*, 535–551. [CrossRef]
31. Kim, K.-J.; Choi, K. Bridging the Perception Gap between Management and Customers on DINESERV Attributes: The Korean All-You-Can-Eat Buffet. *Sustainability* **2019**, *11*, 5212. [CrossRef]
32. Markovic, S.; Komsic, J.; ve Stifanic, M. Measuring Service Quality in City Restaurant Settings Using DINESERV Scale. 2013. Available online: http://www.wseas.us/elibrary/conferences/2013/Dubrovnik/MATREFC/MATREFC-27.pdf (accessed on 14 April 2017).
33. Adeinat, I. Measuring Service Quality Efficiency Using Dineserv. *Int. J. Qual. Res.* **2019**, *13*, 591–604. [CrossRef]
34. Knutson, B.J.; Stevens, P.; Patton, M. DINESERV. *J. Hosp. Leis. Mark.* **1996**, *3*, 35–44. [CrossRef]
35. Kim, H.J.; McCahon, C.; Miller, J. Assessing Service Quality in Korean Casual-Dining Restaurants Using DINESERV. *J. Foodserv. Bus. Res.* **2003**, *6*, 67–86. [CrossRef]
36. Bougoure, U.-S.; Neu, M.-K. Service Quality in the Malaysian Fast Food Industry: An Examination Using DINESERV. *Serv. Mark. Q.* **2010**, *31*, 194–212. [CrossRef]
37. Wang, C.H.; Lin, I.H.; Tsai, J.Y. Combining fuzzy integral and GRA method for evaluating the service quality of fast-food restaurants. *J. Interdiscip. Math.* **2018**, *21*, 447–456. [CrossRef]
38. Keith, N.K.; Simmers, C.S. Measuring Service Quality Perceptions of Restaurant Experiences: The Disparity Between Comment Cards and DINESERV. *J. Foodserv. Bus. Res.* **2011**, *14*, 20–32. [CrossRef]
39. Polyorat, K.; Sophonsiri, S. The influence of service quality dimensions on customer satisfaction and customer loyalty in the chain restaurant context: A Thai case. *J. Glob. Bus. Technol.* **2010**, *6*, 64–76.
40. Hansen, K.V. Development of SERVQUAL and DINESERV for Measuring Meal Experiences in Eating Establishments. *Scand. J. Hosp. Tour.* **2014**, *14*, 116–134. [CrossRef]
41. Kukanja, M.; Planinc, T. Toward cost-effective service excellence: Exploring the relationship between managers' perceptions of quality and the operational efficiency and profitability of restaurants. *Qual. Manag. J.* **2020**, *27*, 95–105. [CrossRef]
42. Kim, W.G.; Ng, C.Y.N.; Kim, Y.-S. Influence of institutional DINESERV on customer satisfaction, return intention, and word-of-mouth. *Int. J. Hosp. Manag.* **2009**, *28*, 10–17. [CrossRef]
43. Tan, Q.; Oriade, A.; Fallon, P. Service quality and customer satisfaction in Chinese fast food sector: A proposal for CFFRSERV. *Adv. Hosp. Tour. Res.* **2014**, *2*, 30–53.
44. DiPietro, R. Restaurant and foodservice research. *Int. J. Contemp. Hosp. Manag.* **2017**, *29*, 1203–1234. [CrossRef]
45. Marinković, V.; Senic, V.; Mimovic, P.M. Factors affecting choice and image of ethnic restaurants in Serbia. *Br. Food J.* **2015**, *117*, 1903–1920. [CrossRef]
46. Oyewole, P. Multiattribute Dimensions of Service Quality in the All-You-Can-Eat Buffet Restaurant Industry. *J. Hosp. Mark. Manag.* **2013**, *22*, 1–24. [CrossRef]
47. Wu, H.-C.; Mohi, Z. Assessment of Service Quality in the Fast-Food Restaurant. *J. Foodserv. Bus. Res.* **2015**, *18*, 358–388. [CrossRef]

48. Ulkhaq, M.M.; Nartadhi, R.L.; Akshinta, P.Y. Evaluating Service Quality of Korean Restaurants: A Fuzzy Analytic Hierarchy Approach. *Ind. Eng. Manag. Syst.* **2016**, *15*, 77–91. [CrossRef]

Article

Towards Sustainable Food Services in Hospitals: Expanding the Concept of 'Plate Waste' to 'Tray Waste'

Nouf Sahal Alharbi *, Malak Yahia Qattan and Jawaher Haji Alhaji

Department of Health Sciences, College of Applied Studies and Community Service, King Saud University, Riyadh 11451, Saudi Arabia; mqattan@ksu.edu.sa (M.Y.Q.); jalhejjy@ksu.edu.sa (J.H.A.)

* Correspondence: noufsahal@ksu.edu.sa; Tel.: +96-65-0415-1280

Received: 29 July 2020; Accepted: 20 August 2020; Published: 24 August 2020

Abstract: Early debates on the sustainability of food-plating systems in hospitals have concentrated mostly on plate waste food served, but not eaten. This study aims to address the need for more comprehensive studies on sustainable food services systems by expanding the concept of plate waste, to that of tray waste (organic and inorganic materials), through a case study of a hospital in Saudi Arabia. Tray waste arising at the ward level was audited for three weeks, covering 939 meals. It was found that, on average, each patient threw away 0.41, 0.30, 0.12, and 0.02 kg of food, plastic, paper, and metal, respectively, each day. All this equated to 4831 tons of food, 3535 tons of plastic, 1414 tons of paper, and 235 tons of metal each year at hospitals across Saudi Arabia. As all of this waste ends up in landfills, without any form of recycling, this study proposes the need for a more comprehensive, political approach that unites all food system stakeholders around a shared vision of responsible consumption and sustainable development.

Keywords: sustainability; food production and consumption; sustainable food systems; sustainable menu; food catering practices in the public sector

1. Introduction

Over the last few years, a large number of international organizations have recognized the economic and environmental impact of the waste generated by food systems [1,2]. According to the Food and Agricultural Organization (FAO), 1.3 billion tons of food are wasted every year, which costs around USD 936 billion [3–5]. At the international environmental level, it has been reported that food waste accounts for a portion of global carbon emissions, equivalent to that of a medium-sized country [6].

Within the various food sectors, hospital food waste has been estimated as being two to three times higher than other sectors, such as restaurants, work places, and schools [7]. Moreover, hospital food service waste can contribute to as much as half of the total waste generated in a ward [8,9]. Actually, from an economic and environmental perspective, in places, like the UK, Portugal, Brazil, and Saudi Arabia, the estimated hospital food waste costs ranged from USD 90,960 to USD 342,449 per year, while the average emission of CO_2 was estimated as 1.8 kg per patient, per day in Portugal [10–13]. As a consequence of this economic and environmental drain, early debates on the food industry and its sustainability have mostly concentrated on waste elimination and recycling, which were seen as critical strategies for creating a food system that promotes environmentally friendly practices. This is the main objective of the United Nations Sustainable Development Goals (UNSDGs) [14].

Previous international studies have shown that food service waste is mostly generated from production, cooking, and, lastly, at the point of the serving stage or plate waste [1,7,15]. Several studies carried out in the last decade have addressed plate waste by trying to quantify the amount of food waste arising from meal delivery services at hospitals [12,16–19]. However, none of these studies have attempted to quantify both the food and its combined solid waste. In order to fill the gap, this paper sets out to extend the concept of plate waste to also include that of tray waste by analyzing, for the first time, the food and solid wastes arising within the hospital.

In order to do this, the paper focuses on a general hospital catering system in Riyadh, a city in Saudi Arabia, which is deeply committed to reviewing the status of the UN sustainable development goals and the country's alignment with Vision 2030. The Saudi government is aiming to achieve environmental sustainability, by preserving the natural resources and increasing the efficiency of waste management [20]. In this respect, the selected hospital in Riyadh, the country's capital city, provides an excellent research context to explore the extent to which the sustainability objectives were being translated into practice. Our study aims to provide new insights into the multiple diminutions of a sustainable food catering system by asking the following research questions: (1) What are the types and quantities of waste resulting from the catering services in hospitals? (2) What kind of sustainable measures do the food catering systems provide?

2. Materials and Methods

2.1. Case Study Description

The contemporary study was carried out at one of the biggest governmental tertiary hospitals in Riyadh city with 1200 beds, serving the various medical departments. For confidentiality purposes, the hospital has not been identified. The catering food system in the hospital offers 20 menu categories, aimed at meeting the nutritional requirements of the patients, according to their different health status. For example, apart from the normal diet menu, there are customized menus for patients with diabetes, renal disease, and other health conditions. Under each menu category, there are seven menus with a variety of food options for each day of the week, according to patient preferences. For example, in addition to the patient's selection of a cold or hot drink, breakfast also consists of packaged pita or toast bread and a choice from an array of main dishes, including beans, corn flakes, lentils or eggs, and two pieces of cheddar or cream cheese, honey and jam. Lunch and dinner include either a portion of rice or pasta with chicken, meat or fish, and mixed vegetables as the main course. Table 1 provides an example of a Sunday menu with the amounts of each meal for the diet of a typical patient. Every day, the menu is circulated to all in-patients between 1:30 and 2 p.m. for them to order their food for the next day. Using a computerized system, based on the list of patients according to their ward, room and diet requirements, the three meals are freshly cooked at a central kitchen in the hospital, and plated according to the patient's requirements. The patients are offered three main meals a day—breakfast at 6:30 a.m., lunch at 11 a.m., and dinner at 5:45 p.m. In addition to this, there are three refreshments snacks at 9 a.m., 1:30 p.m. and 5:45 p.m.

In the hospital, in some cases, patients can choose to share a room, or stay in one by themselves. However, patients with infectious diseases, or those who lack proper immunity are usually isolated. The main course for all isolated patients is served on plates made out of foil, and placed on cardboard trays. On the other hand, the meals for the non-isolated patients are served on a ceramic plate, placed on a reusable plastic tray. For all the patients (isolated or non-isolated), water, dairy drinks, sweets, fruits, bread and salads are served in plastic containers or packaging. Hot drinks are served in paper cups, and soup in foil plates, Figure 1. The meal is placed on a tray covered by a sheet of paper along with the paper menu. In addition, all the cutlery and cups are made from plastic or paper. The trays are transported to the wards in trolleys, and served to the patients at scheduled times.

After the meal, the trays are collected and transported back to the central kitchen, and all the tray waste is, at first, placed together, without any form of sorting, into waste bins (NAPCO Sanita, G.B.70 G SASO BIO, high-density polyethylene) and deposited into 800 L containers, located in the basement. They are then transported twice a day to the hospital's waste depot. From there, the waste is collected by a special private company, without any form of recycling procedures, it is then carried directly to the landfill sites.

Table 1. Inpatient normal diet Sunday menu (2800–3000 kcal).

	Breakfast	Lunch	Diner
Water	600 mL	600 mL	600 mL
Diary drink *	200 mL Milk or Butter milk or 170 mL Yogurt	200 mL Butter milk or 170 mL Yogurt	200 mL Butter milk or 170 mL Yogurt
Hot drink *	200 mL Tea or Coffee	200 mL Tea or Coffee	200 mL Tea or Coffee
Sweet and fruits *	30 gm Honey or Jam	An Orange or 150 gm Pineapple or 150 gm Custard	A Banana or 150 gm Cream Caramel
Bread *	125 gm (white or brown): Toast Bread or Pita Bread or Bun Bread	125 gm (white or brown): Toast Bread or Pita Bread	125 gm (white or brown): Toast Bread or Pita Bread
Main course *	Main course *: 50 gm Corn Flakes or 170 gm Lentil or 50 Shakshuka (scrambled egg with tomato) Cheese *: 50 gm Slice Cheese or Cream Cheese	Starches *: 200 gm Rice or 200 gm Pasta Meat *: 150 gm Grilled Chicken or 150 gm Grilled Fish or 100 gm Grilled Meat * Vegetables *: 150 gm Mixed vegetables or Cauliflower with Carrot sauté	Main course *: 150 gm Grilled Chicken + 200 gm Biryani Rice or 100 gm Grilled Lamb + 200 gm Biryani Rice or 100 gm Grilled Fish + 200 gm Biryani Rice or 2 Pieces Tuna Club Sandwich Vegetables *: 150 gm Cooked Bean or 150 gm Cooked Zucchini
Soup *	NA	150 gm Mushroom Soup or Barley soup	150 gm Mushroom Soup or Vermicelli Soup
Salad *	NA	150 gm Green Salad or 150 gm Mixed Salad	150 gm Green Salad or 150 gm Coleslaw Salad

* based on patient preferences.

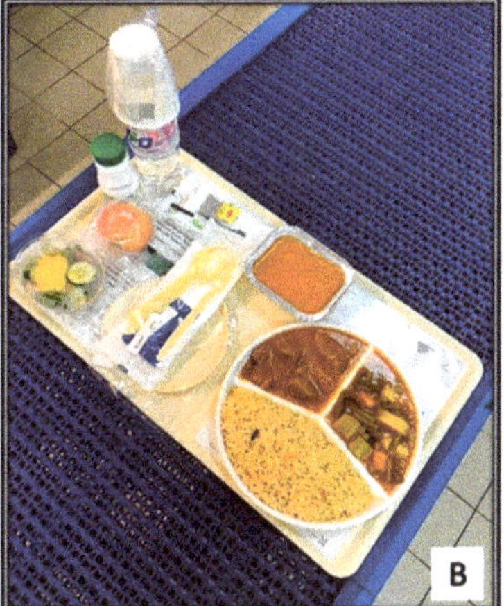

Figure 1. (**A**) A typical dinner for an isolated patient, including rice, meat, vegetables, soup, an apple and salad. (**B**) A meal for a non-isolated patient.

2.2. Waste Audit: Examination and Categorization

The waste was audited during the period of 15 September 15 to 6 October 2019 in eight wards, and consisted of 939 trays for the main meals. In this study, we included only the patients with solid diets, who represented approximately 89% of the total number of patients admitted to the hospital [21]. Patients with tube feeding, liquid diets and supplements were excluded. The data collection was carried out in two stages. Before the meal was served, all data about all the tray components and the weight of the meals by the ward name and bed type were obtained from the electronic food services system. In order to increase the data accuracy, a random sample from each meal was weighed on a digital scale (MOTEXT weight scale ML 30 N). At this first stage, we recorded the data on an Excel sheet, Figure 2. At the second stage, the overall food waste from each ward for the different medical departments was first sorted daily and weighed separately; then empty packaging and other tray waste were sorted and weighed separately, according to the waste type as follows: plastic, paper, and metal. The waste containers (NAPCO Sanita, G.B.70 G SASO BIO, high-density polyethylene) with tags were set aside in a specified area in the main kitchen to be weighed. All of the waste was weighed three times a day—the morning sample included waste arising from breakfast, the afternoon sample included waste arising from lunch, and the evening sample included waste arising from dinner. Since food is consumed, unlike the other waste made of plastic, paper and metal, we determined the plate waste by dividing the amount of food waste by the amount of food served, using the following equation:

$$\text{Plate waste } \% = \text{Food waste}/\text{Food served} \times 100$$

non-ISO	ISO	Patient No.	Ward	Meal	Day	Date
14	12	26	ADULT ACUTE CARE	Lunch	Friday	11/10/19

#	PATIENT MEALS										
	Water	Dairy Pro.	Drink	Sweet	Fruit	Bread	Starches	Meat	Vegetables	Soup	Salad
1	1 (600)ML	L (200) ML	T		B 150 G	125 GM	R (200) GM	C (150) GM	150 G	150 G	150 G
2	1 (600)ML	L (200) ML	T		B 150 G	125 GM	R (200) GM	C (150) GM	150 G	150 G	150 G
3	1 (600)ML	Y (200) ML	T		B 150 G	125 GM	R (200) GM	C (150) GM	150 G	150 G	150 G
4	1 (600)ML	L (200) ML	T		B 150 G	125 GM	R (200) GM	C (150) GM	150 G	150 G	150 G
5	1 (600)ML	L (200) ML	C		B 150 G	125 GM	R (200) GM	M (100) GM	150 G	150 G	150 G
6	1 (600)ML	L (200) ML	C	CC 150		125 GM	R (200) GM	M (100) GM	150 G	150 G	150 G
7	1 (600)ML	Y (200) ML	C	CC 150		125 GM	R (200) GM	C (150) GM	150 G	150 G	150 G
8	1 (600)ML	Y (200) ML	T	CC 150		125 GM	R (200) GM	M (100) GM	150 G	150 G	150 G
9	1 (600)ML	L (200) ML	C	CC 150		125 GM	R (200) GM	M (100) GM	150 G	150 G	150 G
10	1 (600)ML	L (200) ML	C	CC 150		125 GM	R (200) GM	C (150) GM	150 G	150 G	150 G
11	1 (600)ML	L (200) ML	T	CC 150		125 GM	R (200) GM	C (150) GM	150 G	150 G	150 G
12	1 (600)ML	L (200) ML	T	CC 150		125 GM	R (200) GM	F (150) GM	150 G	150 G	150 G
13	1 (600)ML	L (200) ML	C	CC 150		125 GM	R (200) GM	F (150) GM	150 G	150 G	150 G
14	1 (600)ML	L (200) ML	T	CC 150		125 GM	R (200) GM	C (150) GM	150 G	150 G	150 G
15	1 (600)ML	L (200) ML	C	CC 150		125 GM	R (200) GM	C (150) GM	150 G	150 G	150 G
16	1 (600)ML	L (100) ML	T		A 150 G	125 GM	R (125) GM	C (120) GM	150 G	100 G	150 G
17	1 (600)ML	L (100) ML	C		A 150 G	125 GM	R (125) GM	M (100) GM	150 G	100 G	150 G
18	1 (600)ML	L (100) ML	C		A 150 G	125 GM	R (125) GM	F (100) GM	150 G	100 G	150 G
19	1 (600)ML	Y (100) ML	C		O 150 G	125 GM	R (125) GM	F (100) GM	150 G	100 G	150 G
20	1 (600)ML	L (100) ML	T		O 150 G	125 GM	R (125) GM	M (100) GM	150 G	100 G	150 G
21	1 (600)ML	L (100) ML	T		A 150 G	125 GM	R (125) GM	C (120) GM	150 G	100 G	150 G
22	1 (600)ML	Y (100) ML	T		O 150 G	125 GM	R (125) GM	M (100) GM	150 G	100 G	150 G
23	1 (600)ML	L (100) ML	C		O 150 G	125 GM	R (125) GM	M (100) GM	150 G	100 G	150 G
24	1 (600)ML	L (100) ML	T		O 150 G	125 GM	R (125) GM	C (120) GM	150 G	100 G	150 G
25	1 (600)ML	L (100) ML	T		A 150 G	125 GM	R (125) GM	C (120) GM	150 G	100 G	150 G
26	1 (600)ML	L (100) ML	T		A 150 G	125 GM	R (125) GM	M (100) GM	150 G	100 G	150 G
Total	15.6	4.3		1.5	2.4	3.25	4.375	3.23	4.5	3.23	3.9

Shortcut Details (Drink)	
Tea	T
Coffee	C
Juice	J

Shortcut Details (Dairy)	
Milk	M
Laban	L
Yogurt	Y

Shortcut Details (MAET)	
Maet	M
Chicken	C
Fish	F

Skortcut Details (Starches)	
Rice	R
Macaroni	M

Shortcut Details (Sweet)	
Jello	J
Custard	CU
Cream caramel	CC

Shortcut Details (Fruit)	
Apple	A
Orange	O
Fruit salad	FS
Pineapple	P
Grape	G
Banana	B

Figure 2. An example of the data collection sheet for lunch meal.

Untouched main meals were individually counted and their weight was included in the waste. Protective clothes were worn during the categorization and quantification of the waste.

2.3. Statistical Analysis

Data entry and analysis were conducted using the IBM Statistical Package for Social Science (SPSS) version 20.0 (Armonk, NY, USA). Statistical analysis procedures included a descriptive analysis of the total amount of each type of waste, and the means and the confidence intervals of each type for the three main meals per patient per day were computed separately. Finally, after verification that the data were normally distributed, we examined the association between tray waste and bed type using the *t*-test. The statistical significance level was assumed for all estimations as p value ≤ 0.05. Values are presented as means and confidence intervals.

3. The Results

The average tray waste of the food, paper, plastic, and metal were 0.41, 0.30, 0.12, and 0.02 kg per patient per day, respectively. A comparison of the tray waste showed that the paper and metal waste levels were significantly higher among isolated patients—0.21 vs. 0.08, and 0.034 vs. 0.016–kg per patient per day, respectively, with no statistical significance for other tray waste types, Figure 3.

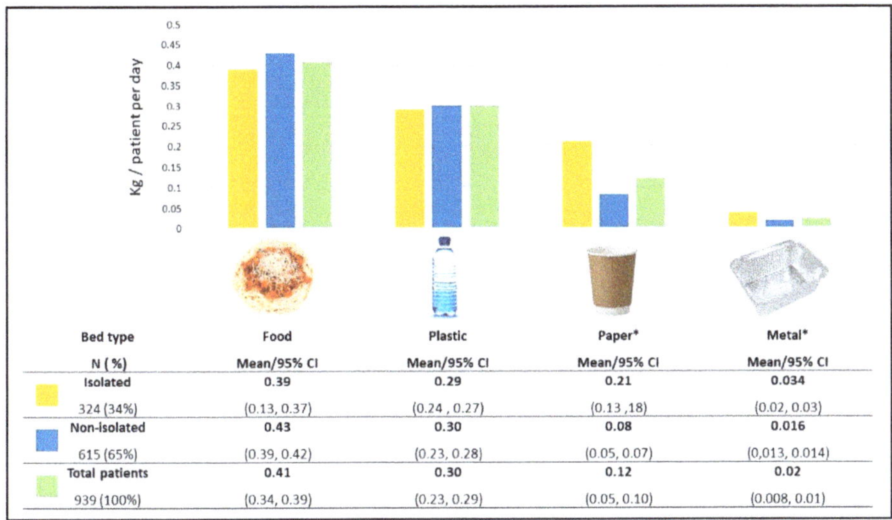

Figure 3. A comparison for tray waste by bed type for all meals per patient per day.

With regard to the plate waste across the main meals, although statistical relevance was not found, the lowest plate waste recorded for lunch was 15.5%, while the highest plate waste was 22% for dinner. On average, each patient threw away 412 g of food each day, representing 18.2% of the total food served. Waste, in accordance with the different main meals, is presented in Table 2 and Figure 4.

Table 2. Summarized mean weight tray waste (kg) per patient per day in accordance with meal time.

Waste	Breakfast	Lunch	Dinner
Food	0.08	0.15	0.16
Paper	0.042	0.051	0.039
Plastic	0.0974	0.0991	0.097
Foil	0.0063	0.01	0.0072

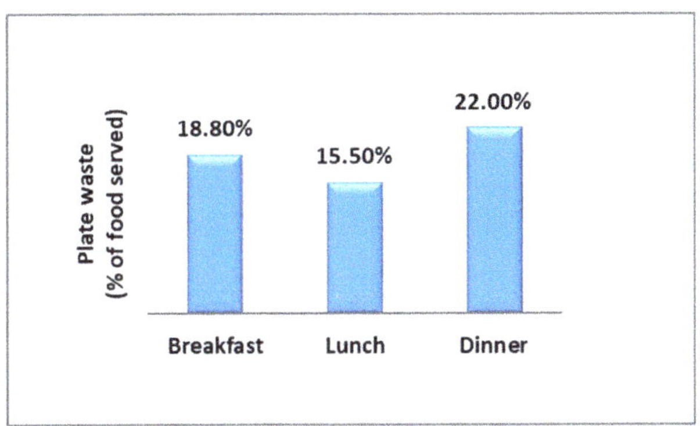

Figure 4. Plate waste per patient across the main meals.

4. Discussion

To the best of our knowledge, this was the first study that yielded a comprehensive picture about the extent of sustainability in food catering services, by using the tray as an assessment unit. In this study, we audited the tray waste (939 trays in all) at the ward level for three weeks, using a general hospital as a case study. Our results revealed that the overall food waste was 412 g per patient per day, and this figure was nearly similar to the average net of all inorganic wastes (plastic, paper, and metal) of 441 g per patient per day. However, according to the patient's bed type, this study found that the total amount of inorganic waste exceeded that of the food waste, where the average estimated inorganic waste was 534 g per patient per day. This figure is statistically higher by 34% than the waste generated from the patients who stayed in shared rooms.

The plate waste values of previous international studies conducted in the UK, Brazil, Portugal, the Netherlands and Australia, applying a similar assessment method in the context of hospital food services, ranged from 29% to 42%. Compared to our study, the results showed a lower value of plate waste at 18% [12,16–19]. In addition, the amount of plate waste arising from this Saudi case study was almost 40% lower (412 g) than that estimated in another study conducted in a general hospital in Portugal, where it was 953 g. This was so for these two studies, the food services department applied the same food serving system of "plating, not bulk". However, the dissimilarity in the results of the plate waste might be due to the differences in the food services systems. In the Saudi study, the meals were freshly cooked every day and the plating was according to the patients' preferences, while in the Portugal study, the food preparation was based on the cook-chill method, and the patients had limited options when choosing from the menu [12].

According to the latest national official statistics in Saudi Arabia in 2018, there were 284 government hospitals with a total of 43,690 beds. During that year, it was estimated by the Saudi Ministry of Health that, for these hospitals, a total of 35 million solid meals was provided [22]. Taking into consideration these numbers and our findings, this equates food tray waste in governmental hospitals in the country amounting to the discarding of about 4831 tons of food, 3535 tons of plastic, 1414 tons of paper, and 235 tons of metal each year. Thus, these indicators represent both a challenge to, and an opportunity for, the Saudi government.

From a sustainability point of view, by comparing retrospectively our food waste results with those from previous studies, it seems that the Kingdom of Saudi Arabia has made a remarkable move in achieving food security—one of the sustainable development goals [10,14]. However, on the other hand, there are still opportunities for stakeholders to meet the challenges of responsible consumption and production, which is another main sustainable development goal [14]. From a political point of view, in order to ensure that the Saudi government minimizes the carbon emissions associated with healthcare waste landfill, they can consider recycling to help reduce the depletion of plastic, paper and metal. Furthermore, the sustainable handling of food waste can return nutrients to the soil [23]. To achieve this, our study emphasizes the importance of developing a more integrated strategy to manage the waste—organic and inorganic—generated by the food systems in Saudi Arabia. This can be achieved by creating a legislative organization that mobilizes and unifies the practices of all the actors in the food industry, in order to create a shared vision for sustainable development in the country.

5. Conclusions

This was the first study that explored the extent of food service sustainability practices in Saudi hospitals. All the estimated tons of food, paper, plastic and metal transformed into waste equate to an environmental impact and economic losses. Indeed, the figures presented in our study highlight the opportunity for financial and environmental savings that can accrue to the Saudi health system by tackling this challenge. However, despite the contribution of this paper, it was challenging to compare our results with those obtained in other studies, due to the novelty of the research method that was characterized by our adopting of the tray as a new assessment unit. The plate waste results drew attention to the analysis of all waste, resulting from the different food preparation stages. Therefore, it would be interesting to gain insights into the reasons for the amount of plate waste generated in Saudi hospitals by conducting an in-depth analysis that included the perspectives of both the food service staff and the patients. In addition, further research is needed to evaluate the long-term environmental costs to society, and the possible measures to be adopted for cost-saving with regards to the food service budget. Other areas for investigation could also include water and energy usage and carbon emissions.

Author Contributions: N.S.A.; Conceptualization; Project administration; Methodology, reviewing and editing the final reaft M.Y.Q.; Data collection; Writing first draft, J.H.A.; Data collection; Validation; Reviewing and editing. All authors have read and agreed to the published version of the manuscript.

Funding: This research received no external funding.

Acknowledgments: The authors extend their appreciation to the Research Center for Humanity, Deanship of Scientific Research at King Saud University for funding this Research Group No. (HRGP-1-19-05).

Conflicts of Interest: The authors declare that they have no competing interests.

References

1. Gustavsson, J.; Cederberg, C.; Sonesson, U.; Van Otterdijk, R.; Meybeck, A. *Global Food Losses and Food Waste*; Food and Agriculture Organization of the United Nations (FAO): Rome, Italy, 2011.
2. Schneider, F. The evolution of food donation with respect to waste prevention. *Waste Manag.* **2013**, *33*, 755–763. [CrossRef] [PubMed]
3. Food and Agricultural Organization. *Food Wastage Footprint: Full-Cost Accounting (Final Report)*; Food and Agriculture Organization of the United Nations (FAO): Rome, Italy, 2014.
4. Scherhaufer, S.; Moates, G.; Hartikainen, H.; Waldron, K.; Obersteiner, G. Environmental impacts of food waste in Europe. *Waste Manag.* **2018**, *77*, 98–113. [CrossRef] [PubMed]
5. Schiavone, S.; Pelullo, C.P.; Attena, F. Patient Evaluation of Food Waste in Three Hospitals in Southern Italy. *Int. J. Environ. Res. Public Health* **2019**, *16*, 4330. [CrossRef] [PubMed]
6. Food and Agricultural Organization. *Food Wastage Footprint: Impacts on Natural Resources Food and Agriculture*; Food and Agriculture Organization of the United Nations (FAO): Rome, Italy, 2013.
7. Williams, P.G.; Walton, K. Plate waste in hospitals and strategies for change. *e-SPEN Eur. e-J. Clin. Nutr. Metab.* **2011**, *6*, e235–e241. [CrossRef]
8. Alam, M.; Sujauddin, M.; Iqbal, G.M.A.; Huda, S.M.S. Report: Healthcare waste characterization in Chittagong Medical College Hospital, Bangladesh. *Waste Manag. Res.* **2008**, *26*, 291–296. [CrossRef] [PubMed]
9. Mattoso, L.H.; Schalch, V. Hospital waste management in Brazil: A case study. *Waste Manag. Res.* **2001**, *19*, 567–572. [CrossRef] [PubMed]
10. Al-Shoshan, A.A. Study of the Regular Diet of Selected Hospitals of the Ministry of Health in Saudi Arabia: Edible Plate Waste and Its Monetary Value. *J. R. Soc. Health* **1992**, *112*, 7–11. [CrossRef] [PubMed]
11. Barton, A.; Beigg, C.; Macdonald, I.A.; Allison, S. High food wastage and low nutritional intakes in hospital patients. *Clin. Nutr.* **2000**, *19*, 445–449. [CrossRef] [PubMed]
12. Dias-Ferreira, C.; Santos, T.; Oliveira, V. Hospital food waste and environmental and economic indicators—A Portuguese case study. *Waste Manag.* **2015**, *46*, 146–154. [CrossRef] [PubMed]
13. Nonino, C.B.; Rabito, E.I.; Da Silva, K.; Ferraz, C.A.; Chiarello, P.G.; Dos Santos, J.S.; Marchini, J.S. Desperdício de alimentos intra-hospitalar. *Rev. Nutr.* **2006**, *19*, 349–356. [CrossRef]
14. United Nations (UN). Transforming Our World: The 2030 Agenda for Sustainable Development. Available online: https://sustainabledevelopment.un.org/content/documents/21252030%20Agenda%20for%20Sustainable%20Development%20web.pdf (accessed on 7 March 2020).
15. Bond, M.; Meacham, T.; Bhunnoo, R.; Benton, T. *Food Waste Within Global Food Systems*; Global Food Security: London, UK, 2013.
16. Barrington, V.; Maunder, K.; Kelaart, A. Engaging the patient: Improving dietary intake and meal experience through bedside terminal meal ordering for oncology patients. *J. Hum. Nutr. Diet.* **2018**, *31*, 803–809. [CrossRef] [PubMed]
17. Schueren, M.V.B.-D.V.D.; Roosemalen, M.M.; Weijs, P.J.; Langius, J.A.E. High Waste Contributes to Low Food Intake in Hospitalized Patients. *Nutr. Clin. Pract.* **2012**, *27*, 274–280. [CrossRef] [PubMed]

18. Etchepare, M.; Buzatti, N.; Martini, C.; Mesquita, M.; Bilibio, M.; Schramm, P. Relacao de aceitabilidade de dieta hospitalar com oindice de resto-ingesta. *Higiene Alimentar* **2015**, *29*, 50–53.
19. Sonnino, R.; McWilliam, S. Food waste, catering practices and public procurement: A case study of hospital food systems in Wales. *Food Policy* **2011**, *36*, 823–829. [CrossRef]
20. United Nations (UN). High-Level Political Forum. Available online: https://sustainabledevelopment.un.org/hlpf/2018 (accessed on 27 March 2020).
21. Elia, M.; Micklewright, A.; Shaffer, J.; Wood, S.; Russell, C.; Wheatley, C.; Scott, D. The Annual Report of the British Artificial Nutrition Survey (BANS). Available online: https://www.bapen.org.uk/pdfs/bans_reports/bans_98.pdf (accessed on 2 March 2020).
22. Ministry of Health (MOH). Statistical Yearbook. Available online: https://www.moh.gov.sa/en/Ministry/Statistics/book/Documents/book-Statistics.pdf (accessed on 2 March 2020).
23. Duane, B.; Ramasubbu, D.; Harford, S.; Steinbach, I.; Swan, J.; Croasdale, K.; Stancliffe, R. Environmental sustainability and waste within the dental practice. *Br. Dent. J.* **2019**, *226*, 611–618. [CrossRef] [PubMed]

Article

Sustainable Food Consumption in Nursing Homes: Less Food Waste with the Right Plate Color?

Kai Victor Hansen * and Lukasz Andrzej Derdowski

Norwegian School of Hotel Management Ullandhaug, University of Stavanger, NO-4036 Stavanger, Norway; lukasz.a.derdowski@uis.no

* Correspondence: kai.v.hansen@uis.no

Received: 24 June 2020; Accepted: 10 August 2020; Published: 12 August 2020

Abstract: The problem of unsustainable food consumption among vulnerable residents of nursing homes who suffer from dementia is often multifaceted. From an individual perspective, people with dementia who do not finish their meals are likely to encounter serious health issues associated with malnutrition. Moreover, at the institutional level, nursing homes generate tons of nonrecoverable food waste each year, impairing not only their economic position but also the natural and social environment at large. The purpose of this study is to explore the possibility of reducing food waste in Norwegian nursing homes by appraising how large this reduction could be as one replaces traditional dining white porcelain with plates with diverse color combinations. A quasi-experimental method was adopted. The results of the pilot study were extrapolated to the annual amount of food wasted at the national level. The findings indicate that, on average, 26% of food was thrown away when served on white plates compared to only 9% when served on one of the colored plate options tested. Nationally, approximately 992.6 tons of food per year could potentially be saved with only a single change, ultimately ameliorating the unsustainable food consumption problem among residents of nursing homes.

Keywords: food waste; sustainability; nursing home; plate colors; pilot study

---

1. Introduction

Generally, food waste is perceived as an ecological, economic, and social problem. Existing estimations of global food production and consumption indicate that every year roughly 1.3 billion tons of food are lost or wasted [1,2]. The gravity of the situation has been recognized by the United Nations, which issued a list of Sustainable Development Goals (SDGs) that included a 50% reduction per capita in global food waste at the retail and consumer levels by 2030 (Goal 12: "Responsible Consumption and Production" [3]). Similarly, Borzan [4] articulates the concern that "for every kilogram of food produced, 4.5 kg of CO_2 are released into the atmosphere; whereas in Europe the approximately 89 Mt of wasted food generate 170 Mt CO_2 eq./yr" (pp. 4–5). When one considers the complexity of the entire food supply chain (i.e., production/procurement, distribution, preparation, consumption, and waste management/disposal [5]), it becomes apparent that, next to food waste and emissions, vast resources (e.g., energy, water, and land) are also being dissipated every year. Thus, from a global perspective, the waste of edible food appears to have far-reaching implications for environmental, social, and economic conditions of individuals and society at large.

Moreover, existing research supports the assertion that the foodservice sector accounts for a considerable percentage of the total food wasted within the confines of the food supply chain. For instance, it has been reported that the European Union (EU) foodservice sector produces approximately 12,263,210 tons of food waste per year, accounting for 14% of the total food waste generated [6]. Furthermore, several authors [7] argue that the level of meal waste tends to vary according to the type of foodservice setting being investigated, such as schools and universities, workplace restaurants and canteens, or hospitals and nursing homes. As an illustration, Engström and Carlsson-Kanyama [8] report 9–11% meal waste in some school foodservices. Norton and Martin [9] find 17% waste in a university dining hall environment and waste in elderly nursing care centers between 20% and 27% of the food produced [10,11].

Having recognized the previously mentioned arguments, this study aims to contribute to the line of research examining the unsustainable food waste problem in institutional settings. Specifically, we scrutinize through a (quasi-) experimental manipulation of how plates with different color combinations influence the amount of food wasted among people with dementia living in nursing homes in Norway. For the sake of specificity, as the definition of "food waste" is not universally shared, [12] in this article, it refers to waste from food that is not eaten from the plates on which they are served. Throughout this paper, the terms "food waste" and "plate waste" are used interchangeably.

2. Food Waste Problem in Nursing Homes

It has long been advocated that consumer behavior is critical in today's society in the fight for a more sustainable future. This includes meeting organizers such as the World Food Summit in Copenhagen, where various organizations and groups meet every year to discuss, among other things, food waste as part of sustainability in the world. [13] Residents of nursing homes that offer care and services for those no longer able to live independently represent a group of people who do not have the same opportunities to decisively affect their consuming patterns. In fact, people with dementia living in nursing homes receive all their care, including all meals, from staff members working there. According to WHO Dementia [14], dementia is chronic and progressive; cognitive function deteriorates beyond what might be expected from normal aging. Most people affected by dementia are 65–90+ years. It affects memory, thinking, orientation, comprehension, calculation, learning capacity, language, and judgment and has major consequences when sufferers perform activities in everyday life [15]. Regular and balanced food and fluid intake represent a case in point. Indeed, malnutrition is found between 10% and 60% of patients in Norway's hospitals and nursing homes [16,17]. Undernutrition increases this rate up to 70% [18]. For persons with dementia, this risk is increased due to physical and psychological changes, which lead to lower food intake, combined with age-related malnutrition risks [19,20]. One can argue that these conditions present a challenge for most countries that have an aging population.

Worldwide, an estimated 21 million people suffered from dementia in 2009, and experts predict that this number will increase to 81.1 million by 2040 if no treatment methods for dementia are found [21]. In the Norwegian context, people with long-developed dementia often live in sheltered wards. According to the latest estimates, approximately 70,000 individuals have dementia, and this number is expected to increase to 140,000 by 2040 [15,18,22]. That number is significant in the whole world, and the estimates and projections today indicate considerably large numbers [21].

Norway had 942 nursing homes in 2018 [23], providing about 40,000 places with capacity close to 100%. Every day of the year, dinner is served to all these nursing home residents. An estimated 80% of long-term residents in nursing homes have dementia [24], which is equivalent to approximately 32,000 people. The number of dinners served throughout the year is then approximately 11.6 million and appropriate dinnerware, such as plates, cups, and mugs, needs to be in place and preferably in a shape that encourages people to eat.

A 1997 study indicated a reduction in food waste in long-term care homes when using dinnerware instead of tray service [25]. This may be related to residents feeling as if they were in their own home when they ate dinner [17]. According to Hackes et al. [25], family-style service produced the least food waste among three different serving methods.

Food waste accounts for a large part of waste in industrialized parts of the world and contributes to the fact that important nutrients are not used for human consumption, but instead go to waste and create environmental problems of great magnitude [26]. Food waste in hospitals, nursing homes, and other health facilities contributes to the total food waste, and is often referred to as plate waste [27]. The reduction in food waste in nursing homes affects vulnerable groups of people. Today, there may be several strategies for reducing food waste, as suggested in the article by Williams and Walton, [27] such as clinical, food and menu, service, and environmental issues.

A previous study that focused on dinnerware color versus the traditional white dishes on which the vast majority of Norwegian nursing homes served dinners in 2015/16 concluded that people with dementia less often ate all the food on the white plates compared to food served on different colored plates [28]. Although the authors of this study focused on several aspects related to porcelain, dementia, and colors, they did not offer an assessment of food waste per se [28]. In a similar vein, a study by Rossiter and associates presents an alternative of a completely blue crockery used among elderly patients in an acute setting and concludes that colored porcelain was associated with increased food intake, [29] yet again no food waste amount was estimated.

Sustainability in the health sector and food service has been discussed and recommended in a Danish study [30] conducted over an 8 year period. The results showed little progress in public hospitals when it came to sustainable food systems for elderly people.

Therefore, given these arguments, we endeavor to explore the possibility of reducing food waste in Norwegian nursing homes by appraising how large this reduction could be by simply replacing traditional white porcelain dinnerware.

3. Method

To achieve the overall estimates of meal intake from estimates of individual food ingestion, it is necessary to perform some sort of calculation [31]. Therefore, this scholarly endeavor employed the following quantitative methodological approach.

3.1. Research Design and Data Collection Procedure

Whereas a given research design is often seen as a blueprint for a study, this project utilized an exploratory approach where no *a priori* (theory-driven) hypotheses were tested [32]. Instead, our primary objective was to explore the subject in question so as to provide ideas and insights that could potentially serve as an initial step for future investigations. Furthermore, in order to go beyond the frequently employed, yet not uniformly appreciated, survey method, [33] we put to use the quasi-experimental method instead [34]. Thus, for the purpose of this investigation, four plate types with different color compositions were designed (see Figure 1), where the white plate (A) option was treated as a baseline for comparisons. Of relevance, plates where all food was eaten were counted and marked as fully consumed food. The different plate combinations are described outside each image, and the interventions were carried out in that particular order. The surveys were conducted on random days over a three-week period.

Concerning the data collection procedure, this project was conducted at a nursing home in Rogaland County in Norway among people with dementia. It was carried out on two wards, with altogether 12 residents (five females) between 65 and 85 years; all residents had been diagnosed with dementia. The departments received the dinner meal in food containers directly from the communal kitchen. The staff put the food on the plates, which they then served. Plates subsequently collected by the staff were analyzed to determine whether they were with or without food residue. Some of the residents needed special diets, which were laid out ready-made from the communal kitchen and delivered together with the other meals. The staff served the ready-made plates to each resident. All photography was performed in a separate post kitchen to avoid disturbing residents before, during, and after the meal. Altogether, 88 pictures were taken, which resulted in 44 pairs of photographs (pre- and postconsumption) viable for further analysis.

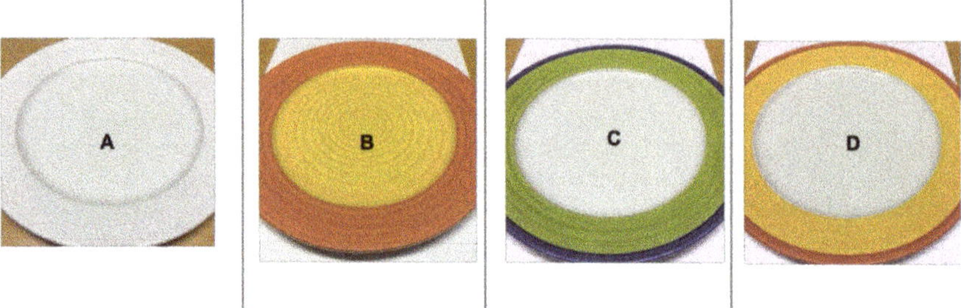

Figure 1. The different plate types used in this research: photo by author. Explanation of the different plates: Plate (**A**): white porcelain (a baseline for comparisons). Plate (**B**): yellow well, red lip, and red ring around the edge. Plate (**C**): white well, green lip, and blue rim on the edge. Plate (**D**): white well, yellow rim, and red ring around the edge.

3.2. Photo Analysis

In general terms, food waste can be measured directly by weighing the amount of food waste or estimated indirectly by, for example, visual estimations [35]. Given the diverse data collection techniques, this pilot project used data extracted through an indirect approach. Specifically, each dish was photographed before the serving occurred and later on when the plates were returned to the post kitchen. To keep the track of incoming/outcoming dishes, all plates were marked with a unique code. Finally, when all paired pictures were in place, an estimate of the percentage of actual food consumed was derived for each meal. The following formula was adopted to capture the amount of food wasted in a single meal: FW_i (in %) = 100% − FC_i (in %), where FW_i is food waste from *i*th single meal and FC_i is food consumed in *i*th single meal.

3.3. Ethical Concerns

As this scholarly endeavor focuses on individuals who belong to so-called vulnerable populations [36], the following processes were established to ensure ethical conduct during the study. Information and consent forms were provided to the nursing home before the intervention. Separate forms were provided for staff and residents and adjusted for their role in the study. Nursing home managers distributed the consent forms, provided information to residents and their families, and brought in the consent forms. When residents were not able to give consent, a relative's vicarious consent was used. The project was submitted for research approval to the Norwegian Centre for Research Data (NSD) and was approved (No: 44818/3/LT).

4. Results

4.1. Experimental Results

The results presented in this quasi-experimental pilot study are based on the calculation of uneaten food during dinner by an individual with dementia who lives in a nursing home. The next section offers extrapolated calculations of how much food waste would occur during a year among people with dementia in Norway if one used different types of plates designed with a focus on color.

As for the results derived from the pilot project, the food waste when using traditional white porcelain (option A, see Figure 1) was on average equal to 26% (N_A = 11, M = 0.26, SD = 0.28, range from 0 to 0.80). This estimate is in line with approximations found in existing literature, as several authors have asserted that 20% to 27% of food produced in nursing homes is being wasted [10,11]. Plate B manipulation (yellow well, red lip, and red ring around the edge) resulted in an average food waste of 10% (N_B = 10, M = 0.10, SD = 0.17, range from 0 to 0.40). Furthermore, the second manipulation (Plate C, white well, green lip, and blue rim on the edge) led to an average food waste of 22% (N_C = 12,

M = 0.22, SD = 0.22, range from 0 to 0.60). The last condition (Plate D, white well, yellow rim, and red ring around the edge) recorded the least waste at 9% (N_D = 11, M = 0.09, SD = 0.14, range from 0 to 0.40).

4.2. Extrapolated Results

Table 1 presents the extrapolated results of our study, taking into account general statistics unique to the context of Norwegian nursing homes (as described in Section 2) and the findings obtained from the pilot project.

Table 1. Estimates for plate waste and potential saving of food waste depending on different plate color.

1	2	3	4	5	6	7	8
Plate Type	Nursing Home Beds in Norway	Average Number of People with Dementia	Average Weight of a Dinner (in kg)	Weight of Dinners Served in 365 Days (in tons)	Average Percentage of Plate Waste	Dinner Plate Waste in Norway (per year, in tons)	Saving of Food Waste (Plate A as a Baseline, per year, in tons)
	40,000	80%	0.5 kg	365 days	Pilot test		
A	40,000	32,000	16,000	5840	26%	1518.4	-
B	40,000	32,000	16,000	5840	10%	584	934.4
C	40,000	32,000	16,000	5840	22%	1284.8	233.6
D	40,000	32,000	16,000	5840	9%	525.6	992.6

Our data reveal that, if dinners served to elderly people with dementia continue to be delivered on traditional white porcelain, it will produce approximately 1518.4 tons of food waste (per year) from this single meal (column 7 in Table 1). However, introducing color-based variants into the design of pottery can possibly lead to saving a nonnegligible amount of meal waste. That is, our crude estimates imply that around 934.4 tons of food per year (column 8 in Table 1) could potentially be spared by introducing Plate B's design in nursing homes in Norway. Plate C's design could save up 233.6 tons of meal waste, whereas Plate D's design could save the biggest amount of food, equaling 992.6 tons per year.

Available literature on food waste has long recognized the distinction between edible food and nonedible food as well as between recoverable and nonrecoverable food [37]. The recoverable food category includes surplus food from restaurants, grocery stores, and cafeterias, whereas nonrecoverable food consists of animal bones, shells, and skin as well as uneaten food prepared/served by institutions. The latter argument is of particular relevance here. Food waste in nursing homes represents a major challenge in that food served to residents is more difficult to reuse for further human consumption. The number of kilos per day that go to waste from residents who do not eat all of the food served is not desirable, but when the amount of food waste throughout the year is considered, it becomes a challenge. The estimates presented imply that the savings are the greatest between Plate A and Plate D, which could save 992.6 tons per year for all of Norway. Regardless, the results clearly indicate that, of all plates, food served on white plates comes out the worst compared to the other three plates with color combinations.

4.3. Sources of Bias

The estimated results are based on several different calculated figures. Different norm numbers were extracted and approximated including dinner portion weight in kilogram per person, number of nursing home places in Norway, number of individuals with dementia in nursing homes in Norway, and percentage of dinners not fully eaten from different color combinations in the pilot project. Thus, it is necessary to point out several sources of errors that might confound our estimations. Regardless of their magnitude, in our view, the presented evidence should still be a cause for concern when considering sustainable food consumption among residents with dementia in Norwegian nursing homes.

One potential source of error in the study is the small number of individuals who participated in the pilot project, meaning our conclusions may not necessarily be transferred directly to the whole country. The small sample size also did not allow us to statistically test and verify differences in food waste production across the four investigated quasi-experimental conditions (e.g., using an ANOVA test).

A second possible source of error is that various dishes were served on the different intervention days. The residents may have liked the food better on one day than the other day. In addition, the same staff did not serve meals every single intervention day. This may have led to a greater variation in how the food was served and added up. There might have been large differences between consistently adapted food from the kitchen and how the food was plated. A last source of error was that the kitchen manual that provided pictures to show what the dinner plates should look like was oftentimes not used by the staff.

5. Discussion

According to Borzan [4], "[t]o reduce food waste, improve food safety as well as enhance the overall sustainability of food production, research and development have a highly important role to play in all sectors of the food supply and consumption chain" (p. 29). Of relevance, past research points to the fact that food waste in developing countries occurs primarily in the postharvest stages, whereas food waste in developed countries (such as Norway) occurs primarily in the consumer and postconsumer stages [38]. Thus, this project focuses on (un)sustainable food consumption patterns observed among residents of Norwegian nursing homes who suffer from dementia.

The discussion is based on the original findings published that showed how much food was eaten in terms of the colors of dinnerware on which the food was served. This was the relevant issue when the project was conducted. The result was that all types of dinnerware with colors showed a greater effect in the number of dinners eaten among people with dementia. The data indicated that the white plate used to serve dinner at the nursing home resulted in the fewest residents eating all the food on the plate.

In this article, any food not eaten by the residents was considered food waste. This means that any measures that can increase food intake and reduce food waste will be important in the fight for the better utilization of food for the sake of ensuring a sustainable future (as emphasized by the United Nation's Sustainable Development Goals).

However, at a more fundamental level, food that is eaten provides the necessary energy, protein, minerals, vitamins, and other nutrients humans need for a good life. There are many areas that can stimulate increased food intake, but the focus of this article was how changing color combinations on a plate would affect the appetite of people with dementia.

The group that participated in this research project (often described as a vulnerable population) included people with a diagnosis of dementia. They need a regular supply of food to prevent malnutrition, from which many elderly people suffer. It is important that food is eaten to prevent malnutrition and that people with dementia receive natural nutrients and the building blocks for their body. An equally important part is that food waste needs to be reduced. A reduction in food waste among people with dementia in nursing homes faces several challenges, but the benefits of the elderly eating more food affect the individual occupant, the nursing home, the community, and finally, the natural environment.

As previously noted, food served on a plate to people in nursing homes cannot be reused and must be considered as food waste. The food is then treated as if it were contagious and cannot be destined for any further human consumption. In general terms, it is important that the risk of any possible infection is reduced, and that the food follows current laws and legislations such as, e.g., hazard analysis and critical control points (HACCP) and other national regulations [39,40]. Thus, food leftovers produced in nursing homes cannot be reused in any way that would resemble, for instance, several hotel chains and restaurant strategies (i.e., where they offer mobile apps such as "Too Good To Go" where one can purchase food at a discounted price that would otherwise be thrown away). From a sustainability perspective, it is, therefore, important that as many residents of nursing homes as possible eat the food being served. The basic premise is that everyone should be fed on white porcelain. However, our findings suggest that white dinnerware can be associated with the largest amount of food waste, while Plate D can save more than 992.6 tons of food per year. The amount of food served on average on one plate was 0.5 kg per person per day [41]. Different interventions with white plates versus colorful plates showed different outcomes in terms of food waste from the various servings. The figures are based on estimates documented through various sources and research [41–43]. The lowest estimate is 525.6 tons of food waste if all nursing homes switched to the colored plate that showed the best result (Plate D) rather than continuing to use white plates. As a final thought, it is important to note that the provided estimations relate to only one meal (i.e., dinner). If other meals are considered when determining the amount of food waste avoided, even greater prosustainable changes could occur that would perhaps make a considerable difference to the environment and the people directly affected.

Taken together, improving the efficiency of food production and consumption, as well as changing the general diet in Western countries, appear to be vital for securing the sustainable future of food supply. Along this line, several authors contend that food waste occurring in particular at the end of the food supply chain (i.e., consumption) is especially harmful to the environment and economy due to the resources invested and emissions produced while growing, transporting, and retailing foodstuffs [44,45]. Having considered the gravity of the situation, this scholarly endeavor brings in a possible solution for ameliorating the unsustainable food consumption problem specifically among residents of nursing homes.

6. Conclusions

The findings of this article indicate that with only a single change (such as redesigning the colors of dinnerware), a lot of food can be eaten instead of it turning into waste in institutions such as nursing homes. Although this was just a pilot project, the estimated number of tons of food saved from waste is tremendous (i.e., up to 992.6 tons of food per year). It seems that sustainable consumption at institutions such as nursing homes has not received adequate attention thus far, and we believe that this stream of research holds the potential to benefit individuals (e.g., residents' health), institutions (e.g., their general food expenses), and/or the natural environment at large (e.g., by producing less nonreusable food waste).

Therefore, further research should look at conditions that encourage nursing home residents to eat more of the food being served, especially people with dementia, who make up a large proportion of these residents. Several measures can be adopted and (e.g., experimentally) manipulated to comprehend how, for instance, the size of the plate, atmosphere in the dining room, brightness in the food room, or diverse plate colors encourage or discourage overall food consumption and waste production among the elderly with dementia living in nursing homes.

Author Contributions: Conceptualization, K.V.H.; Formal analysis, L.A.D.; Investigation, K.V.H.; Methodology, L.A.D.; Project administration, K.V.H.; Software, L.A.D.; Validation, K.V.H.; Writing—original draft, K.V.H.; Writing—review & editing, L.A.D. All authors have read and agreed to the published version of the manuscript.

Funding: This study was funded through the public policy system VRI (policy instrument for regional innovation and development). The authors would like to thank everyone at the nursing home where the study was conducted.

Acknowledgments: This project would not have been possible without the close cooperation of the nursing home in Rogaland County.

Conflicts of Interest: The authors declare no conflict of interest.

References

1. Gustavsson, J.; Cederberg, C.; Sonesson, U.; Van Otterdijk, R.; Meybeck, A. Global Food Losses and Food Waste. In *Save Food Congress*; Interpack: üsseldorf, Germany, 2011.
2. *State of Food Insecurity in the World 2013: The Multiple Dimensions of Food Security*; FAO: Rome, Italy, 2014.
3. United Nations Sustainable Development Goals; United Nations. 2016. Available online: https://www.un.org/sustainabledevelopment/sustainable-consumption-production/ (accessed on 10 June 2020).
4. Borzan, B. Report on Initiative on Resource Efficiency: Reducing Food Waste, Improving Food Safety. 2017. Available online: https://www.europarl.europa.eu/doceo/document/A-8-2017-0175_EN.html (accessed on 21 June 2020).

5. Carino, S.; Porter, J.; Malekpour, S.; Collins, J. Environmental Sustainability of Hospital Foodservices across the Food Supply Chain: A Systematic Review. *J. Acad. Nutr. Diet.* **2020**, *120*, 825–873. [CrossRef] [PubMed]
6. *European Commission's Preparatory Study on Food Waste Across EU 27*, European Commission: Paris, France, 2010; p. 213.
7. Ofei, K.T.; Werther, M.; Thomsen, J.D.; Holst, M.; Rasmussen, H.H.; Mikkelsen, B.E. Reducing Food Waste in Large-Scale Institutions and Hospitals: Insights from Interviews With Danish Foodservice Professionals. *J. Foodserv. Bus. Res.* **2015**, *18*, 502–519. [CrossRef]
8. Engström, R.; Carlsson-Kanyama, A. Food losses in food service institutions Examples from Sweden. *Food Policy* **2004**, *29*, 203–213. [CrossRef]
9. Norton, V.; Martin, C. Plate waste of selected food items in a university dining hall. *Sch. Food Serv. Res. Rev.* **1991**, *15*, 37–39.
10. Nichols, P.J.; Porter, C.; Hammond, L.; Arjmandi, B.H. Food intake may be determined by plate waste in a retirement living center. *J. Acad. Nutr. Diet.* **2002**, *102*, 1142. [CrossRef]
11. Silvennoinen, K.; Katajajuuri, J.M.; Hartikainen, H.; Jalkanen, L.; Koivupuro, H.-K.; Reinikainen, A. Food waste volume and composition in the Finnish supply chain: Special focus on food service sector. In Proceedings of the 4th International Symposium on Energy from Biomass and Waste, Venice, Italy, 12–15 November 2012.
12. Garrone, P.; Melacini, M.; Perego, A. Opening the black box of food waste reduction. *Food Policy* **2014**, *46*, 129–139. [CrossRef]
13. Summit, W.F. Better Food for more People. Available online: https://bfmp.dk/servicemenu/communications-material/ (accessed on 31 January 2019).
14. WHO. Dementia. Available online: http://www.who.int/topics/dementia/en/ (accessed on 22 June 2016).
15. *Demensplan 2015—Delplan til Omsorgsplan 2015*, Helse- Og Omsorgsdepartmentet: Oslo, Norway, 2015; p. 25.
16. Forbrukerrådet. Ernæringsforbundet, K.-o. In *Appetitt på Livet*; Forbrukerrådet og Kost- og ernæringsforbundet: Oslo, Norway, 2015; p. 39. Available online: https://fil.forbrukerradet.no/wp-content/uploads/2015/11/Appetittpaalivet.pdf (accessed on 21 October 2019).
17. Carrier, N.; Ouellet, D.; West, G.E. Nursing Home Food Services Linked with Risk of Malnutrition. *Can. J. Diet. Pract. Res.* **2007**, *68*, 14–20. [CrossRef]
18. Aukner, C.; Eide, H.D.; Iversen, P.O. Nutritional status among older residents with dementia in open versus special care units inmunicipal nursing homes: An observational study. *BMC Geriatics* **2013**, *13*, 1–7.
19. Nazarko, L. Maintaining good nutrition in people with dementia. *Nurs. Resid. Care* **2013**, *15*, 590–595. [CrossRef]
20. Rognstad, M.-K.; Brekke, I.; Holm, E.; Linberg, C.; Luhr, N. Malnutrition in elderly people living at home with cognitive impairment and dementia. *Sykepleien Forskning* **2013**, *8*, 298–308. [CrossRef]
21. van Hoof, J.; Schoutens, A.M.C.; Aarts, M.P.J. High colour temperature lighting for institutionalised older people with dementia. *Build. Environ.* **2009**, *44*, 1959–1969. [CrossRef]
22. *Demensplan 2020*; Helse- Og Omsorgsdepartmentet: Oslo, Norway, 2015.
23. Sjukeheimar, heimetenester og andre omsorgstenester—Plasser og rom i sykehjem og aldershjem. In *KOSTRA*; SSB: Oslo, Norway, 2019. Available online: https://www.ssb.no/helse/statistikker/pleie (accessed on 22 April 2020).
24. Ranhoff, A.H.; Vollrath, M.E.M.T.V.; Skirbekk, V.F. *Dementia in Norway*; FHI: Oslo, Norway, 2019.
25. Hackes, B.L.; Shanklin, C.W.; Kim, T.; Su, A.Y. Tray Service Generates more Food Waste in Dining Areas of a Continuing-Care Retirement Community. *J. Am. Diet. Assoc.* **1997**, *97*, 879–882. [CrossRef]
26. Griffin, M.; Sobal, J.; Lyson, T.A. An analysis of a community food waste stream. *Agric. Hum. Values* **2009**, *26*, 67–81. [CrossRef]
27. Williams, P.; Walton, K. Plate waste in hospitals and strategies for change. *e-SPEN* **2011**, *6*, e235–e241. [CrossRef]
28. Hansen, K.V.; Frøiland, C.T.; Testad, I. Porcelain for all—A nursing home study. *Int. J. Health Care Qual. Assur.* **2018**, *31*, 662–675. [CrossRef] [PubMed]
29. Rossiter, F.F.A.; Shinton, C.A.; Duff-Walker, K.; Behova, S.; Carroll, C.B. Does coloured crockery influence food consumption in elderly patients in an acute setting? *Age Ageing* **2014**, *43* (Suppl. S2), ii1–ii2. [CrossRef]

30. Mikkelsen, B.E.; Beck, A.M.; Lassen, A. Do recommendations for institutional food service result in better food service? A study of compliance in Danish hospitals and nursing homes from 1995 to 2002–2003. *Eur. J. Clin. Nutr.* **2007**, *61*, 129–134. [CrossRef]
31. Castellanos, V.H.; Andrews, Y.N. Inherent flaws in a method of estimating meal intake commonly used in long-term-care facilities. *J. Am. Diet. Assoc.* **2002**, *102*, 826–830. [CrossRef]
32. Churchill, G.A.; Iacobucci, D. *Marketing research: Methodological foundations*; Dryden Press: New York, NY, USA, 2006.
33. Buchanan, D.A.; Bryman, A. 'Not another survey'. The value of unconventional methods. In *Unconventional Methodology in Organization and Management Research*, 1st ed.; Bryman, A., Buchanan, D.A., Eds.; Oxford University Press: Oxford, UK, 2018; pp. 1–24.
34. Shadish, R.; Cook, D.; Campbell, D. *Experimental and Quasi-Experimental Designs for Generalized Causal Inference*; Wadsworth Publishing: London, UK, 2002.
35. Boschini, M.; Falasconi, L.; Giordano, C.; Alboni, F. Food waste in school canteens: A reference methodology for large-scale studies. *J. Clean. Prod.* **2018**, *182*, 1024–1032. [CrossRef]
36. Sutton, L.B.; Erlen, J.A.; Glad, J.M.; Siminoff, L.A. Recruiting vulnerable populations for research: Revisiting the ethical issues. *J. Prof. Nurs.* **2003**, *19*, 106–112. [CrossRef]
37. Kantor, L.S.; Lipton, K.; Manchester, A.; Oliveira, V. Estimating and addressing America's food losses. *Food Rev. Natl. Food Rev.* **1997**, *20*, 2–12.
38. Parfitt, J.; Barthel, M.; Macnaughton, S. Food waste within food supply chains: Quantification and potential for change to 2050. *Philos. Trans. R. Soc. B Biol. Sci.* **2010**, *365*, 3065–3081. [CrossRef] [PubMed]
39. FDA HACCP Principles & Application Guidelines. Available online: https://www.fda.gov/food/hazard-analysis-critical-control-point-haccp/haccp-principles-application-guidelines (accessed on 5 August 2020).
40. Lovdata, Forskrift om næringsmiddelhygiene (næringsmiddelhygieneforskriften). In *FOR-2008-12-22-1623*; Nærings- og fiskeridepartementet, L.-o. m., Helse- og omsorgsdepartementet (Ed.) Nærings- og fiskeridepartementet, Landbruks- og matdepartementet, Helse- og omsorgsdepartementet: Oslo, Norway. Available online: https://lovdata.no/dokument/LTI/forskrift/2008-12-22-1623 (accessed on 1 July 2019).
41. Matprat Porsjonsberegning i Hverdagen. Available online: https://www.matprat.no/artikler/matsvinn/porsjonsberegning-i-hverdagen/ (accessed on 12 May 2020).
42. *Nasjonal handlingsplan for bedre kosthold (2017–2021)*; Departementene Handlingsplan: Oslo, Norway, 2017; Vol. I-1177 B.
43. *Kosthåndboken—Veileder i Ernæringsarbeid i Helse- og Omsorgstjenesten*; Helsedirektoratet: Oslo, Norway, 2016; p. 280.
44. Priefer, C.; Jörissen, J.; Bräutigam, K.-R. Food waste prevention in Europe–A cause-driven approach to identify the most relevant leverage points for action. *Resour. Conserv. Recycl.* **2016**, *109*, 155–165. [CrossRef]
45. Silvennoinen, K.; Nisonen, S.; Pietiläinen, O. Key concepts, measurement methods and best practices. In *Routledge Handbook of Food Waste*, 1st ed.; Reynolds, C., Soma, T., Spring, C., Lazell, J., Eds.; Routledge: London, UK, 2020.

 *sustainability*

Article

Creative Food Cycles: A Cultural Approach to the Food Life-Cycles in Cities

Manuel Navarro Gausa, Silvia Pericu *, Nicola Canessa and Giorgia Tucci

Department Architecture and Design (DAD), Università di Genova, 16126 Genoa, Italy; gausa@coac.net (M.N.G.); n.canessa@go-up.it (N.C.); tucci.giorgia@gmail.com (G.T.)

* Correspondence: silvia.pericu@unige.it

Received: 9 June 2020; Accepted: 7 August 2020; Published: 11 August 2020

Abstract: The new contemporary multi-city needs the landscape as a proactive eco-systemic infrastructure in order to rethink the whole food system, from the design of public spaces to domestic spaces. In this direction, Creative Food Cycles (CFC) is an EU project that, according to the Sustainable Development Goals (SDGs), addresses the topic of food as a cross-cutting factor and powerful accelerator toward the co-design of sustainability in cities. Design culture today has begun to question and innovate production, distribution, and recycling models of food cycles. In the post-consumption and disposal phase illustrated herein, making the most of food means conceiving waste as a resource for the creation of new sustainable materials or prototypes. The concept of food waste and food losses has been shown to be not only a topic at the center of the debate but also a powerful tool for raising awareness of sustainable development at the community level. The CFC actions shown here were developed with the objective of persuading consumers to change their behaviors, while at the same time exploring cultural and social perceptions. With the aim of making cities more sustainable, this paper describes tools to engage different stakeholders, such as architects, product designers, and citizens, from a cultural point of view. The ongoing research has turned in the end into an educational campaign and an open platform where prototypes, new materials, and products are developed as inspiration for change.

Keywords: resilient and sustainable cities; food waste; design culture; food cycles

---

1. Introduction

1.1. Land Links: Fractal Multi-Cities, Meshed Territories, and Operational Landscapes

During recent decades, increasing anthropization and the competitive positioning of cities and territories in a global economic framework [1], associated with the growing increase in mobility and internationalization of the soil market and the appearance of a new cultural and environmental sensitivity, have led to the need to think about new urban reformulation processes and initiate significant, innovative, and qualitative operations within these global "circuits of flow and exchange." The definitions of possible "multi-inter" strategies—multi-level and inter-network, but also multi-urban and inter-territorial [2]—for the great challenges that arise in this exchange scenario oblige us to contemplate some of the great transversal themes associated with the "re" factor (re-naturalization, re-cycling, re-structuring, re-activation, and re-information) that today mark the goals of the new urban–territorial agendas in the beginning of this century [3–5].

The new urban and territorial approach today appeals to a new mutable, dynamic, complex, evolutionary, and networked "systematiCity," which is more relational (transversal), intelligent (holistic), and imaginative (creative) and leans toward a new conceptual logic (more strategic and informational), a logic in which the ancient "urban swing" or "urban needlework" would be based not only on the continuity of building plots but on the capacity of new integrated network models [6]. These models are associated with the more active importance of a natural and, above all, semi-natural (agro-productive) landscape capable of promoting an interlaced linkage of large "meta-politan" development areas [7], coordinated synergistically with different territorial mobility links. This type of new multi-urban governance [8] obviously requires a reinforcement, an enhancement, and a qualitative (re)definition of its main nodal tissues and centers, and therefore the reuse and recycling of pre-existing urban structures, through strategies aimed at favoring programmatic and social diversity, but also a more effective relationship with the landscape and between landscapes.

We have used, on several occasions, the terms "land links," "land grids," and "recycling" [9] associated with these new dynamics, which are open to define possible integrated strategies intended to ensure local and global development, coordinated qualitatively at the large (territorial) scale and the intermediate (urban) scale—developments in which the new multi-city [10] would no longer interpret itself as a large "building extension" linked to a single mono-central reality, but as a possible polycentric structure [11]. Today, it is a question of interpreting landscapes as infrastructures (and even infrastructures as landscapes) or infra-structures such as eco-structures [12,13].

1.2. Agro-Cultures, Agro-Cities: Potential for New "Rurban" Proactive Development

In this sense, the evolution of these new urban territorial cities and the mutation of our environments has produced in recent decades in Europe (particularly in the Mediterranean area) a complex set of questions and research topics going beyond the traditional relationships—city–landscape, landscape–nature, and nature–city. Consistent parts of the work in urban disciplines and territorial sciences have been dedicated to reinterpreting the role of open spaces (free, semi-natural, and in-between spaces) closely related to agricultural production and how they can become (re)generative elements for defining new paradigms in the construction of the urban forms [3].

The transfer from an oppositional reading between city and countryside to an integrated and intertwined reading, in which the peri-urban territory can assume a vital and active role, with new productive functions associated with creative and complex added value, supposes a new kind of holistic approach to land-use governance in this new geo-urban definition [2]. Challenges that require new types of structural land spaces, or "rurban" spaces [14], call for combining primary and tertiary activities: agricultural and technological production, environmental sensitivity and tourist attractions, private spaces and public spaces, etc.

The role of agriculture in this interpretative framework is hence fundamental as one of the most decisive and transcendent uses of the soil, linked to the concept of landscape and basic for its conservation and the efficiency of the new urban territorial dynamics [15]. In the most paradigmatic zones of the Mediterranean Latin Arch, agriculture generally represents an average of 35% to 65% of the geographical area, occupied by only 1% to 5% of the working country's active population [16]. It is important to understand agricultural spaces as being no longer conceived solely as primary spaces but as complex spaces (green infrastructure, ecological corridors, natural matrices, wellness environments, innovative production scenarios, agro-touristic attractions, etc.), spaces that can foster an understanding of the landscape as a "system of ecosystems" [17].

A condition linked to the basic agricultural food component [18], but also connected to social well-being, economic development, the environmental and resilient urban quality, and a new technological and operational dimension, is smart landscapes [19] or advanced landscapes [20]. The smart planning concept alludes to a set of integrated systems and subsystems (safety, resilience, water, health, infrastructure, economy, environment, food, etc.) calling for orienting and managing the development and sustainable growth of these new scenarios [21]. In this "smart" framework, urban and interurban agriculture can not only contribute to ensuring healthier and more efficient nutrition processes related to algorithmic data optimization of environmental and economic parameters, but also promote new energy and waste cycles, reduce water consumption, and improve and manage resilient answers for the environment.

In this sense, some basic research questions can be formulated around this new prospect linked with agricultural spaces, their local traditions, and their ability to survive and adapt their role and characteristics according to the current transformational trends of this "glocal" and "rurban" scenario in which rural and urban are no longer strictly separated, as it is possible to see in the proposed schemes in Figure 1, as follows:

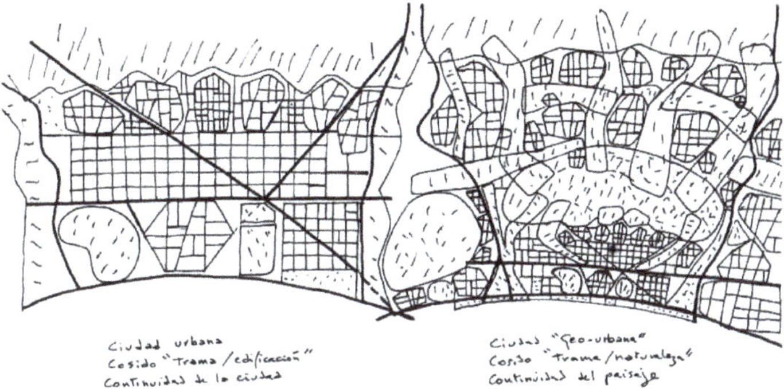

Figure 1. Patchworks of urban building plots and geo-urban landscape grids. By building continuity, old connections of the old urban fabric are replaced by a natural–artificial interlacement in the networked city where buildings and the landscapes in between are meshed. Image: M. Gausa.

1. To what extent can urban agriculture become a form and structure of the city, considering the new technological possibilities linked to production and distribution systems as well as the interest in a quality food chain and the processes of urban renaturalization by citizens?

2. How can diverse neighboring zones (functional, residential, commercial, eco-recreational, and industrial) that exist today along the edges of these areas be rethought to encourage new positive interactions among agriculture, social activities, leisure and innovative production, new mixed operations, and users?

3. How can we reformulate and reinterpret the old notion of food as a primary product, combining it with secondary and tertiary levels of definitions related to the recycling of waste but also with its reuse oriented toward innovating research in pharmaceutical applications, cosmetics, chemistry, and new bio-materials?

The research units and partners in the Creative Food Cycles (CFC) project have tried to answer these questions, starting from the ideas of previous studies associated with the prospective planning and social design laboratories of the University of Genoa, for example, AC+, Agri-culture, and Agro-cities [15]; Albenga GlassCity [22] and MedCoast AgroCities [23]; and in 2018 innovative actions linked to the CFC framing a set of urban perspective projects on the contemporary multi-city and its relationship with equations of city–territory–landscape–architecture and resilience in the Mediterranean coastal territory.

The main actions presented in these pages related to new approaches to food and its creative transformation and reinterpretation call for a recognition of the current context of these potentially hyper-agricultural scenarios in new polycentric and meta-metropolitan contexts and their strategic repercussions from the point of view of the high territorial and environmental value, a value connected to our "living–working–resting … enjoying and visiting" our own habitats [24]. New urban and territorial systems need to be stimulated to propose holistic solutions to multi-level problems related to society, the environment, health, food, and cities. This new agricultural transformation of the city, not bucolic but functional, renews the whole system, from the design of public spaces to domestic ones, with new needs and new opportunities.

In this direction, the CFC project, particularly with the work of the University of Genoa (UNIGE) team on reducing waste generation through recycling and reuse in everyday life, aims, on one hand, to test within the academic design community new products and materials derived from food waste, and, on the other hand, to organize creative events to raise awareness of the impact of food cycles in our cities. The CFC research starts by detecting good practices from the urban to the productive scale and goes on to analyze and experiment with the social impact of this paradigm shift. The experimentation related to the agricultural supply chain is a vessel and a stage that can show how much the scientific and creative communities are already working hard on these issues.

The project focuses on identifying those tools and methods related to the production and recycling of food waste that are innovative and can be either combined or simplified for non-industrial use. To do this, the project mainly targets training and education as capacity-building tools for architects and product and event designers in order to widen the interfaces between creativity, places, and public awareness through active engagement and co-creation events. In this framework, this paper presents the structure of the CFC research, illustrating the different phases and events and describing the materials and methods of the network and its possible implementations but also opening a possible discussion on awareness campaigns for issues related to food cycles and their impact in our cities.

2. Methodology

2.1. Urban Cultural Revolution in the CFC

As early as the late 1990s, Pothukuchi and Kaufman argued that food systems need a place in planners' concerns so that planning can be oriented toward the future and the public interest in an effort to improve the livability of the community through community systems and their interconnections [25]. The next step means understanding how holistic agro-cultural and social systems intercept the spaces, actors, resources, and dynamics present in a city, moving from the food system—understood as a chain of activities related to production, processing, distribution, consumption, and post-consumption, including related institutions and regulatory activities—to a new kind of integrated agro-urban system where innovative food and multi-scalar approaches are combined. The same CFC project follows this sequence by addressing the theme of food in 360 degrees, from production to disposal, structuring the project into three main phases.

The production phase is demonstrated in the city in the experiences of urban and peri-urban agriculture (producing in or around the city) and in the approach of commercial farms and agricultural parks, the heterogeneous set of horticultural experiences (social gardens, vegetable garden collectives, private gardens, school gardens, regulated or abusive gardens, guerrilla gardening practices, etc.). With a view of the food system at the city–region scale, it is equally important to know the characteristics of production, analyzing the agricultural sector in terms of quality and quantity. Specifically, the CFC project in this first phase aims to demonstrate how the use of technology can help produce food in urban environments or in close proximity and enhance city resilience. Urban agriculture can contribute to enhancing the resilience of cities, making available inexpensive healthy food for citizens. With the use of digital fabrication and control interfaces, the aim is to create hydroponic and aquaponic systems in a closed loop, teaching citizens, architects, and product and event designers how to manage self-sufficient cultivation. The use of digital fabrication allows custom-designed gardens to be built, and the use of sensors helps in controlling the performance. If soil cultivation is not practicable in many urban conditions, especially in dense city cores, hydroponic cultivation can represent a practical solution where the main limitations are lack of space or farming knowledge.

The distribution phase (large-scale food distribution, retail stores, markets, alternative food networks, online commerce) is the service activity aimed at transferring food products from producers and processors to consumers. In general, food distribution intercepts urban dynamics in spatial (affecting the way in which space is lived, designed, and consumed), social (in the relationships between actors), and environmental (generating impacts in terms of air and soil pollution, energy consumption, etc.) aspects. In the CFC project, the concept of this phase is to focus on new models of distributing, marketing, and processing, as well as cooking, displaying, and sharing food and regional products from a collective aggregation point (place-making effect). This can be an "urban food hotspot" characterized by a multipurpose stage connecting different places to a single manifestation of material and nonmaterial open public activities, trends, and movements. The aim is to collect into movable pieces of urban furniture different sensory experiences, augmented reality data processing, and art installations, offering interactive ways for audiences to participate in products or services and address extended audiences to ensure that the goods and commodities are difficult for customers to resist. A sense of originality and unparalleled creativity is a critical aspect that buyers take into consideration when shopping, consuming, and interacting in the urban foodscape.

The phase of urban consumption, combined with the last disposal phase, is complex and difficult to analyze since it includes a multiplicity of issues, ranging from the spaces in which items are consumed (public and private collective catering, domestic catering) to the social and cultural implications related to habits, traditions, consumer choices, ways and times of consumption, food accessibility, the relationship between food and health, etc. The last phase of disposal addresses the issue of waste and scraps, which the Food and Agriculture Organization of the United Nations (FAO) distinguishes between food loss (in the production, collection, distribution, and transformation phases) and food waste (produced in the final stages of sale and consumption), and it is becoming increasingly important in relation to issues such as global climate change, social justice, and food education. In particular, within the CFC project, this phase explores the process that brings food from consumption to disposal by not only offering options for new uses of discarded products (from waste to resources) but also defining new potential meanings and spatial combinations in an art–design reinterpretation (from scrap to art). It proposes a series of actions and performances based on a combination of projects and research that explore a new way of rethinking and reinterpreting food after consumption or discarded products for art, material, or reuse; the creation of ephemeral and flexible installations to define new configurations of public spaces (urban and artistic scenography) in order to attract the attention of target groups and stakeholders in the framework of public events; and the reuse of abandoned heritage buildings in order to promote civic participation and a convivial dimension in urban settings.

It is therefore these elements and their integration that the analysis of qualitative and quantitative aspects and local relationships and those with higher levels are concentrated on in a multiscale approach, with the aim of constituting an effective support tool for future territorial policies. An important challenge for the future will be to strengthen collaboration and knowledge sharing between users in the food sector (groups, organizations, businesses, individuals, etc.), research, and companies by combining the technological capacity of companies and their practical, operational, and market visions with conceptual capacity, or the experimental and creative role of research, in order to launch proactive exchange platforms on the theme of food and its expressive capacity as a cultural vehicle of identity, innovation, and social integration.

2.2. Tools

The CFC research is thus configured in three steps with different methods and tools.

The first step, food interactions, was a call for the creation of a database of good practices, already existing at a global level, of innovative food production processes and the exploitation of food waste. The idea to start from best practices came from the academic field in order to involve the research units in the three cities in the first two phases and to spread the outputs to architecture and design schools.

The second step, food crossovers, follows the research activity by proposing three creative workshops, one for each partner city, meant as open co-creation labs to empower thematic skills and engage diverse audiences. It was intended both to test some of the catalog experiences and interview some of the subjects who had made them, but above all, within the didactic university laboratories, to also experiment with new combinations, productions, and materials. This phase did not stop there but went further by designing new containers for food production and new objects produced from food waste. We consider this the most important phase for two reasons: on one hand, it is experimental and innovative for the results achieved, and on other hand, everything that has been prototyped from the point of view of both manufacturing and chemical processes is easily replicable in a fabrication laboratory and often simply at home.

This has been very important for us because the third step, Food Cycles in action, is the step of dissemination, and as mentioned, since this is a creative project, it is linked to dissemination targeted to wide audiences. This part included the following kinds of activities:

- The development of three art installations in Hannover, Barcelona, and Genoa aimed at connecting professionals and citizens with creatives through the co-production of art installations and place-making laboratories;
- Itinerant exhibitions on best practices and learned experiences held in Barcelona, Ljubljana, and Genoa;
- An international symposium to present the project's results to selected international representatives (experts and creatives); and
- An international festival, aimed to explore, through prototypes, art installations, and art performances, the process that brings food from consumption to disposal, by offering new potential meanings and spatial combinations in design reinterpretation.

Festivals, shows, performances, or even playful activities become the output of the project because all of these activities should not only provide information about production processes but also be real activities involving people in order to have an effective impact so that the processes are then replicable by the participants autonomously. This was very important because we believe that understanding the ease of the process and its replicability allows us to increase interest in a whole chain of food cycle processes and greater awareness.

Food Cycles in action displayed co-produced art installations and place-making events in the three cities and an itinerant exhibition traveling to other places, ending up with the final festival and symposium that will be held in 2020 to present prototypes and proofs of concept to target groups and stakeholders.

For this reason, the three units involved divided the phases of the main cycle of the food chain for research and experimentation, always maintaining cross-over on objectives, content, and methods. Figure 2 represents the exchanges set up by the three city partners, each one taking care of a food cycle phase. The Institute for Advanced Architecture of Catalonia (IAAC) developed the food production phase through the use of new technologies or new production processes, but also by experimenting on new foods and containers mainly for domestic food production. The UNIGE with the Department Architecture and Design instead developed the phase in relation to the reuse of food waste and the prototyping of new products from the materials obtained but also decided to go further, especially with the popular model for the use of methods of food consumption, and the idea was to organize real banquets to consume experimental food with supplies produced by food waste in an atmosphere of conviviality. The Leibniz Universitat Hannover (LUH) worked on the intermediate phase, distribution, imagining pop-up markets that could allow small or spontaneous producers to easily commercialize or exchange their products but at the same time create spaces and multi-level objects to be both new vessels and new platforms for exchange and generation of new sociality.

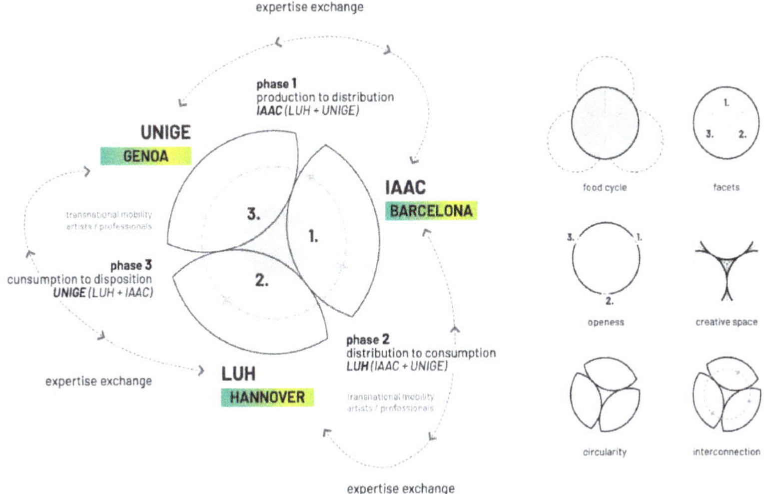

Figure 2. Food cycle phases and partners: international exchanges. Image: LUH Regionales Bauen und Siedlungsplanung.

Every action of the project is conducted by extending open calls for action and for projects in order to allow wider participation by professionals and encourage local organizations to deepen their audiences and experience international exchange at the same time. Calls for projects are meant as essential tools to collect ideas on the topic of food but also to allow the creation of a network of proposals [26] with a high innovation level that can be spread at a bigger scale.

The collaboration is evident, as every partner worked with the same tools. Calls for projects/papers/actions helped to reach a wider audience. These calls represent a wide-reaching dissemination program aimed at increasing the visibility of the research through social channels and a dedicated website.

During the pandemic period in spring 2020, other tools were tested, such as webinars, where participants were asked to take an active role and to experiment with materials derived from food waste in the domestic environment.

2.3. CFC Activities, Impact, and Network Dissemination

The CFC approach merges new ways of design and digital interaction in a transdisciplinary way, exploring cultural, social, and economic innovations through the activities. The research structure was inspired by a previous two-year Creative Europe project developed by IAAC in 2017, a current CFC research partner, entitled Active Public Space (APS). The purpose of APS is to develop knowledge of public spaces, fostering people's interaction with flows of energy, materials, services, and finances to catalyze sustainable economic development, resilience, and high quality of life. Thanks to the use of smart urban technologies, the project was able to demonstrate how they were essential for the change of public city space, allowing real-time data capture, energy generation, storage and reuse, material adaptability, management of time use, and citizen–space interaction.

Similar to the structure of APS, the CFC research has been structured through a series of activities to bring research closer to the social context and encourage the cultural dissemination of project results. Activities have been organized by the three international partners, LUH, UNIGE, and IAAC, with an open and inclusive approach and targeted communication, deepening the interconnection among architects, designers, cultural operators, institutional stakeholders, and active urban society, and combining the concept of food resilience with the cultural sphere.

Institutions, local experts, artists, cultural operators, and stakeholders of the three partner cities were involved in an open co-creation work program.

The workshops, exhibitions, and festival strongly address target groups such as cultural operators and local stakeholders, as well as active urban society in the three partner cities, while the symposium is aimed at dissemination among academic and institutional stakeholders.

These target groups are considered integral parts of the project work and are incorporated into a specific audience development strategy based on creative workshop experiences and open co-creation moments for building art installations and international biennial festival exhibitions in order to extend the audience. Thanks to the intense digital presence of the project (website, streaming and social channels) and major publications (CFC catalogue, international festival experiences, and symposium proceedings), the project's results have also been transferred to other cities and available online to everyone.

The main results obtained in terms of impact and dissemination are assessed on the basis of the number of participants (citizens, creatives, stakeholders, etc.) involved in the activities detailed in Table 1.

Table 1. Activities carried out in the period 2018–2020 involving the three partners, LUH, IAAC, and UNIGE.

Activity ID	Type of Activity	Activity Description
A	Call for the Food Interactions Catalogue	The Call for the Food Interactions Catalogue was spread via social networks and mailing lists and asked the participants to send realized projects related to three types of classification: typological, readiness, and performance categories. https://creativefoodcycles.org/food-interactions-catalogue/
B1	Creative Urban Farming Workshop, Barcelona, IAAC	The Creative Urban Farming Workshop, Barcelona, 4–6 May 2019, was aimed at developing innovative urban food production system prototypes. Each participant group made one 1:1 scale prototype for food production in the urban environment. The event involved 59 participants/creatives. https://creativefoodcycles.org/workshops/workshop-barcelona/
B2	Food Cycles Pop-Up Workshop, Hannover, LUH	The Food Cycles Pop-Up Workshop, Hannover, 23–25 May 2019, was about new models of distributing, marketing, and consuming food, as well as cooking, displaying, crafting, and sharing, in a collective "urban food hotspot." Workshop participants learned about how to conceive and effectively communicate innovative concepts for pop-up market prototypes based on instant urban design principles. The workshop involved 66 participants/creatives. https://creativefoodcycles.org/workshops/workshop-hannover/
B3	Food Shakers | Food Remakers Workshop, Genoa, UNIGE	Food Shakers | Food Remakers Workshop, Genoa, 17–21 June 2019, was about food waste as new material, from organic food waste to the creation of new industrial materials, or food waste and packaging for new products, from organic food waste to real products for consumers. Workshop participants had the opportunity to showcase the designed prototypes at the 2019 SUQ Intercultural Food, Art and Craft, and Music Festival. The event involved 49 participants/creatives, and about 5000 citizens on the closing day, and was open to the public, within the SUQ Festival. https://creativefoodcycles.org/workshops/workshop-genova/
C1	Myco-scape Installation, Barcelona, IAAC	The CFC myco-scape installation, Barcelona, 27–29 June 2019, staged a modular system supporting the growth of edible mushrooms in the urban environment, producing both food and construction materials. The installation involved citizens of different education levels and ages. More than 1200 people participated. https://creativefoodcycles.org/installations/installation-barcelona
C2	PorTable Installation, Hannover, LUH	The CFC "PorTable" pop-up installation, Hannover, 15–17 October 2019, staged a modular and movable unfolding table covered by a raised cultivating bed in which culinary or wild herbs were grown. The Installation involved 63 participants/creatives and about 220 citizens. https://creativefoodcycles.org/installations/installation-hannover/

Table 1. *Cont.*

Activity ID	Type of Activity	Activity Description
C3	Food (re)makers Installation, Genoa, UNIGE	At the food (re)makers installation, Genoa 29–31 October 2019, the prototypes designed during the Food Shakers | Food Remakers workshop were implemented through educational laboratories organized as open days during the Festival della Scienza 2019 program. The event involved 30 participants/creatives and more than 70 middle and high school students, who conducted some workshops with the creatives. https://creativefoodcycles.org/installations/installation-genova/
D1	Responsive Cities Expo, Barcelona, IAAC	The CFC itinerant exhibition was hosted on 15–27 November 2019 in Barcelona at the IAAC main exhibition hall in parallel with the work of Responsive Cities 2019 Symposium. It involved about 680 citizens. https://creativefoodcycles.org/exhibitions/cfc-exhibition-barcelona/
D2	Future Architecture Fair, Ljubljana, LUH	The CFC itinerant exhibition was hosted on 12–13 March 2020 in Ljubljana during the Future Architecture Fair, part of the Creative Exchange 2020 event, an international gathering organized by the Future Architecture Platform, involving more than 550 citizens. https://creativefoodcycles.org/exhibitions/cfc-exhibition-ljubljana/
E1	International Online Symposium, Hannover, LUH	The CFC International Online Symposium will be organized online by LUH on 17–18 September 2020. https://creativefoodcycles.org/symposium/
F1	International Festival Food interAction!, Genoa, UNIGE	The International Festival Food interAction! will be hosted on 9–11 December 2020 in an abandoned heritage building, Albergo dei Poveri, in Genoa. The aim is to explore the process that brings food from consumption to disposal by offering new potential meanings and spatial combinations in design reinterpretation.

3. Results

3.1. The Consumption to Disposal Phase

The UNIGE team explored the consumption to disposal phase and proposed new ways of recycling and reusing food waste as a resource for the creation of new environmentally friendly materials or prototypes, as can be seen in Figure 3. The question that emerges at this point is how to evaluate this approach, which uses food as an accelerator of disruptive change toward the co-design of sustainability in our cities and implement it in everyday life. Food is certainly a powerful medium because it is related to our emotional processes [27], even if it has become a product for mass consumption and a true industrial product linked to profit. Food can be combined with creativity as a lever of innovation, redesigning its entire life-cycle from production to disposal, in an attempt to anticipate what will happen in the short term but also to subvert what we are used to and broadly raise awareness.

Table 2. Some of the prototypes and new materials developed by the designers within the CFC project.

Reused Food	Type of Processing	Product Description
Coffee grounds	Compression material	1. Mooka is a circular product, it is a pot for planting that becomes fertilizer. Presented in a setting that offers visitors a visual and olfactory experience.
Coffee grounds	Bioplastic combination	2. DishBratta line is made by mixing coffee ground and a biological resin. It consists of a set of two dishes, a dinner plate and a deep dish, a fork, a spoon, and chopsticks.
Chamomile infusion	Bioplastic combination	3. BioPlastic was born from the desire to create a line of packaging for chamomiles and infusions starting from the classic internal waste of the bags once used.
Fennel and walnut waste	Bioplastic combination	4. Fennut light is a lamp that combines two materials borne from food waste.
Eggshell, pasta, lentils, etc.	Bioplastic combination	5. Bis Bioresina and Bis Compostable, are tableware with different uses: the first can be re-used, and the other one is single-use and biodegradable.

Table 2. *Cont.*

Reused Food	Type of Processing	Product Description
Rice husk	Bioplastic combination	6. V.pot is a dish made from the waste of rice husk compressed in a mold with the addition of bio-resins.
Fish bones	Bioplastic combination	7. BOFISH is an innovative material obtained from bone and cartilaginous waste from fish sourced locally.
Peanut shell	Cooking chemistry	8. Hanging Plates from peanut shells into bowls.
Honey	Cooking chemistry	9. Miellow is a honey-based bioplastic with a high resistance to water. The semi-transparency given by honey gives it a glass-like appearance.
Milk	Cooking chemistry	10. Galalith is a natural plastic material manufactured by the interaction of casein and formaldehyde. It is odorless, insoluble in water, biodegradable, non-allergenic, antistatic, and inflammable.
Soybean	Drying and weaving	11. S.D.S. The skin made of soybean, combined with the weaving process, makes healthy and environmentally friendly coasters and placemats.
Loofah	Drying and weaving	12. The mission of the Loofah fiber is to completely reuse decayed and inedible loofah and combine the good physical properties of the loofah.

Figure 3. Some of the products and materials designed by University of Genoa (UNIGE) students on the Creative Food Cycles (CFC) project, which follow the list in Table 2.

This becomes even more evident if we talk about food waste. While consumers' awareness of the issue is growing, it remains a significant barrier to achieving a sustainable food system. Even if technologies are ready to make a new sustainable lifestyle possible with new products and techniques, the disappearance of unsustainable practices is not yet on the agenda. The goal for food waste, followed in the CFC project, is to halve per capita global food waste at the retail and consumer levels by 2030 (SDG 12.3). This cannot be achieved without raising awareness of the topic at the community level. Currently, a third of all food produced globally is thrown away each year. Food loss and waste represents one of the most significant environmental and economic issues, and it is generally recognized that if it were a country, it would be the third-largest greenhouse gas emitter behind China and the United States [28]. It is a well-known issue that also fully involves European and Italian cities, which have to take an active role by making the most of food by redistributing surplus edible food while turning inedible byproducts into new products, ranging from food products to organic fertilizers and biomaterials [29]. Rather than being seen as final destinations for food, cities and communities have to be seen as places and environments where food byproducts are transformed by emerging technologies and innovations into a broad array of valuable materials. This is a philosophy that is contextualized within the circular economy, using the material energy of food, but it is also conceptual, for better safeguarding of resources.

Communities are at the center of these experiences and in a way also drive the food industry and large-scale distribution. Creative communities, accustomed to social innovation practices in cities, can design and make visible new ways of recycling and reusing food waste, as a resource for the creation of new environmentally friendly materials or prototypes. These projects are developed as actions aiming to persuade people to change their behaviors around food waste, at the same time exploring cultural, social, and economic perceptions. The experiences proposed in the following section are intended to answer the question of how to configure new design and creative activities related to food and food waste, engaging the public to design by and for themselves [30] and making sustainable habits and behaviors more compelling and attractive [31]. In Genoa, product designers, researchers, students, and local urban activists presented and implemented activities in which design played a role as a form of culture and a major driving force for envisioning and realizing processes of social innovation toward resilience, where people change their behavior and act collaboratively [32]. This educational campaign targeted to this group of stakeholders in the food system can make these projects effective and successful, because consumers' knowledge is integral to reducing food waste and recycling in similar activities [33].

The main output of the UNIGE activities, previously described, was to create real products and prototypes derived from food waste that were displayed in a way that implied an active role for visitors through artistic performances and co-creation workshops.

These prototypes were useful for fully understanding the relationship between ethical elements and the way we produce, consume, and recycle in our cities, and even more because this topic represents an interesting field of investigation for design that has a "reparative role" [34] with respect to these kinds of environmental and social issues in which the system at a general level and the choices of individual consumers are intimately intertwined.

"Designers and artists are able to formulate, through artifacts and concepts, urgent political questions that cannot rely solely on regular processes to enter public discourse. In regards to the environment and all associated concerns, in particular, state policy is driven to make reformations by the priorities that researchers, designers, activists, scientists, architects, and citizens set forth" [34] (p. 18).

To do this, we must refer to the circular economy for food as a natural system of regeneration, in which waste again becomes food, transforming itself into a new resource. "Making the most of food" [35] means involving local communities, stakeholders, and active urban society, developing a cultural and holistic approach, and joining all aspects of food cycles, but also stimulating with an open and inclusive approach a deeper interconnection of all disciplines dealing with the urban environment to reduce food waste and co-design a new concept of waste.

3.2. Outputs

The first result of the CFC project is represented by the catalog available as an open-access digital publication, which gives access to information through social media, webpages, online videos, and interactive resources to increase the framework of knowledge. This database includes both industrial and start-up projects, as well as university and other research institutes. All this information has been incorporated in a catalog, showing how innovative production or processing of food waste can be done by showing the characteristics that make it possible, for example, to transform orange peels or pineapple leaves into fabric. This is interesting because it allows us to understand, for example, the logic and chemistry behind a transformation in order to make it accessible and replicable with different food waste but similar in substance. The catalog expresses sustainable food cycles that emerge from digitalization, advanced technological implementation, digital manufacturing, sharing and informal economies, innovative participatory processes, increased awareness of climate change, and advanced strategies for urban and territorial resilience.

Besides this, the other important research results include the development and prototyping of new materials deriving from food waste. In fact, thanks to the CFC project, the research unit of Genoa has developed numerous design products made with the use of new materials derived from food waste and recyclable together with students and designers. These projects always work with the main idea of making the population aware of recycling, looking for easy-to-reproduce procedures and daily use of the products made. There have been different ways of processing waste, but we can say that the three most recurrent macro-categories are the addition of bio-resins or homemade processes that we could define as "cooking chemistry" or situations of drying and weaving of food waste. An excerpt of the processes and products produced can be found in Table 2 and Figure 3.

The results of the research can be measured through the number of prototypes developed, the people physically involved in the activities and the knowledge produced and disseminated through the online channels; the real impact of these objects must in fact still be evaluated, because the project is still in progress, and it is believed that the prototypes can act as demonstrators of the principles that originated them.

The topic of food waste should involve each one, and it is actually at center stage, so the approach of CFC can become a guideline to create events open to the public to involve citizens in the production of prototypes and materials derived from food waste. Schools, universities, and municipalities with waste management companies are the main stakeholders of this process. From the project, a strategy could be extrapolated that brings together collections of best practices, co-creation workshops, and installations open to the public that involve different levels of education. This could be a format that allows schools to involve students in project activities. Putting together best practices with co-designing activities has proved to be very effective in raising people's awareness and involving them directly in the creative and realization phase, allowing them to come into contact with food waste as a material that takes on new meaning.

In addition, the many prototypes coming out of the project activities of the workshops also have value at the level of innovative proposals, which should be explored and evaluated separately. The involvement of young architects and designers in the design processes of food waste allowed the research teams to introduce new ideas and generate a high rate of innovation.

From a conceptual point of view, the research is aimed at defining where we can act with consumer awareness as a priority, but also, and above all, at the possibility of putting in place possible actions that make everyone's impact effective, in order to understand how to amplify this message that the project activities started from an analysis of best practices, followed by creative workshops to develop ideas and installations aimed at spreading and testing alternative practices in the three cities involved. This educational campaign has the ambition to build an open platform where everything is designed and developed in the research; i.e., prototypes and new materials and products are available as inspiration for change in different communities and in other contexts. Physically, the project has also become an exhibition that circulates, spreading its message in cities related to the CFC network, but at a digital level, it is also a website (www.creativefoodcycles.org), conceived as an open platform where every single action that has been produced can be viewed and therefore replicated by using the tools, i.e., workshops, calls to action, calls for projects, and webinars. This modus operandi is particularly effective when referring to food waste, an issue for which mixing inspirational best practices with artistic installations and projects of new materials, products, and services can have a major impact on generating new ideas and approaches.

This can be illustrated as an output, as proposed in this paper, by analyzing the phases of the food life-cycle, starting from the one that is generally at the end of the process, the food waste phase. Putting food waste at the beginning of the life-cycle, as a new starting point in this case, takes on significance as a radical change of perspective.

The food life-cycles guide us backward on a journey through the activities carried out in CFC, in which the raw material is the waste to produce food and zero-mile items, to distribute and consume the surplus, up to food reuse processes and packaging made from the food itself. The scheme in Figure 4 provides a guide to descriptions of the project findings by the actions proposed in the activities with stakeholders in the three cities.

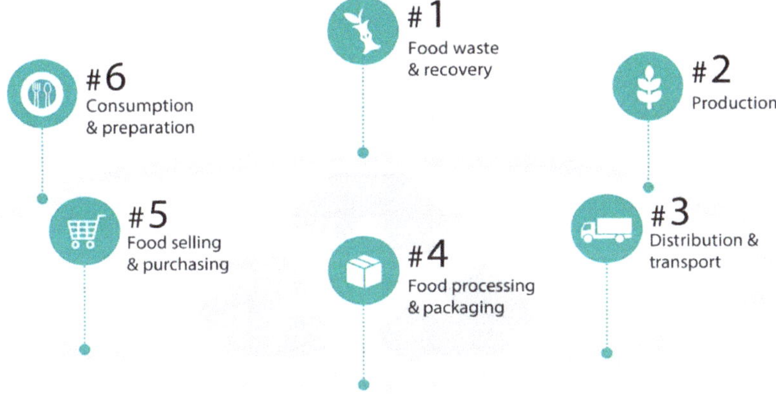

Figure 4. Food life-cycle phases starting from food waste.

Phases 1 and 2: From Food Waste to Production—Urban Environment

"Waste equals food." In nature, everything has a purpose: each organism's process contributes to the health of the whole ecosystem, and when it becomes waste, it is food for other organisms. Designers can optimize products and services, creating closed-loop material flows that are sustainable [36]. The Food Cycles in action installation presented in Barcelona in 2019 by the IAAC displayed a myco-scape, a modular wooden system with an external surface supporting the growth of edible mushrooms in the urban environment, producing both food and construction materials. After harvesting the mushrooms, the material contained in the cultivation area can be used as construction material. This prototype project acts as a real food life-cycle demonstrator, creating a culture of caring for locally sourced and produced food and raising awareness of sustainable development and lifestyles in harmony with nature according to the SDGs. It should be noted that the concept that a basic element of architecture, a wall, can function as a prototype of these possibilities brings interesting developments as a means of communicating to a wide audience.

Phases 1–4: From Food Waste to Food Processing and Packaging—Products

Climate change demands original and radical thinking, and if, as Papanek and Fry argue [37,38], design is a vital form of political action, designers play a major role as powerful agents of change who can imagine long-term freedom. Freedom from plastic packaging, for instance, is a necessity for designs that not only can serve the market but also can realize alternative dreams. To spread this message, the CFC UNIGE team, in summer 2019, organized as a food crossovers workshop "Lay the Table," a performance that combined an exhibition of objects made from food waste conveying a message with a stage show to explore new ways of rethinking and reinterpreting post-consumer food as everyday objects and packaging. The workshop/performance took place at a summer festival in Genoa, an event that enhances mutual knowledge exchanges and artistic collaborations across the Mediterranean area, combining food, music, and other cultural activities.

Phases 1–6: From Food Waste to Food Selling—Services and Food

Services also have a major role in supporting communities of citizens as users and companies, by creating a virtuous circle in which everyone actively interacts for sustainability with a positive impact on the territory and the quality of life of all those involved. With proper service and interaction design strategies, companies can promote their sustainable actions and behaviors, while consumers can lead sustainable lives. The example of Too Good to Go pushes in this direction. Designed as a free app by a movement against food waste, it allows the purchase of unsold food to prevent it from becoming waste and ending up in a landfill. Following this strategy, as part of the CFC project, the Food Shakers | Food Remakers installation explores the topic of food surplus by experimenting with food to become new material as packaging, but also as real products for consumers. The installation was a part of the Festival della Scienza program, an annual science event in Genoa at an international level. In the installation, waste becomes a means for education but also a possible and desirable answer to problems we all face, in which the true essence of contemporary design is expressed not merely as an intellectual exercise. In this sense, the aesthetic and emotional dimension represents a fundamental theme that, together with the ethical emergency, can become leverage for persuading final consumers to change their habits. Based on the idea of experiencing beauty and related to the consumption of food, in the project, discarded food, such as dry bread and vegetable waste, was cooked according to the idea that ethics and aesthetics become one thing. Food thus becomes an artistic experience, in which art makes the invisible visible and generates a sense of responsibility, which in turn is a social act in the form of creativity.

4. Main Findings

The CFC project, funded by the European Commission within the Creative Program, started with the intent to combine research and dissemination through the use of tools that can reach citizens of different ages and cultural profiles. It is therefore a mixture of investigations into the current panorama of innovative techniques of production, distribution, consumption, and reuse of food; workshops with students from universities; and presentations of the various results at events suitable for citizens, integrated with artistic performances and open festivals.

This structure of the project, on the one hand, allows high scientific rigor in research and experimentation and, on the other hand, combines educational and creative playful aspects that help to pass on the message about the importance of food cycles and their potential within people's own houses and urban environments.

The research starts from refining and improving dissemination actions already addressed in another creative project carried out by the same network, APS, to improve and develop a new format based on the previous one, oriented to involving a wide audience, whose impact can be measured by the numbers of participants in the project activities.

In the CFC research, the wide topic of food and cities was divided and deepened in the activities of the three partners, allowing possible implementation of new research clusters.

The aim of this paper is to present the CFC research in its structure, illustrate the different phases and events, and describe the working methods of the network as a format that can be implemented in other contexts. As we have seen, the project with its phases is linked by moments of research and experimentation and moments of dual dissemination to the academic and scientific world and to citizens, as foreseen by the European reference project. This openness to citizens through cultural events makes its effects on the territory easier, but it also opens a discussion on the importance of raising public awareness of issues related to food cycles in daily life. The CFC network is constituted today by the three research groups, UNIGE, LUH, and IAAC, and in addition to collaborating with each other, they have built a network of small businesses, start-ups, creative groups, and visualization activities, combining potential and developing new prototypes. The various actors involved had the opportunity to interact with each other, often working in direct contact with university students in a mutual exchange. The ensuing events allowed these interactions to be shown to citizens in a process where they were not only consumers but active participants.

The research, especially the part followed by UNIGE, moves toward the capacity of self-sufficiency, understood not as a survivalist scenario, but as the capacity of self-production and, above all, awareness of the potential and richness of food waste. The current situation has also led us to reflect further on how much the social capacity of food processing in all its cycles can be important within cities, to create better habitats and facilitate production in certain urban contexts [39] and within homes, in single or associated form, for new models of production in daily life.

Hannover, Genoa, and Barcelona are the three cities that have had a direct impact from CFC research, because these cities are where the research groups organized events and workshops for dissemination. However, today, with the website full of content and itinerant exhibitions, the research interest has expanded to other cities and stakeholders. The next step will be about better integrating local administrations, in terms of dissemination and practice, not only to develop or incentivize new materials generated by, e.g., second-life food waste but also to make them a real option.

Since the research is not linked to a precise geographic area and does not require close interaction with the administrative world for all its phases, even if this would allow great facilitation, the project, and above all the scenario presented, is replicable in other geographic contexts. Certainly those territories that already have a deep-rooted agricultural culture can be facilitated, although perhaps they are in decline and far from cities, but with an active and young entrepreneurial capacity, especially if there is interest in the rebirth of the territory and traditions related to innovation.

Author Contributions: M.N.G.—Introduction; N.C.—2. Methodology: 2.2 Tools; G.T.—2. Methodology: 2.1 Urban cultural revolution in the CFC, 2.3 Activities, impact and network dissemination, S.P.—3. Results. All authors have read and agreed to the published version of the manuscript.

Funding: This research was funded by CREATIVE EUROPE PROGRAMME (2014–2020), EACEA 32/2017, 22.11.2017 CREATIVE FOOD CYCLES 01.10.2018 – 30.09.2020.

Conflicts of Interest: The authors declare no conflict of interest.

References

1. Muñoz, F. *Urbanalización*; Gustavo Gili: Barcelona, Spain, 2008.
2. Gausa, M. *Multi-Barcelona, Hiper-Catalunya. Estrategias Para Una Nueva Geo-Urbanidad*; ListLab: Trento, Italy, 2010.
3. Ricci, M. *Nuovi Paradigmi*; ListLab: Trento, Italy, 2012.
4. Fabian, L.; Munarin, S. *Re-Cycle Italy, Atlante*; Lettera Ventidue: Siracusa, Italy, 2017.
5. Carta, M.; Lino, B.; Ronsivalle, D. *Re-Cyclical Urbanism*; ListLab: Trento, Spain, 2017.
6. Gausa, M.; Guallart, V.; Muller, W. *Hipercatalunya, Territories of Research*; Actar Publishers: Barcelona, Spain, 2004; pp. 1–704.
7. Asher, F. *Métapolis Ou L'Avenir Des Villes*; Éditions Odile Jacob: Paris, France, 2010.
8. Puig Ventosa, I. Polítiques Econòmiques Locals Per Avançar Cap a Formes Més Sostenibles D'habitatge I D'ocupació. In *Cap A Un Habitatge Sostenible*; A.A.V.V., Cads, Consell Assessor Per El Desenvolupament Sostenible, Generalitat De Catalunya: Barcelona, Spain, 2011.
9. Gausa, M.; Ricci, M. *Aum 01, Atlante Urbano Mediterraneo, Med.Net.It.1.0 Ricerche Urbane Innovative Nei Territori Della Costa Italiana*; ListLab: Trento, Italy, 2014.
10. Gausa, M. *Open. Espacio-Tiempo-Información*; Actar Publishers: Barcelona, Spain, 2010.

11. Nel.Lo, o. *Ciutat De Ciutats, Reflexió Sobre El Procés D'Urbanització a Catalunya*; Empúries Editorial: Barcelona, Spain, 2011.
12. Llop, C. Paisatges Metropolitans: Policentrisme, Dilatacions, Multiperifèries i Micro Perifèries del Paisatge Clixé al Paisatge Calidoscopi. *Papers: Regió Metropolitana De Barcelona* **2008**, *47*, 8–13.
13. Deleuze, G.; Guattari, F. *Mille Plateaux*; Éditions De Minuit: Paris, France, 1980.
14. Guallart, V. *The Self-Sufficient City*; Actar Publishers: New York, NY, USA, 2014.
15. Gausa, M.; Canessa, N.; Tucci, G. *Agri-Cultures, Agro-Cities, Eco-Productive Landscapes*; Actar Publishers: Barcelona, Spain, 2018.
16. Ministerio De Agricultura, Pesca y Alimentación/Nipo: 251-06-115-0. 2006. Hechos Y Cifras De La Agricultura, La Pesca Y La Alimentación En España. Available online: https://www.mapa.gob.es/es/desarrollo-rural/temas/programas-ue/00_tcm30-84598.pdf (accessed on 7 December 2019).
17. Buonanno, D. Nuovi Paesaggi, interventi di rinaturalizzazione Urbana. *Planum* **2012**, *2*, 76–78.
18. Sommariva, E. *Cr(Eat)Ing City. Agricoltura Urbana. Strategie Per La Città Resiliente*; ListLab: Trento, Italy, 2015.
19. Carrabba, P.; Di Giovanni, B.; Iannetta, M.; Padovani, L.M. Città E Ambiente Agricolo: Iniziative Di Sostenibilità Verso Una Smart City. *Eai Energ. Ambient. Innov.* **2013**, 21–26. [CrossRef]
20. Gausa, M. City Sense: Territorializing Information. In *City Sense, 4th Advanced Architecture Contest*; A.A.V.V. Actar Publishers: Barcelona, Spain, 2012; pp. 6–13.
21. Ratti, C. *City of Tomorrow (The Future Series)*; Yale University Press: New Haven, CT, USA, 2016.
22. Tucci, G. *Albenga GlassCity. From the Glass City to the GreenCity*; ListLab: Roma, Italy, 2016.
23. Tucci, G. Resili(g)ence and Agro-Urban Approach, from the global framework to the Genovese Dimension. In *Resili(G)Ence: Goa Resili(G)Ent City*; Canessa, N., Ed.; Actar Publishers: Barcelona, Spain, 2020.
24. Gausa, M. *Resili(G)Ence, Intelligent Cities, Resilient Landscapes*; Actar Publishers: Barcelona, Spain, 2020.
25. Pothukuchi, K.; Kaufman, J. The Food System: A Stranger to the Planning Field. *J. Am. Plan. Assoc.* **2000**, *66*, 113–124. [CrossRef]
26. Fassi, D.; Sedini, C. Design Actions with Resilient Local Communities: Goals, Drivers and Tools. *Strateg. Design Res. J.* **2017**, *10*, 36–46. [CrossRef]
27. Di Lucchio, L. Design to feed the world. *Diid Design* **2010**, *46–47*, 140–152.
28. Stenmarck, Å.; Jensen, C.; Quested, T.; Moates, G.; Buksti, M.; Cseh, B.; Jull, S.; Parr, A.; Politano, A.; Redlingshöfer, B.; et al. *Estimates of European Food Waste Levels*; Fusions: Stockholm, Sweden, 2016.
29. BCFN; Fixing Food 2018. Best Practices Towards the Sustainable Development Goals. The Economist. Intelligence Unit. 2018. Available online: https://www.barillacfn.com/it/magazine/cibo-e-sostenibilita/chef-e-aziende-in-prima-linea-contro-lo-spreco-alimentare/ (accessed on 14 February 2020).
30. Brown, T. *Change by Design: How Design Thinking Transforms Organizations and Inspires Innovation*; Harper Business: New York, NY, USA, 2009.
31. Mau, B.; Leonard, J. *Massive Change*; Phaidon: London, UK, 2004.
32. Manzini, E. *Design, When Everybody Designs: An Introduction to Design for Social Innovation; Design Thinking, Design Theory*; The MIT Press: Cambridge, MA, USA; London, UK, 2015.
33. Wunderlich, S.M.; Martinez, N.M. Conserving natural resources through food loss reduction: Production and consumption stages of food supply chain. *Int. Soil Water Conserv. Res.* **2018**, *6*, 331–339. [CrossRef]
34. Antonelli, P.; Tannir, A. *Triennale di Milano. Broken Nature*; La Triennale di Milano Electa: Milano, Italy, 2019.
35. Ellen MacArthur Foundation. Cities and Circular Economy for Food. 2019. Available online: https://www.ellenmacarthurfoundation.org/assets/downloads/Cities-and-Circular-Economy-for-Food_280119.pdf (accessed on 14 February 2020).
36. McDonough, W.; Braungart, M. Design for the Triple Top Line: New Tools for Sustainable Commerce. *Corp. Environ. Strategy* **2002**, *9*, 251–258. [CrossRef]
37. Papanek, V. *Design for the Real World: Human Ecology and Social Change*; Thames & Hudson: London, UK, 2019.
38. Fry, T. *Design as Politics*; Berg Publishers Ltd.: Oxford, UK, 2010.
39. Cockrall-King, J. *Food and City: Urban Agriculture and New Food Revolution*; Prometheus Book: New York, NY, USA, 2012.

Article

Sustainable Use of the Littoral by Traditional People of Barbados and Bahamas

Brent Stoffle [1], Richard Stoffle [2,*] and Kathleen Van Vlack [3]

1 NOAA Fisheries, Miami, FL 33149, USA; brent.stoffle@noaa.gov
2 School of Anthropology, University of Arizona, Tucson, AZ 85721, USA
3 Living Heritage Research Council, Cortez, CO 81321, USA; kvanvlack82@gmail.com
* Correspondence: rstoffle@email.arizona.edu; Tel.: +1-520-907-2330

Received: 7 April 2020; Accepted: 5 June 2020; Published: 11 June 2020

Abstract: This paper is about the traditional people of Barbados and The Bahamas, in the Caribbean and their sustainable adaptations to the littoral, which included both marine and terrestrial components. Traditional people are defined as having lived in a sustainable way in an environment for five generations, the littoral is described here as an ecological zone at the sea's edge, which is composed of hundreds of medicine and food plants and animals, and resilient adaptations are understood with the environmental multiplicity model. The analysis is based on more than a thousand site intercept interviews conducted by the authors and their research teams. These data argue that culturally based patterns of sustainable food use and environmental preservation can be understood from generations of successful adaptations of traditional people.

Keywords: traditional people; coastal littoral; Barbados; Bahamas; environmental co-adaption; Caribbean

---

1. Introduction

This essay contributes to discussions occurring worldwide that have crystalized in the United Nations Agenda for 2030, which calls for protecting the planet by promoting sustainability in food production and consumption by informing both policy and practice. This lofty goal is being acted on in many places, at many scales, and with various intervention strategies. While some have argued for new patterns of sustainable adaptation to be uniquely developed with specific reference to a place, a people, and an environment, the present analysis argues that a better starting point is to draw upon the sustainable environmental adaptations of traditional people. Instead of creating sustainability patterns out of whole cloth, the time-tested cultural adaptations of traditional people should be drawn upon first.

This analysis assumes that it is essential to understand the diachronic foundations for the development of sustainable lifeways and food consumption before considering how to share these insights with other people. Two case studies serve this purpose. They involve the traditional coastal people of Barbados, West Indies (Figure 1), and The Bahamas, Caribbean (Figure 2) who have lived in their environment for hundreds of years during which time they have come to understand and adapt to their littoral, which includes both terrestrial and marine resources occurring at the sea's edge (Figure 3). Traditional people are defined for this essay as having lived in a sustainable way in their environment for five generations. The littoral is described as composed of hundreds of medicine and food plants and animals. Resilient adaptions are understood within the environmental multiplicity model. The analysis, therefore, is focused on traditional coastal people who have resided in their communities for more than 150 years. These are people who have learned about their environment to a level that can be termed traditional ecological knowledge (TEK) and so they can be called traditional. The people are associated with littorals that have never been fully disrupted by development but are not considered pristine. The environmental multiplicity model is argued as a useful intellectual frame for understanding adaptations where the people, their ways of life, and the environmental resources they utilized have mutually changed in sustainable ways. We also argue that the environmental multiplicity model, which has both social/economic and natural resource components, can be used to extrapolate elsewhere and up-scale study findings.

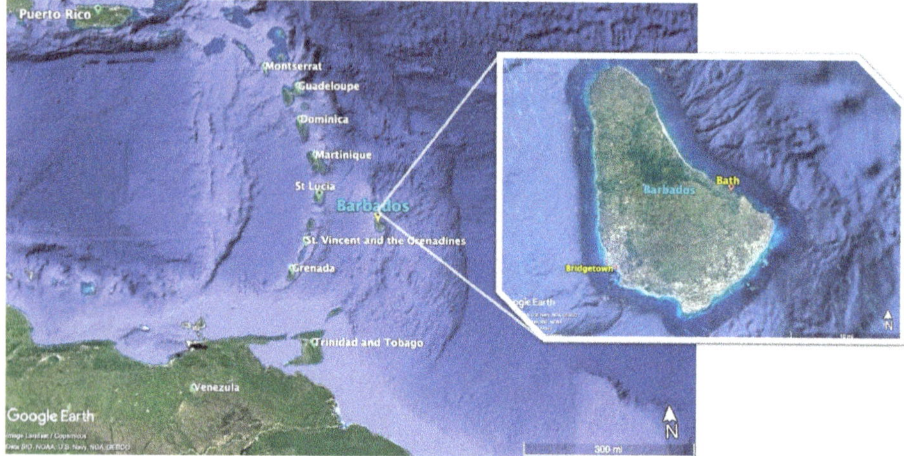

Figure 1. Barbados.

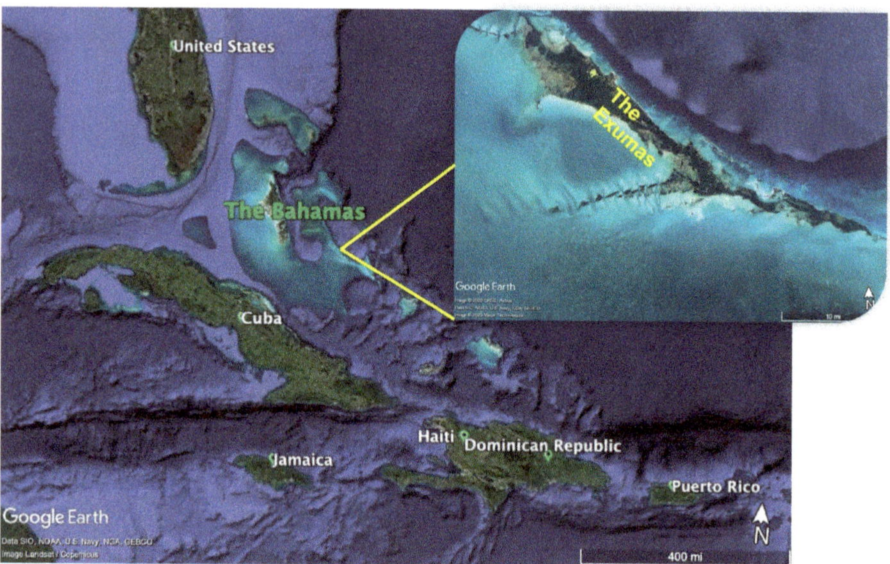

Figure 2. The Bahamas.

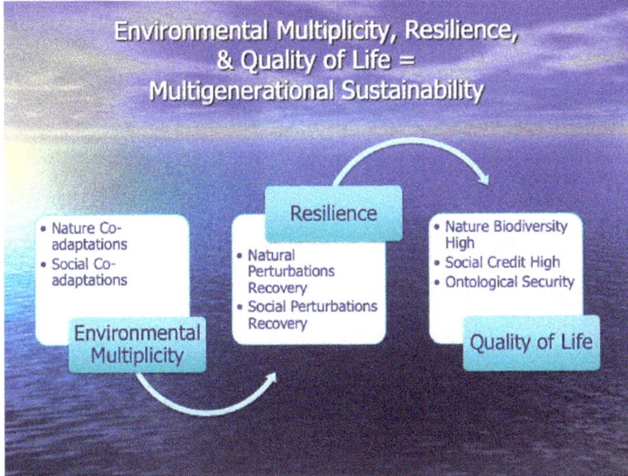

Figure 3. Environmental multiplicity.

Research that is potentially used to develop policy and influence public decisions must be understood in terms of clear limits if they are to be extrapolated to other societies and environmental situations as well as up-scaled to different societies and places. Sustainable food use and production policies can be set at the community and national levels. The authors believe that the findings are useful for understanding other coastal people who share similar histories, locations, adaptations, and live elsewhere in the Caribbean. This conclusion is based on having conducted similar studies in the Dominican Republic [1,2], Antigua [3], St. Thomas [4], and St. Croix [5].

The analysis argues that documenting and protecting the complex adaptions of coastal people is essential in order to protect them from disruption. Furthermore, we argue that some findings can be cautiously up-scaled to dissimilar social and environmental situations by using basic sustainability principles identified in the studies. In this analysis we select urban farming as an example of where new communities can be established based on a shared commitment to stable food production.

The primary power (agency) of the slave and post enslaved peoples was that their littoral use areas and nearby gardens were unwanted by more powerful people and corporations. Initially they then had only themselves as a threat to the littoral and so self-management with conservation norms developed and persisted for hundreds of years. In more recent times threats from interior non-coastal peoples and corporations have posed a threat to the littoral. We suggest that their nations declare them heritage communities in recognition of their sustainability practices and in so doing afford them higher levels of protection. All-inclusive hotels who excavate the mangroves to make boat docks and golf courses are the main threat to The Bahamas and elsewhere in the Caribbean.

Urban gardeners in the US and Cuba have a similar agency problem even after they become a community of farmers. They must establish shared conservation norms to coordinate production and to prevent their products from being taken by other community members; which is less of a problem because they share in the bounty of the gardens. The threats by outsiders who must somehow be policed by the broader community. Successful urban gardeners will be threatened by the potential sales of their farms to larger scale non-locally controlled commercial business.

2. Conceptual Background

Two concepts frame this analysis: environmental multiplicity and the littoral. Both are described as critical components of societies produced by and for slave-based and post-colonial industrial agricultural plantations in the Caribbean. The colonial societies of the Caribbean were designed to provide their European owners (which in this analysis is England) with profits through the production of tropical cash crops. Most of these plants were imported from the East Indies and established on new lands in what would be called the West Indies. The people who farmed, processed, and shipped these cash crops did so as unfree laborers [6] and continued as underpaid laborers after the end of English colonial slavery in 1833 well into modern times [7]. The cash crop exporting societies that made Europe wealthy did not share this abundance with the workers who often had to engage in subsistence activities just to survive. Thus, the sustainable cultural adaptations to this suppressive economic situation and to the littoral resources that were not disrupted or destroyed by industrial production are key for understanding these coastal Caribbean peoples and their small island ecology today.

2.1. *Environmental Multiplicity*

This analysis builds on the environmental learning model suggested argued by Berkes and Turner [8], which maintains that learning and adaptation is based on an accumulation of ecological knowledge and how to protect people from the influence of social and natural perturbations. Learning and adaptation can become the foundation for self-organizing and developing conservation-orientated practices [8]. Common property, and in this case common destiny as a community, can be key ingredients in the elaboration of resource management practices at the local level.

People begin to learn about nature as soon as they arrive in a new environment that has unique (to them) fauna and flora and ecosystem functions [9]. Such knowledge is often termed local knowledge, and it may be useful in terms of proper environmental behavior within a generation [10]. To move from simple observations to deeper ecological understandings of food webs and trophic levels takes many generations. This case supports the co-adaptation model of learning, which argues that within five generations or about 150 years in a new environment a community can acquire extensive and complex ecological understandings and become what is termed traditional begin to build a resilient way of life [9]. As local knowledge is tested, becomes shared and integrated into the culture it can lead to adaptive behaviors including conservation and resilience [11].

Connell's research [12] documented that natural disturbances occurring at an intermediate scale can cause positive changes biodiversity and biocomplexity. Traditional people use their knowledge of ecosystems and make intermediate changes that have positive benefits by clearing spaces in forests [13], moving seeds to new habitats [14], digging tubers [15], changing behavior of herding animals [16], pruning wild nut trees [17], and designing agricultural fields to stimulate animals and plant populations as well as provide sustainable farming [18]. Especially important for this analysis is Turner's study of seaweed collection among the native peoples of Northwestern America where harvesting of intertidal zones improved littoral habitats for both plants and animals [19]. Gifting and trading the foods cemented social relationships, developed economic sufficiency, and built the reliance of families and communities [19].

When a people learn about the fauna and flora of their ecosystem, they can adjust their adaptive strategies to protect themselves from natural and social perturbations. When they do this and live in a sustainable way, they can be said to have developed a resilient way of life [20,21].

In The Bahamas and Barbados the African-ancestry people have made a resilient way of life by building a series of social and ecological redundancies, which we have termed environmental multiplicity (Figure 3).

This concept builds on Lambros Comitas's [22] theory of occupational multiplicity, which is widely recognized as a foundation of social adaptation and resilience in the Caribbean. He documented that Jamaicans acquire skills, invest in resources, and hold many jobs at one time. Even though it makes more economic sense to invest all of their work time in the highest paying job, they spread their efforts across a range of jobs because these come and go due with economic booms and busts. Environmental redundancies occur when people have multiple places to fish and gather the same fauna or flora. They also have agricultural fields that are left fallow for five or more years, restoring nutrients in the soil and serving as a buffer to environmental damage. These redundant and rotation use patterns restores the soil and reduces fishing and gathering pressures. These use patterns only cause intermediate environmental disruptions. In this analysis we combined natural and social redundancies into the environmental multiplicity model so that it describes these adaptations as both parallel and functionally interrelated.

2.2. Littoral

Edges are special places for human ecology [13,23] because critical natural and human life cycle events occur there. In the Caribbean the edge of the sea is especially important for fish nurseries, mangrove wetland nutrient exchanges, and people [24]. This is a place where freshwater touches saltwater, birds' nest, and amphibians thrive. Here too people begin and end their lives.

Unlike other important areas in the terrestrial and marine environment, the human dimensions of the Caribbean littoral tend to be underrepresented or misunderstood in ecological studies and in the social impact assessment research. Research data from former slave communities in Barbados and The Bahamas illustrate that over hundreds of years the people learned about their littoral environment and adaptation to it through sustainable uses. This explains why the littoral has become culturally central to the people of these communities and why they established a sense of customary ownership and responsibility for protecting it and treating it as "family land" [25,26].

The term littoral is used in many different ways. Some scholars restrict it to the wet portion of the coast between high and low tides [27], whereas others view it as a general term of reference for socially and biologically integrated portions of the seacoasts. These ideas are combined to operationally define littoral to mean that a portion of the sea immediately adjacent to the land but no deeper than the waist of an adult at low tide. It includes places on land that are socially and biologically connected with the sea.

The littoral extends onto the land through food webs that critically depend on both salt and freshwater habitats. Thus, it extends up estuaries into mangrove wetlands, and as far as amphibious animals like crabs travel inland. Minimally the littoral involves the following kinds of places: shallow coral reefs, sea grass beds, exposed beach rock, foreshore, backshore, sand dunes, sea cliffs, mud flats, estuaries, mangrove swamps, brackish ponds (anachialine), freshwater deltas, springs, and streams.

These case studies involve a number of kinds of littorals, which are presented at various points in the essay. Figure 4 is a leeward side shallow marine sand bank littoral with slow tides. Figure 5 is a tall resistant cay with steep banks and shallow sea and fast tides. Elsewhere is a photo of a windward rain-driven littoral pressured by persistent northeast trade winds with fast tides. Also later in the article is a mostly enclosed mangrove littoral with a mud bottom and slow tides. Each of these littorals represents a different econiche. The biodiversity and biocomplexity of these types of littorals have been documented through our marine and land mapping studies [11,28,29].

It is essential to understand the biocomplexity and biodiversity of the littoral because only this explains how these coastal peoples survived periods of industrial agriculture and why they are so strongly attached to these areas today [24]. Tables A1–A3 (Appendix A) highlights some of the culturally important species by their cultural uses and names. These use species were selected from among hundreds of species identified in research interviews and published reports, in order to illustrate the many sources of traditional food, medicine, and construction.

Socially and culturally the littoral extends into the contemporary activities, history, and culture of traditional communities. So, the littoral also exists where it participates in the lives of coastal people. The littoral is more than a physical place; it is a part of the social fabric of coastal people. It is a place of teaching and learning. It is a place where knowledge is passed from generation to generation and where it is commonplace for an elder to pass cultural knowledge regarding appropriate fishing techniques and strategies to younger people. It is where lessons are taught regarding different species and ways to monitor environmental changes, be they monthly, annual, or an aspect of climate change. Here the younger generations are taught lessons regarding types of plants that can be collected and used for both food and medicinal purposes. This type of knowledge sharing has gone on for generations and according to oral history began with the forced arrival of their ancestors. Some knowledge was brought from Africa and other types were learned over time. Environmental learning occurred from their need to sustain family and group health.

Figure 4. A leeward sand bank littoral.

Figure 5. Littoral on small key with sea grass beds at edge.

Today these activities in the littoral provide subsistence and small-scale commercial (informal economy) opportunities for young men and women who engage the littoral, gather, and collect to offset unemployment and as a means of contributing to the household income and welfare. Young adults between 16 and 25 tend to experience high levels of unemployment. Their ability to go to the littoral allows them to harvest fish and other marine products for sale or trade. It is not uncommon to see young adults on their way to the beach stopping by various homes to see if individuals want a certain species. Upon return that species is sold or given to the individual. This allows for either cash payment or the ability to call on a future favor. Dinner stews often include marine species and plants gathered on the way home.

3. Methods

This paper presents an analysis of traditional communities situated along the coasts of the Exumas islands in the central Bahamas and the island of Barbados. Data from the two studies were collected with different funding sources and each study had it own specific purpose. The studies are comparable because each included the systematic gathering of data regarding use of littoral by community members. In both studies the members of the communities had been dependent for food, medicine, and construction for more than 150 years on the animals and plants from the littoral. Residing in place for this period and sustainably using the littoral argues that the people should be defined as traditional, which is an analytically important stage of adaptation. In fact, the communities in the Exumas, Bahamas were established 235 years ago (post-1785) and the Bath plantation was established more than 227 years ago (post-1793) [30].

Social science researchers use mixed methods [31,32] and triangulation [33]. Mixed methods involve collecting qualitative and quantitative data, and where there is convergence, confidence in the findings grows considerably [34]. Participant observation was an important component of each case study. The Exuma study used seven instruments (a) sea attachment, (b) quality of life, (c) grubbing, (d) tourism, (e) ethnobotany, (f) land mapping, and (g) sea mapping. The Barbados study used two instruments one focused on occupational multiplicity and other on household finance and micro-credit. Both studies used oral histories, which were both structured and open, ended and were comprehensive given the dozens of hours each required. Most data collection instruments were diachronic in order to contextualize contemporary life ways in short (30 years) to long (200 years) adaptation time frames. All formal interviews were systematic in that they were administered using a structured data collection instrument, thus permitting direct comparisons from instrument to instrument and person to person.

Informal interviews are an important tool for collecting data when formal interviews are not possible because of either time or interest of the interviewee [32]. They are often the best way of listening to people about subjects not currently contained within the formal interview instruments. Informal interviews permit topics to emerge that may become critical to the study, perhaps eventually requiring their own formal instrument. In-depth understanding of some topics like ethnobotany required dozens of hours of informal interviews. All informal interviews were recorded in bound field notebooks and logged into a data base.

It is important to recognize, however, that confidence in these findings derives from an overall triangulation of comparable findings from any of the instruments and oral histories. The triangulation of data thus involves comparing responses generated with divergent instruments. When two or more instruments provide the same answer to a research question then the confidence in the accuracy of the answer is increased. Confidence in the accuracy of responses also increases to the extent that most interviewees provide the same answers.

3.1. Barbados

The Barbados interviews were conducted during a 3-year study of rotating savings and credit associations known in the Caribbean as meeting turn, sou sou, asu, box hand, and partner [35]. During five field sessions a single researcher [36] conducted 500 formal and informal interviews in Barbados; of these 120 were with the people of the Bath plantation area on the northeastern coast. Data were collected on microeconomic systems, which are a creole (or informal) economic system [7] and community lifeways. Responses regarding direct production from the sea provided data on patterns of littoral use.

The research is built on the findings from earlier studies, especially questions generated by previous interviews. The research methodology included: literature review; participant observation; formal interviews with instruments; and informal interviews using a memorized interview schedule.

The Barbadian analysis assumes that broader patterns of Caribbean life, especially ones occurring over the life cycles of individuals and traditional patterns established over many generations can be understood through systematic interviews as well as through a few typical stories (oral history) from a single settlement. The 500 plus interviews established the cultural centrality of the littoral in the lives of people in coastal Barbados. The Barbados analysis situates these findings through the diachronic story of the Bath settlement and one fisherman.

3.2. The Bahamas

The Bahamas research focused on how proposed marine protected areas (MPAs) could impact six local communities in the Exumas. Community perceptions of potential MPA impacts were assessed as predictors of local responses to the MPAs. Beginning with an open-ended ethnographic approach, the study sought to elicit variables rather than test them.

The Bahamas case involved 572 interviews conducted with 193 people from six coastal communities in the Exumas islands and cays. Various data collection instruments were used, often with the same people. There were 352 formal and 221 informal interviews. An overall sample size of 34% of the census recorded population was achieved for each community.

The research is (a) inductive, (b) iterative, (c) mixed methodologically, (d) collaborative, and (e) consultative. Eight field sessions occurring over a six-year period permitted an iterative cycle of collecting data, analyzing findings, and returning with both new and revised data collection instruments.

The sea attachment, land mapping, and sea mapping instruments were central to this analysis. It is important to recognize that confidence in these findings derive from an overall triangulation of comparable findings from any of the seven instruments and oral histories. A fourteen-page sea attachment instrument was developed to explore the widest range of marine uses and cultural meanings. It has 208 questions distributed across seven knowledge and use domains such as: material arts, sea biology, underwater landscapes, land biology, expressive arts, identity symbols, and settlement stories. Land mapping and sea mapping interviews were used to define resource use patterns across space and through time from slavery until the time of research.

The oral history accounts, some of which involved hundreds of hours of interview time, describe environmental learning and the subsequent behaviors of ancestors during slavery and just beyond. These are up to 235-year-old heritage memories (post-1785). It is important to remember that many of the people today and their ancestors have continuously lived in or near their initial slave village, have taken and kept the last name of the original planters, and for most of this time have relied upon the same marine and terrestrial ecosystems that their ancestors faced. Continuity of people and place is illustrated by the fact that many people remember when the first commercial pharmaceuticals became available in the Exumas in the 1950s. So, bush teas and medicines were relied upon throughout this period and are used extensively today. Memory timelines, given these criteria, are well within the accepted standards of accuracy [37]. Elsewhere, Stoffle and Zedeño [38] document accurate oral history accounts going back thousands of years. From an ethnological perspective the slave-period interpretations also are robust because many people interviewed in the Exumas similarly describe the environmental learning and adaptations of their ancestors.

4. Cases

These cases have the common theme of the diachronic development of sustainable life ways in former enslaved communities who have maintained a core traditional population. Both cases are primarily about of African-ancestry people, but each involves traditional European-ancestry neighbors who for hundreds of years have worked alongside them on the plantations. Neither Barbados nor the Bahamas colonial society provided a social/economic safety net for its rural poor; thus, individual security was established by the people themselves as members of small-scale communities located near the edge of the sea.

Both cases are focused on a marine littoral activity, but it is important to understand that complex sustainable use patterns exist for terrestrial fauna and flora. Oral history studies of plant specialists, for example, have documented 264 species of traditional use plants for the Exumas [39] and almost as many for Barbados [40–42].

4.1. Case One: Barbados

The east coast of Barbados can be characterized by its rugged coastline where prevailing easterly wind and wave action carved out a variety of locations in which flora and fauna thrive. One impressive area is the former Bath plantation where in the nearby shallow sea is a predominant feature known as the Great Rocks. The Great Rocks serves as a defining marker between deeper ocean and land interaction (Figures 1 and 6). The area outside of the Great Rocks tapers off to mixed patches of shallow and deep-water coral reefs that create an environment where fish and underwater plant life flourish. The Great Rocks littoral supports land and marine based plants and animals.

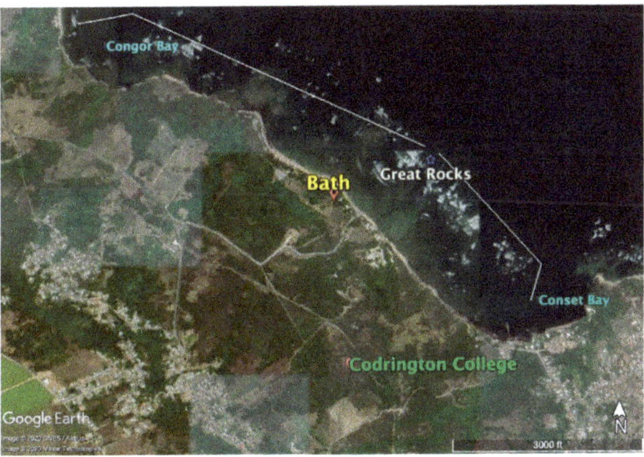

Figure 6. The Bath area and littoral zone.

An extensive littoral zone extending from the Great Rocks north to Martins Bay and south to Consett Bay (Figure 7) has been and continues to be a socially, culturally, and economically important part of the lives of two distinct ethnic groups of people; Irish indentured servants brought over in the latter 1600s and subsequently African enslaved people in the late 1600s. Both of these groups continue to inhabited the area and have done so ever since the first plantations were established in this part of Barbados. The littoral provided food and medicine in addition to creating social and economic security and stability. The collection of plants and animals in this area was an important strategy for offsetting the cruel and harsh conditions of slavery and allowed the people to sustain themselves, when provisions from the plantation were grossly inadequate.

Figure 7. The Bath littoral with sea swells made by northeast winds.

The north east coastal littoral was used from the earliest times of English settlement (1627) but it seems that fishing was primarily the responsibility of 40 Arawak families many recently from Dutch Guiana. Richard Ligon [43], who lived in Barbados from 1647 to 1650 provided a map of the Indian area located just inland near Bath on the north east coast. He observed:

> "that Indian women were primarily, if not solely, employed in household tasks. On the other hand, the men...were use for footmen, and killing of fish, which they are good at. With their own bows and arrows they will go out, and in a day's time kill as much fish as will serve a family of a dozen persons two or three days, if you can keep the fish so long". [41,43]

This observation is an intriguing insight into the pristine condition of the littoral. It is not clear how long Indian men were employed in the fishing industry with their six-foot bows and long arrows, but it is unlikely that African-ancestry people were permitted to make and use of such weapons.

Ligon [43] recorded on an inland plantation that African-ancestry males were allowed two mackerel a week and each woman one. African ancestry people were not observed fishing except under the direction of a coastal plantation owner who had a seine net. This may have occurred because they were restricted to industrial plantation labor and thus their visits to the littoral were at odd hours and on Sunday and conducted with little equipment. Some later evidence of this comes from the analysis of the nearby Codrington Plantation (pre-1710 to 1782) where about 300 people were enslaved [44]. The Codrington plantation records of 1776 list the occupations of 51 skilled men, and 21 women and 9 boys who were not on the field gangs, but the occupation of a fisher is not listed [44]. This is especially interesting because Codrington plantation became a major English Religious College occupied by students and ministers of the Anglican Church all of whom consumed fish on a regular basis. Perhaps the lack of the fisher job was unique to Codrington because 12 years later in 1788 the island's governor, reported that the number of fishermen may be about 500 out of an enslave population of about 62,000 [41].

Crab fishing during slavery may be a better analog for patterns of littoral use than fishing. They are caught at night by fishers using torches that both provided light for movement and to attract the crabs [41]. Similar patterns occurred in the shallow water for collecting lobsters. Torches combined with nets were used to catch flying fish along the shore. McKinnen in 1802 recorded that in Barbados the local people are very successful at taking flying fish. At night they spread their nets before a light and disturb the water at a small distance. The fish rise and eagerly fly toward the light and are intercepted by the nets [45]. This pattern of using torches and nets to catch flying fish was observed again by another visitor in 1818 [46]. Use of torches to collect in the littoral is in keeping with the need for enslaved people to be out of sight when they are away from the plantation.

There are many key features and resources of the littoral utilized by the enslaved people and their descendants in the area today. The collecting of marine resources was an important strategy for providing much needed protein in a diet that was inadequate at best when based on provisions provided by the plantation owner or overseer. These littoral collection strategies are much the same as they were in the past using hands to grab or hook, small/light fishing poles or lines, and self-made nets (cast nets) as seen in Figure 8.

The Bath case is centered on the life of one outstanding, but in many respects, a typical fisher. His is the story of the people of Bath and many other coastal communities in Barbados. Fred Watson is a 94-year-old fisherman (in 2020) who was born in a little house across from the beach in Bath. He has lived his whole life in and around this area, fishing as a means of providing food and money for his family. As a youth Fred was trained to fish and collect marine and terrestrial resources such as seacat (*Octopus brareus*), conch (*Strombus gigas*), spiny lobster (*Panulirus argus*), Red Sea crabs (*Careilius corallinus*), whelks (*Buccinum undatum*), and curbs (*Polyplacophora* ssp.) from the sea, and seaweeds, sea grapes (*Coccoloba uvifera*) and white swampee crabs (*Cardisoma guanhumi*) from the land. These lessons were a common part of the informal environmental education of young boys as older relatives and community members often felt obligated to pass on what they knew to the younger generation. As he grew older, he attended school and upon completion of his formal education was made a primary school teacher. During that time Fred continued to live and support his family, especially his 15 family members. The problem was that even with Fred's salary there still was not enough food to feed everyone.

RED LEG FISHERMEN AT MARTIN'S BAY CIRCA 1910

Figure 8. Irish fishermen with littoral fishing equipment in Barbados near Bath [47].

Fred stated, at the age of 13:

> "I had to give up being a teacher even though I enjoyed it. Even though I had a steady paycheck I couldn't always make enough money to make sure that everyone was fed. I would still fish at that time but because of my commitment to the school I wasn't able to do it as often as I needed to. So, I gave up being a teacher and went to the sea full time. I would fish all day and all night, sometimes inside Great Rocks and sometimes outside. Even though I made a lot less money, I was able to make sure that everyone ate. My brothers and sisters never went hungry because the sea always provides."

Fred comes from a fishing family and his fisheries knowledge and success in training fishermen from the area comes from his over 85 plus years of fishing plus the generations of knowledge acquired from those that came before him. His knowledge of the sea has garnered a great deal of local as well as national respect and attention. In 2016, the Prime Minster awarded Fred the Honor of Oldest Active Fisherman in Barbados building on his previous award from 2002 for his lifelong service in fishing and recognizing him as the Best Fisherman in Barbados [48]. This service includes not only providing food for the people of Barbados but also includes his role in assisting researchers and policy makers in creating laws that benefit the creation of sustainable fishing practices. The young boys and young men still turn to him for advice and even at 94 he spends his days mending nets and holding "classes" for those that need assistance.

The littoral is full of marine and terrestrial resources that can be sustainably used for consumption and sale. Many plants are needed for making medicinal tonics. "Bush teas" are made from various combinations of plants and used for a variety of ailments, from sickness related to colds and flu to detoxifying internal organs. People bathe in the saltwater to cleanse the body from wounds as well as consume saltwater to cleanse congestion from head and lung ailments. These folk cures are passed down from generation to generation and are often selected over the use of "western" chemical medicines.

Fred is a person who acquired generations of knowledge about the littoral zone from elder family and friends and he continues to be an important conduit of this information. He is quick to fulfill his role as teacher for those who wish to carry on the responsibility to share and build upon the lessons of the past. While he is no longer able to fish in the same ways as he did in the past, he still is a regular fixture at Bath where young men come to improve upon their fishing knowledge and practices. He is revered as the "best fisherman in all of Barbados" and has made sure that the youth are educated as to the best practices for sustainable use.

In 1999 he demonstrated how to use a cast net in the littoral can provide enough for a single person to provide adequate amounts of food to multiple feed families. In this instance he went out across the shallows targeting small fish known as fray (a small fish) and sardines. With two net throws and over 30 min he was able to fill a bucket of these small fish providing enough food for two days of meals for both his family and that of his helper (Figure 9).

Figure 9. Fred Watson on the right with a younger fisherman.

When asked why he did not continue to throw and catch more fish his response demonstrated a conservation strategy. Fred stated, "Why would I continue to throw the net when I have all that I need for now (Figure 10). Sure, I could catch much more but why would I hurt the fish. I do not need more than I have and by taking more than I need all I do is hurt them (the fish). Now I know that they will be there for me for the future.".

This strategy of creating a balance between conserving the fish while supplying food for the family ensures that resources will be there for future use. This knowledge stems from generations of day-to-day involvement with the resources and is a lesson that when passed to future generations of fishermen will ensure that the practice is sustainable for both humans and the fish. This is a practice that many coastal people throughout the world have employed because of their recognition of humans' potential impact on the health of the environment.

Figure 10. Fred Watson casting the net.

4.2. Case Two: The Bahamas

The Bahamas case analysis is about how African ancestry people living on the isolated Exumas islands and cays (Figure 10) located in the central Bahamas have learned about and adapted to their environment. The case contains a range of dates for the beginning of environmental learning after individual plantations failed post 1785 and the eventual collapse of most slave plantations in the late 1790s. Two types of TEK, a form of hand fishing and extensive knowledge of ethnobotany, illustrate the complexity of these in-situ knowledge domains. The case documents how TEK has gone beyond understanding species to awareness of trophic levels interactions, ecosystem functions, and eventually to ways to conserve this delicate coastal environment. Their in-situ TEK, co-adaptation, and conservation have produced a lifeway based on environmental multiplicity that is has been resilient for over about 235 years.

4.2.1. History of Bahamian Case

The arrival of the Spanish in the Bahamas in 1492 initiated a period of rapid depopulation and the eventual extinction of the original inhabitants of the Bahamas, the Lucayan people. Most evidence suggests that Bahamian islands and cays subsequently lay unused by humans. The ecology as it had existed under thousands of years of Lucayan farming changed when they became extinct and the Spanish failed to occupy the Bahamas. With no managers, nature went wild so to speak. A new Bahamian state of nature emerged over the next 156 years, until being interrupted in 1648 when English Puritans settled Eleuthera Island in Northern Bahamas.

The Exumas, being more isolated and removed from the centers of the Bahamian colonial economy, remained unoccupied for 293 years, when in 1784 the two large Exuma islands were surveyed and conditionally given as plantations to Loyalist (refugees from the 1776 to 1783 American Revolution). A plat map made in 1792 documents the presence of 115 land grants each of which is a small plantation [49]. Only a few platted areas lacked an indicated owner, most of who resided with their enslaved workers. The Loyalists were required, as a condition of receiving Crown lands, to clear the land and make it into productive cash crop farms. Failure to accomplish this within ten years would result in forfeiture of the land back to the Crown.

Living in a new ecosystem that had been fallow for about 293 years and required by the Crown to rapidly produce a commercial plantation, the Loyalists stripped this long fallow ecosystem. They sold off all commercial timber, moved to define boundaries with coral rock walls, and planted cash crops such as cotton on all suitable land. Tropical rains and hurricanes soon revealed the danger of opening all lands to farming. Keegan and Mitchel [50] estimate that the topsoil of most plantations washed away within three years, exposing the hard-calcariferous bed rock. The chenille bug destroyed much of the cotton grown in the Bahamas in the late 1790s [51]. So, most Exumian slave plantations quickly failed, although a few remained for another 100 years as salt producers. With crop failures, the Loyalists left the Exumas, but their African ancestry slave populations remained in a limbo status because of English laws, which made moving slaves illegal, although slavery continued to be legal and they continued to be considered enslaved people. As an Exumian plantation failed and was abandoned by the Loyalist owner, the workers were left to fend for themselves, which they did by organizing themselves as a community and taking control of the plantation lands. In order to define their common occupation rights (which later would become recognized as usufruct rights or generation lands) the people called their community after the name of the former plantation owner and each person took his name as their last name. This was the beginning of environmental learning and the foundation of contemporary African ancestry communities in the Exumas. Today, the descendants of the former slaves are largely clustered in twenty-six settlements located on or near the post-1784 plantations [51].

Little applicable ecosystem knowledge was brought by the Loyalists or the African ancestry people because neither had lived in environments identical to the Exumas. These peoples neither had access to Native American TEK from the American colonies nor in the Exumas. African ancestry people were both Creole (born in the New World) and arrived directly from Africa. The former came from the revolting mainland colonies, especially South Carolina and Georgia, many of the latter came from the interior of the Senegambia region of West Africa [52].

The English government stipulation that the Loyalists must produce cash crops within ten years, caused the small plantation owners to pressure enslaved people to invest maximum time clearing the land, growing mono crops, and processing for the market. Free time for slaves was not abundant even though Bahamian law required that the slaves have plantation land for their own gardens [53]. The enslaved people were often underfed because most food was grown in another English colony [54]. So, they used small gardens and gathering in the littoral just to survive. Free time during slavery was constrained by limits on permissible distance traveled and time absent. Oral history accounts document that each evening one local planter took his enslaved people by small boat to an isolated cay surrounded by swift tides where they were left to fend for themselves until work the next day.

Most Loyalist plantations were abandoned by the end of the ten year economic viability period, others failed by the early 19th century, and all slaves were freed in The Bahamas in 1834, which defines a point after which all African ancestry people had full access to their own labor, lands, and ocean [51,52]. Like Barbados, by this time most of the land was completely cleared. After the plantations failed there were few natural plants; however, ecologists believe that something resembling the natural ecology did reoccur within a generation due to small island biogeography [55]. Fauna and flora traveled from dozens of undisturbed cays and reestablished a new but not pristine ecology.

The formerly enslaved people began immediately after the failure of their plantation to use the littoral [24]. This area was accessible by foot or by floating on small rafts to the neighboring cays just offshore. Foods for daily consumption, construction materials, and medicines came from the littoral because it was salty and thus never cleared for plantation cash crops. Here people could access a wide variety of protein, while patches of natural vegetation and home gardens were being expanded.

Subsistence farming for African ancestry people occurred on limestone bedrock because the thin soils had been eroded away by unsustainable plantation agriculture; so people developed a system of pot culture in which farming holes were annually expanded by burning small fires in them and supplementing the soil with seaweeds and earth from elsewhere like bat caves. When hurricanes overturned a larger tree, the soil contained within its root ball was eagerly sought after and used in the pot-hole fields. This pattern was observed during this study. Developing this practice, African ancestry people (re)established a form of Native American swidden farming, which seems uniquely adapted to Exumian ecology.

4.2.2. Grubbing: Unique Ecology of Case

During the plantation period, enslaved people stood on the shore and watched ocean currents, learned about the movement of water in the mangroves, and observed how weather patterns such as mid-day storms and hurricanes affected the sea. People used free time to study fish behaviors and to collect marine products. Once on their own they turned to the littoral.

This analysis is based on thirty-four grubbing-specific interviews and half of a dozen family oral histories. Grubbing only occurs in mangroves where the sea is relatively calm and shallow; and when it is effective to use hands and team work to catch fish during low to medium tides. In order to grub, a person must have full knowledge of the littoral and fellow fishers including (1) tides-grubbing occurs during low to medium tides because then it is easy to walk to grubbing locations, which can be up to a mile off shore, and not all grubbers can swim, (2) fish behaviors and types—especially important are life cycles, (3) weather-rapidly changing weather conditions place people at risk, (4) plants—these were used to catch fish and to protect grubbing groups from attack, (5) predators—mangroves are a dangerous place to walk because of sharks, moray eels, and barracudas, (6) mangroves services—a system of regulations was imposed to assure that the key ecological roles were protected, and (7) social relations—normally grubbing involved groups who functioned successfully when there was a shared division of labor, clear communication, mutual commitment, and redundant skills.

4.2.3. Exposed Mangroves and the Grubbing Circle

This description of the grubbing circle comes from Forbes Hill where it normally involved a large group of women, usually one from most households and sometimes her oldest child. The group would venture a mile and a half offshore to the end of a large mangrove covered peninsula (Figure 11). A deep salt-water creek had to be crossed to get to the grubbing area. This mangrove is largely an open system marked by shallow waters and some sheltered areas. One of the best areas is far from shore and surrounded on three sides by ocean thus is especially vulnerable to shifts in tides, adverse weather, and large predators. All areas are open to the sea and shallow where the mangrove dries out completely forming massive mud flats or the mangrove fills rapidly with water making grubbing impossible. Women recount sinking up to their waist in mud fearing the rapidly returning tides, which carry sharks and barracudas. The path taken by the women to the grubbing area is documented in Figure 11.

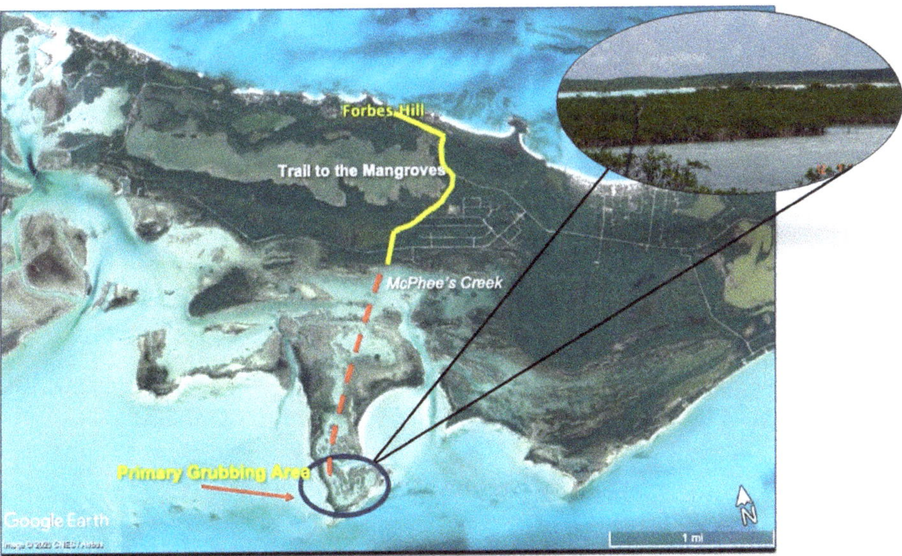

Figure 11. Grubbing area near Forbes Hill and exposed mangrove, with the ocean behind.

The dozen or more women who circle grubbed together had very specialized roles because of the risks and location far away from shore. One woman was charged with watching the whole group; to make sure every woman returned home safely even if she helped them swim across the blue holes and the creek. Another woman was charged with watching unpredictable tides often triggered by strong winds. Those women who did not know how to swim had to be protected if the tide came in suddenly. One woman watched for and fended off dangerous animals like sharks, barracudas, and moray eels. She was fearless. Women interviewed recalled an incident when a big moray eel swam into the group and this woman forced her hand down the moray's throat and strangled it to death.

The majority of the women in the grubbing circle were tasked as fish herders, who were to muddy the mangrove waters by slowly walking in a line, slightly raising a muddy cloud causing the fish to become confused and have a difficult time breathing.

> "They would make the water muddy in the mangrove because when the water get muddy, if any moray in there, they gonna come out [and go away]. They [the fish] gonna get drunk and they gonna keep pushing their head in the mud ... they gonna keep making a noise like a grunt, so you know just where they is."

The herders moved the fish in an ever-tightening semi-circle towards a group of fish catchers each of whom wear a wide flared skirt with the hem tucked tightly underneath their heels. As the fish are driven blindly, they seek refuge under the skirt where they are easily caught and placed in a specially constructed woven grubbing basket with a narrow opening at the top. Children who often hold the baskets are brought to learn grubbing and to not slip in the mud and have the fish swim out of the basket.

The grubbing circle women (Figure 12) developed this unique fishing method using social organization, cooperation, mangrove TEK, a wide flair skirt, and a narrow-mouthed basket. It often was the women who had the responsibility to catch fish for the whole community when the men were gone; a common situation because ships would come seeking laborers and remove all the men in the community for months. When the women returned home, they gave fish to other community members who could not go out into the mangroves to grub or chose other community tasks. Fish was often exchanged for breads or vegetables. Sometimes the fish became part of a large communal meal. The women of Forbes Hill depended on each other and the social networks they created influenced community structure and all other aspects of their life.

Figure 12. Grubbing ladies of Forbes Hill at St. Peter's Union Church.

4.2.4. Sustainability and Grubbing

Grubbing continued to be a primary form of subsistence fishing in the Exumas throughout the post-slavery period, but it declined with new technologies associated with boat building, fishing lines, and nets. Soon people traveled beyond the mangrove system into deeper waters and expanded the fishing territory. As Exumians broadened their knowledge of the sea, people with boats learned about the best distant and deep locations to fish. Deep water fishing teams had multiple places to fish because it was beneficial to rotate fishing areas. By dispersing fishing pressure people did not over-fish the mangroves despite increases in population and more efficient technologies.

Knowing that the mangroves are a vital part of the ecosystem, a system of regulations (conservation ethics) were agreed to and imposed. These regulations involved only taking fish from the mangroves when necessary, not fishing in the mangroves every day, and never taking juvenile fish. The rules were regulated by the family and community, so the mangroves are protected for future generations. A man from Little Farmers Cay explained that in the mangroves, young fish grow before moving to the deeper waters and therefore people learned not to always fish in the mangroves. People have redundant mangrove resource use areas to prevent overexploitation. They fished in two very different mangroves—one close to the settlement and another in the leeward cays. Redundancies are important for conservation, so people have multiple use areas with similar ecology.

Even though people acquired new fishing technologies and used territories away from shore, knowledge of and respect for grubbing persists. People still grub occasionally but speak of it more as security heritage; that is, a traditional way of life that can always be turned to in times of need. It is a source of pride in the resourcefulness of the ancestors and a proven way of fishing to be relied upon during difficult periods.

5. Discussion

Stuart Pimm [56], a foremost ecologist maintains that understandings of world-wide environmental principles have slowly developed because ecology studies are designed to be narrow in time, place-space, and species. Ecologists seek to improve the quality of their findings by carefully focusing their studies; however, Pimm concluded that broader understanding of other places, species, and time frames require a different type of integrative analysis. He, thus, wrote *The World According to Pimm: A Scientist Audits the Earth* to demonstrate the importance for public policy of up-scaling local research findings to the planet level.

Here we cautiously address ways of extrapolating and up-scaling some of the findings from the Barbados and The Bahamas research. We do not assume that just because a community has been in place for hundreds of years and today has clear conservation and sustainable use practices that they have not made mistakes. In fact, if patterns of sustainable food use are to be learned by others, it is necessary to understand how these practices came into being. Some models of environmental learning suggest that people make mistakes (depletion crisis model) and learn what damage they can cause. Over time, they move away from harmful practices and replace them with sustainable uses [8].

One example from both cases is that of learning from mistakes using fish poisoning. In Barbados the practice of poisoning fish with the juices of the manchineel tree (*Euphorbiaceae* sp.), which may have been brought by the Carib or Arawak fishers. It became so widespread and destroyed great quantities of fish in the bays, creeks, and shoals. So, in 1724 the government passed laws against it use. The practice was not eliminated, however, so in 1766 a similar law was passed with more severe penalties [41]. It is understood that the pattern of fish poisoning was largely practiced by non-fishers who were primarily farmers living away from the sea and thus less committed to its sustainability. Fred remembers it was used in his youth about 1950 but there is no evidence of fish poisoning in the Bath area at the time of the study; fishers decided to stop the practice (personal communication Fred Watson 2020).

In the Exumas there is a traditional form of fish poisoning (stunning) called chemical grubbing that uses the bark of dogwood (*Piscidia piscipula*) and joewood (*Jacquinia keyensis*). While still practiced today both for fishing and protection it is closely governed by local customs [39]. Like in Barbados, the poison derives from a native tree bark that is so strong if used while wadding in the water the fisher also can be drugged. Women carry the bark in a bag to protect themselves from shark attacks while rowing small boats. It is only used today by fishers who both know littoral TEK and have experience using it for chemical fishing and it is not widely taught to youth.

Certainly, it is possible to extrapolate findings to other Caribbean coastal people living in traditional post-slavery industrial plantation communities who already use appropriate technology (Schumacher 1973) for littoral fishing and plant harvesting. Based on published research very similar sustainable food use and environmental protection patterns exist elsewhere; such as in the Dominican Republic [2,5,57,58]. The key here is to carefully evaluate through sustainability assessment [59]. Potential impacts of developments would shift the land use patterns or community stability and thus reverse sustainability in food production. Especially critical are coastal littoral impacts from developments that both modify the ecology and exclude the local people, such as national tourism parks [60] and all-inclusive resorts. Traditional coastal communities need to be protected as heritage communities who have learned how to sustainably use and protect the littoral.

The study findings, however, must be up-scaled to more complex societies and their members who are living in different environments, and who must use alternative technologies and change their pattern of food production and consumption in order to meet the UN 2030 goal of increased sustainability. So, where are the findings most likely to be well received and potentially make a useful contribution? Here we consider the example of urban gardening; which has occurred in Cuba, Detroit, and New York (Appendix B). In each of these situations locally controlled urban gardening emerged after the collapse of external support systems and withdrawal of regional and world economies. In other words, like the people of Barbados and The Bahamas the residents were to one degree or another left to adapt on the own.

So what general principles need to be agreed upon to by the people surrounding the urban gardens in order to make a sustainable food producing system? The new gardens, which replace abandoned and removed buildings, must become resilient to climate change, economic, and social perturbations. One strategy is to use environmental multiplicity model whereby different soils, rain shadows, and sunlight distributions can be assessed to in order to establish different places for growing the same plants and to experiment with different plants growing in these same places. Knowledge of outcomes from these micro-experiments needs to be shared among the farmers in order to build a body of knowledge and create a sense of common purpose. New urban farmers will be confronted, like the people of the Caribbean, with shifts in weather and climate. The weather will be dryer or wetter, hotter or colder, and have storms. Learning to adapt to weather shifts and eventually climate changes will require generations. The new farmers will become a component of the urban economy and perhaps be in competition with rural farm systems. If successful the urban gardeners may face capital intrusions whereby larger more powerful business will try to purchase and consolidate the farms in an attempt to profit from past successes and reputation. At these moments urban famers must decide if they were just surviving an economic transition or building a new way of life and community in the city. To survive long term, people in the city must develop a sense of ownership of the gardens and build a system of sustainable rules for their protection and preservation. All of these sustainable principles were developed over long periods by the traditional peoples of Barbados and The Bahamas and elsewhere in the Caribbean and potentially will be needed for urban gardeners.

The Barbados and Bahamas cases are useful for understanding the beginnings of community gardening and exchanges, but more importantly they document the need for such local gardens and exchanges to prepare for perturbations caused by climate change, economic withdrawal, and development intrusion. The Caribbean cases have components proven to be successful for hundreds of years, but new learning methods must be developed to rapidly use old lessons and stabilize past adaptations [61,62]. There is a need to constantly assess current needs and develop adaptive responses for future threats. The key is community ties, thus any threat to natural and human relationships threatens the whole system. As said earlier, common property and common destiny are key ingredients in the elaboration of sustainable resource use practices at the community level. Clearly urban gardening communities and the coastal communities of Barbados and The Bahamas share these challenges.

Author Contributions: B.S. was the Principal Investigator for all of the Barbados research, analysis and write up. He is an equal contributor to this article. R.S. and K.V.V. worked as team members for eight years on The Bahamas research, with Stoffle as Principal Investigator. They equally share responsibility for the production of this article. All authors have read and agreed to the published version of the manuscript.

Funding: The Barbados study was funded by an American Telephone & Telegraph Inc., Community Development Grant, Community Initiative Program. Funding also derived from the Latin American and Caribbean Studies Research Grant at the University of South Florida, Tampa, Florida. Addition funding came from the Wenner-Gren Foundation, New York City, NY through a grant awarded to Trevor Purcell, faculty in the Department of Anthropology, University of South Florida. The Bahamas study was funded by a National Science Foundation grant entitled "Coupled Natural and Human Dynamics in Coral Reef Ecosystems" (or BBP—Bahamas Biocomplexity Project), which was awarded September 2001 as number OCE 0119978 and hosted in the American Museum of Natural History, and directed by Dr. Dan Brumbaugh. A portion of this grant was administered by the University of Arizona, School of Anthropology (UofA).

Acknowledgments: We would like to acknowledge the following people for sharing fishing knowledge, insights into local culture, and providing specific cases on which is based on the littoral discussion and the concept of community. These people are Lorenso Perez (Tuba) from Buen Hombre, Dominican Republic, Fred Watson from Bath, Barbados, Carlos Waldron from Bath, Barbados, Selvin Graeves (Bird) from Bath, Barbados, Guerson Martinez (Nicky) from St. Croix, USVI, and Derke Snodgrass from Tavenier, Florida Keys. The Bahamas Bahamas Biocomplexity Project Study was co-headed by J. Minnis, Chairperson, School of Social Science, at The College of the Bahamas. Her students include K. Arnett, C. Dean, T. McDonald, W. Minnis, T. Smith, and Y. Skinner. R. Stoffle headed the UofA study team. His students include Graduate Assistant A. Carroll and students C. Carroll, F. Chmara-Huff, L. Crowe, J. Dumbauld, H. Fauland, R. Gilmour, B. Goldstein, A. Haverland, C. Jones, S. Kelley, N. Lyell, A. Martinez, A. Murphy, N. O'Meara, K. Payne, T. Pierce, P. Poer, D. Post, and K. Van Vlack.

Conflicts of Interest: The authors declare no conflict of interest.

Appendix A

Table A1. Land animals of the littoral.

Land Animal Name	Scientific Name	Local Name	Location	Interaction
Land crab		Swampy, Land Crab	brackish freshwater springs before ocean	food source, bait for hand line fishing
Marsh Fiddler crab	*Uca pugnax*	Stone Crab	Shoreline	Food source
Bahamaian Nighthawk	*Choredeiles gundlachii*	Nighthawk	Shores, marshes, estuaries, grassy wetland areas	Eats insects that bother people
Blue Crab	*Callinectes sapidus*	Blue crab	Shores, marshes, estuaries, grassy wetland areas	Food source
Blue Heron	*Adrea herodias*	Arsenicker	Shoreline	Food source, prey on crabs
Crescent-eyed Pewee	*Contopus caribaeus*	Pewees	Mangroves, edges of clearings	Eats insects that bother people
Green Heron	*Butorides virescens*	Gaulin birds, Poor Joe	Shoreline	Eat crabs that would come into gardens and eat crops, stories
Hermit Crab	*Paguristes* ssp.	Hermit crab, Solider crab	Reefs, shallows, sand patches	Food source
Killdeer	*Charadrius vociferus*	Killdeer	Shores, marshes, estuaries, grassy wetland areas	National symbol
Laughing Gull	*Larus atricilla*	Sea gulls	Salt marshes, lagoons	Indicates schools of fish
Least Tern	*Sterna antillarum*	Gulls	Shoreline	Fed and cared for by people

Table A1. *Cont.*

Land Animal Name	Scientific Name	Local Name	Location	Interaction
Mangrove crab	*Cardisoma guanhumii*	Land Crab, Cigga	Mangroves	Put nicker bean in crab hole to prevent them from destroying garden, used for crawfish bait
Osprey	*Pandion haliaetus*	Fish Hawk	Nest near the ocean	Eat eggs, story of climbing to the nest of the hawk to get eggs
Pigeon	*Columba leucocephala*	White Crown Pigeon	nests in mangroves	Food source
West Indian Rock Iguana	*Cyclura* spp.	Iguanas	Brush, lagoon areas	Tourist attraction, food source
Zenaida dove	*Zenaida aurita*	Wood dove	low lands	Food sources, hunted and eggs eaten

Table A2. Sea animals of the littoral.

Sea Animal Name	Scientific Name	Local Name	Location	Interaction
Bonefish	*Albula vulpes*	Bonefish	Shallow flats near mangroves	Tourist attraction, food source
Caribbean Reef Octopus	*Octopus brareus*	Octopus, sea cat	In shore reefs	Food source
Caribbean Spiny Lobster	*Panulirus argus*	Lobster, crawfish	Reefs, caves, holes, ledges	Food source, not caught during spawning, spawning crawfish thrown back
Chiton	*Polyplacophora* ssp.	Curb	Rocky shores	Food Source
Chub	*Kyphosus sectatrix*	chub	Sea grass beds	Food Source
Conger Eels, Garden Eel	*Nystactichtys halis*	conga eel	shallows	Food source
French Grunt	*Pomadasyidae*	Grunt	Near reefs, mangroves	Food source
Giant Brain Coral	*Colpophyllia natans*	Coral	Reefs	Made into cement for houses
Great Barracuda	*Sphyraena barracuda*	Barracuda	All, especially reefs	Food source
Green Eel, Green Moray	*Gymnothorax funebris*	green eel	In shore reefs	Food source
Hawksbill Turtle	*Eretmochelys imbricata*	Turtle	Shallow, coastal waters and estuaries	Food source, shells were sold to make jewelry
Jolthead Porgy	*Calamus bajonado*	Porgy	Reefs, sand, coastal interface	Food source
Lemon Shark	*Negaprion brevirostis*	Lemon Shark	Lagoons, estuaries and the shallows	Food source, helps fight cancer, skin used for fertilizer
Mangrove Snapper	*Lutjanus griseus*	Snapper	Near mangroves	Food source
Nassau Grouper	*Epinephelius striatus*	Grouper	Shallow to mid-range reefs	Food source, grouper is not caught during spawning

Table A2. *Cont.*

Sea Animal Name	Scientific Name	Local Name	Location	Interaction
Nurse shark	*Ginglymostoma cirratum*	Nurse Shark	All (shallow and deeper)	Skin used to fertilize gardens
Parrotfish	*Sparisoma viride*	queen parrotfish	Reefs, sea grass beds	Food source
Crevalle	*Caranx hippos*	jack crevalle, rainbow crevalle	located in shallows as well as offshore	Food Source
Queen Conch	*Strombus gigas*	Conch	Sand and eel grass beds	Food source
Queen Triggerfish	*Balistes* ssp.	Triggerfish	Reef tops	Food source
Reticulated Sea Star	*Oreaster reticulatus*	Starfish	Eel grass beds	Tourist attraction
Sea Sponge (1.) Yellow Tube Sponge (2.) Red Cup Sponge	*(1.) Aplysina fistularis (2.) Mycale Laxissima*	sponges		Cleaning, songs are made about going sponging, today spongers uses knives, so sponges grow back
Sea Urchins	*Tripneusts ventricosus*	Sea eggs	Sea grass beds	Food source, shells sold, shells ground and burned, ground into lime to build houses, black ones used for bate for Jacks

Table A3. Plants of the littoral.

Plant Name	Scientific Name	Local Name	Location	Interaction
	Ambrosia hispida	Bay Tansy, Baygereen	Beaches, dune sands or occasionally on rocky shelves along coast	Medicinal properties
Black mangrove	*Avicennia germinans*	Black Buttonwood	Mangrove lagoons and along tidal shore	Medicinal properties
Sea Ox-eye	*Borrichia arborescens*		Coastal sands and rock and margins of brackish water	Medicinal properties
Nicker bean	*Caesalpina bonduc*	Nickers	Native to seacoasts	Children use the seeds as marbles and playing pieces in Wari, a traditional African game widely played in the West Indies, medicinal properties
Seven-year apple	*Casasia clusiifolia*		Coastal rocks but also in coppices	Edible wild plant
	Cassytha filiformis	Love vine	Parasitic on various herbaceous and woody plants	Medicinal properties

Table A3. *Cont.*

Plant Name	Scientific Name	Local Name	Location	Interaction
Cocoplum	*Chrysobalanus icaco*		Coastal swamps and thickets along sea beaches	Edible wild plant
Sea grape	*Coccoloba uvifera*		Coastal thickets and rock outcrops	Edible wild plant
Silver thatch	*Coccothrinax argentata*	Silver Top	In coastal flats along beaches	Plaiting and making straw crafts, thatch material for roofs
Coconut palm	*Cocos nucifera*		Thrives in the low tropics, especially in coastal sands	Food, boat building wood, aesthetic qualities, medicinal properties, plaiting and making straw crafts
Button wood	*Conocarpus erectus*		Coastal mud, savannas and edge of salines	Boat building wood, source of driftwood used for decoration in homes
Lignum vitea	*Guaiacum sanctum*		Rocky slopes and ridges, seaside ledges, palm-shrub associations, and dense coppices	National tree of Bahamas, medicinal properties, boat building wood
Pigeon berry	*Guapira longifolia*		Coppices, scrublands, and on rock flats, often along the coast and on ridges	Recognized as a main food source for wild pigeons. Used during the hunting season to find pigeons
Horse Bush	*Gundlachia corymbosa*		Clayey or rocky saline flats, marshes, dune sands, pinelands, edge of coppices	Medicinal properties
Log wood	*Heamatoxylum campechianum*		Coastal thickets, hillsides and on edge of salinas and periodically flooded places	Medicinal properties
Wild Dilly	*Manilkara bahamensis*		Coppices or scrublands, especially along coastal areas and on rock flats	Edible wild plant, medicinal properties, fruit chewed as chewing gum

Appendix B Urban Gardening

Appendix B.1 Cuba Urban Gardening

The USSR withdrawal from Cuba began about 1990 largely leaving the country on its own for the production of food. Given the crisis Cubans began to clear areas near their homes and plant food [63]. While initially individual efforts soon both the people and the government worked together to feed themselves. Some adaptations were organized into a movement termed Organoponicos, which placed individual gardens into a system using low-level concrete walls filled with organic soils and watered with a drip system. Through time gardeners self-organized and added government knowledge of pest management, alternative forms of fertilization, and crop rotation to make a more effective and sustainable human and natural system.

Appendix B.2 Detroit, Urban Gardens

The collapse of the car industry combined with the urban riots in the 1960 placed the residents of Detroit in a near starvation situation. As hundreds of ruined buildings were removed and a million people left, soil became available where only pavement and bricks were before. People turned to faming to survive and to feed others [64]. Today 23,000 residents participate in urban gardens as even more buildings are removed to eliminate blight and open earthen spaces. Like Cuba, Detroit gardening began as a response to a crumbling economy and eventually the gardeners themselves and the city combined to organize a more efficient system.

Appendix B.3 New York City Urban Gardening

In New York City (NYC), urban agriculture has become integrated throughout the five boroughs and individual neighborhood [65]. Currently, NYC has more than 700 urban agricultural sites with the Department of Parks and Recreation and the Housing Authority running the two largest community gardening programs in the country. These departments oversee with more than 1000 gardens throughout the five boroughs, most located on public land. The NYC Department of Education and the nonprofit GrowNYC support 300 school gardens. Out of the 300 gardens, 117 grow food for a farm-to-cafeteria program in over 50 schools. The food products are used in making healthy lunches for many NYC school children [65].

Even though urban gardening and agriculture span all demographic and geographic categories of people, the city's farms and gardens are clustered in places that were hardest hit by decades of disinvestment, i.e., places abandoned. Residents in these neighborhoods faced a number social and economic challenges such as limited access to healthy food options, underperforming schools, poor health, high unemployment rates, and twice as many vacant lots on average than in the city's wealthy neighborhoods. Urban agriculture gives people a way to address some of their neighborhoods' pressing needs.

References

1. Stoffle, R. When Fish Is Water: Food Security and Fish in a Coastal Community in the Dominican Republic. In *Understanding the Cultures of Fishing Communities: A Key to Fisheries Management and Food Security*; FAO Fisheries Technical Paper: Rome, Italy, 2001; pp. 219–245.
2. Stoffle, R.; Halmo, D.; Wagner, T.; Luczkovich, J. Reefs from Space: Satellite Imagery, Marine Ecology, and Ethnography in the Dominican Republic. *Hum. Ecol.* **1994**, *22*, 355–378. [CrossRef]
3. Stoffle, R. *Caribbean Fisherman Farmers: A Social Impact Assessment of Smithsonian King Crab Mariculture*; Institute for Social Research: Ann Arbor, MI, USA, 1986.
4. Stoffle, B. *The Socio-Economic Importance of Fishing in St. Thomas, USVI: An Examination of Fishing Community Designation*; Technical Memorandum NMFS-SEFSC-623; NOAA: Miami, FL, USA, 2011.
5. Stoffle, B.; Waters, S.; Abbott-Jamieson, S.; Kelley, S.; Grasso, D.; Freibaum, J.; Koestner, S.; O'Meara, N.; Davis, S.; Stekedee, M.; et al. *Can an Island Be a Fishing Community: An Examination of St. Croix and Its Fisheries*; U.S. Department of Commerce, National Oceanic and Atmospheric Administration National Marine Fisheries Service Southeast: Miami, FL, USA, 2009.
6. Williams, E. *Capitalism and Slavery*; University of North Carolina Press: Chapel Hill, NC, USA, 1994.
7. Browne, K. *Creole Economics: Caribbean Cunning under the French Flag*; University of Texas Press: Austin, TX, USA, 2004.
8. Berkes, F.; Turner, N. Knowledge, Learning and the Evolution of Conservation Practice for Social-Ecological System Resilience. *Hum. Ecol.* **2006**, *34*, 479–494. [CrossRef]

9. Stoffle, R.; Toupal, R.; Zedeno, N. Landscape, Nature, and Culture: A Diachronic Model of Human-Nature Adaptations. In *Nature across Cultures: Views of Nature and the Environment in Non-Western Cultures*; Selin, H., Ed.; Kluwer Publishers: Dordrecht, The Netherlands, 2003; pp. 97–114.
10. Olsson, P.; Folke, C. Local Ecological Knowledge and Institutional Dynamics for Ecosystem Management: A Study of Lake Racken Watershed, Sweden. *Ecosystems* **2001**, *4*, 85–104. [CrossRef]
11. Stoffle, R.; Minnis, J. Resilience at Risk: Epistemological and Social Construction Barriers to Risk Communication. *J. Risk Res.* **2008**, *11*, 55–68. [CrossRef]
12. Connell, J.H. Diversity in Tropical Rain Forests and Coral Reefs. *Science* **1978**, *1999*, 1302–1310. [CrossRef] [PubMed]
13. Turner, N.; Davidson-Hunt, I.; O'Flaherty, M. Living on the Edge: Ecological and Cultural Edges as Sources of Diversity for Social-Ecological Resilience. *Hum. Ecol.* **2003**, *31*, 439–461. [CrossRef]
14. Nabhan, G. *Enduring Seeds: Native American Agriculture and Wild Plant. Conservation*; North Point Press: San Francisco, CA, USA, 1989.
15. Wandsnider, L.; Chung, Y. Islands of Geophytes: Sego Lily and Wild Onion Patch Density in a High Plains Habitat. In *Islands on the Plains: Ecological, Social, and Ritual Use of Landscapes*; Kornfield, M., Osborn, A., Eds.; University of Utah Press: Salt Lake City, UT, USA, 2003; pp. 220–244.
16. Anderson, R. Dating Reindeer Pastoralism in Lapland. *Ethnohistory* **1958**, *5*, 361–391. [CrossRef]
17. Fowler, C.; Minnis, P.; Elisens, W. We Live by Them: Native Knowledge of Biodiversity in the Great Basin of Western North America. In *Biodiversity & Native America*; University of Oklahoma Press: Norman, OK, USA, 2000.
18. Atran, S.; Medin, D.; Ross, N.; Lynch, E.; Vapnarsky, V.; Ucan Ek', E.; Coley, J.; Timura, J.; Baran, M. Folkecology, Cultural Epidemiology, and the Spirit of the Commons: A Garden Experiment in the Maya Lowlands, 1991–2001. *Curr. Anthropol.* **2002**, *43*, 421–450. [CrossRef]
19. Turner, N. We Give Them Seaweed: Social Economic Exchange and Resilience in Northwestern North America. *Indian J. Tradit. Knowl.* **2016**, *15*, 5–15.
20. Berkes, F.; Colding, J.; Folke, C. (Eds.) *Navigating Social—Ecological Systems: Building Resilience of Complexity and Change*; Cambridge University Press: New York, NY, USA, 2003.
21. Holling, C.S. Resilience and Stability of Ecological Systems. *Annu. Rev. Ecol. Syst.* **1973**, *4*, 1–23. [CrossRef]
22. Comitas, L. Occupational Multiplicity in Rural Jamaica. In *Proceedings from the American Ethnological Society*; American Ethnological Society: Washington, DC, USA, 1964.
23. McCay, B. Presidential Address Part II: Edges, Fields, and Regions. *Common Prop. Resour. Dig.* **2000**, *54*, 6–8.
24. Stoffle, B.; Stoffle, R. At the Sea's Edge: Elders and Children in the Littorals of Barbados and the Bahamas. *Hum. Ecol.* **2007**, *35*, 547–558. [CrossRef]
25. Besson, J.; Momsen, J. *Land and Development in the Caribbean*; MacMillan: Centre for Caribbean Studies, University of Warwick: London, UK, 1987.
26. Barrow, C. Small Farm Food Production & Gender in Barbados. In *Women and Change in the Caribbean: A Pan-Caribbean Perspectiv*; Momsen, J., Ed.; Indiana University Press: Bloomington, IA, USA, 1993; pp. 181–193.
27. Mokyevsky, O. Geographical Zonation of Marine Types. *Limnol. Oceanogr.* **1960**, *5*, 389–396. [CrossRef]
28. Stoffle, R.; Minnis, J. Marine Protected Areas and the Coral Reefs of the Traditional Settlement in the Exumas, Bahamas. *Coral Reefs* **2007**, *26*, 1023–1032. [CrossRef]
29. Stoffle, R.; Minnis, J.; Murphy, A.; Van Vlack, K.; O'Meara, N.; Smith, T.; McDonald, T. Two-MPA Model for Siting a Marine Protected Area: Bahamas Case. *Coast. Manag.* **2010**, *38*, 501–517. [CrossRef]
30. Unviersity College of London. Legacies of British Slave-Owernship. Available online: https://www.ucl.ac.uk/lbs/estates/ (accessed on 5 April 2020).
31. Tashakkori, A.; Teddlie, C. *Mixed Methodology: Combining Qualitative and Quantitative Approaches*; Sage Publications: Thousand Oaks, CA, USA, 1998.
32. Beebe, J. *Rapid Assessment Process*; Altamira Press: Walnut Creek, CA, USA, 2001.
33. Campbell, D.; Fiske, D. Convergent and Discriminant Validation by the Multi-Trait-Multimethod Matrix. *Psychol. Bull.* **1959**, *56*, 81–105. [CrossRef] [PubMed]
34. Jick, T. Mixing Qualitative and Quantitative Methods: Triangulation in Action. *Adm. Sci. Q.* **1979**, *24*, 601–611. [CrossRef]

35. Gmelch, G.; Gmelch, S. *The Parish behind God's Back: The Changing Culture of Rural Barbados*; Waveland Press: Long Grove, IL, USA, 1997.
36. Stoffle, B. We Don't Put All of Our Eggs in One Basket: An Examination of Meeting Turn, A Rotating Savings and Credit Association in Barbados. Ph.D. Thesis, University of South Florida, Tampa, FL, USA, 2001.
37. Stoffle, B.; Stoffle, R.; Minnis, J.; Van Vlack, K. Women's Power and Community Resilience Rotating Savings and Credit Associations in Barbados and the Bahamas. *Caribbean Stud.* **2014**, *42*, 45–69. [CrossRef]
38. Stoffle, R.; Zedeno, N. Historical Memory and Ethnographic Perspectives on Southern Paiute Homeland. *J. Calif. Gt. Basin Anthropol.* **2001**, *23*, 229–248.
39. Stoffle, R.; Van Vlack, K.; O'Meara, N. Environmental Multiplicity in the Exumas, Bahamas: A Social and Natural Resilience Model. *Appl. Anthropol.* **2017**, *37*, 15–21.
40. Handler, J. The Amerindian Slave Population of Barbados in 17th and Early 18th Centuries. *Caribbean Stud.* **1969**, *8*, 38–64.
41. Handler, J. Aspects of Amerindian Ethnography in 17th Century Barbados. *Caribbean Stud.* **1970**, *9*, 50–72.
42. Watts, D. *The West. Indies: Patterns of Development, Culture, and Environmental Change Since 1492*; Cambridge University Press: Cambridge, UK, 1987.
43. Ligon, R. *A True and Exact History of the Island of Barbados*, 5th ed.; Smith, D.C., Ed.; Wilfrid Laurier University: London, UK, 1657.
44. Bennett, J.H. The Problem of Slave Labor Supply at the Codrington Plantations. *J. Negro Hist.* **1951**, *36*, 406–441. [CrossRef]
45. McKinnen, D. *A Tour through the British West. Indians in the Year 1802 and 1803*; Forgotten Books: London, UK, 1804.
46. Hawthorne, N. *The Yarn of a Yankee Privateer*; Funk & Wagnalls Company: New York, NY, USA, 1926.
47. Yates, A.W. *Bygone Barbados*; Black Bird Studies: Christ Church, Barbados, 1998.
48. Barbados Today. Bajan Gems- Fred Watson. Available online: https://www.youtube.com/watch?v=7_Kdupmmba4 (accessed on 4 April 2020).
49. Craton, M.; Saunders, G. *Islanders in the Stream: A History of the Bahamian People. Volume Two: From the Ending of Slavery to the Twenty-First Century*; University of Georgia Press: Athens, GA, USA, 1999.
50. Keegan, W.; Mitchell, S. Possible Allochthonous Lucayan Arawak Artifact Distributions, Bahama Islands. *J. Field Archaeol.* **1986**, *13*, 255–258.
51. Saunders, G. *Slavery in the Bahamas: 1648–1838*; Nassau Guardian: Nassau, Bahamas, 1995.
52. Wilkie, L.; Farnsworth, P. *Sampling Many Pots: An Archaeology of Memory and Tradition at a Bahamian Plantation*; University Press of Floria: Gainesville, FL, USA, 2005.
53. Johnson, H. *The Bahamas from Slavery to Servitude, 1783–1933*; University Press of Floria: Tallahassee, FL, USA, 1996.
54. Farnsworth, P. From the Past to the Present. In *African Sites: Archaeology in the Caribbeam*; Markus Wiener: Princeton, NJ, USA, 1999; pp. 94–130.
55. Woods, C.; Sergile, F. *Biogeography of the West Indies*; CRC Press: Boca Raton, FL, USA, 2001.
56. Pimm, S. *The Balance of Nature? Ecological Issues in the Conservation of Species and Communities*; University of Chicago Press: Chicago, IL, USA, 1991.
57. Stoffle, B.; Halmo, D.; Stoffle, R.; Burpee, C. Folk Management and Conservation Ethics among Small-Scale Fishermen of Buren Hombre, Dominican Republic. In *Folk Management in the World's Fisheries*; Dyer, C., McGoodwin, J., Eds.; University of Colorado Press: Boulder, CO, USA, 1999; pp. 115–138.
58. Olwig, K.F. *Cultural Adaptation and Resistance on St. John: Three Centuries of Afro-Caribbean Life*; University Press of Floria: Gainesville, FL, USA, 1985.
59. Stoffle, R.; Stoffle, B.; Sjölander-Lindqvist, A. Contested Time Horizons. In *Sustainability Assessment Pluralism, Practice and Progress*; Bond, A., Morrison-Saunders, A., Howitt, R., Eds.; Routledge: London, UK, 2013; pp. 51–67.
60. Olwig, K.F. National Parks. Tourism and Local Development: A West Indian Case. *Hum. Organ.* **1980**, *39*, 22–31. [CrossRef]
61. Chen, W.; Ho, T. The Application of Yantian Cultural Resources in Design Education-Taking the Yantian Community in Tainan as an Example. *Sustainability* **2020**, *12*, 2260. [CrossRef]

62. Strus, M.; Kalisiak-Medelska, M.; Nadolny, M.; Kachniarz, M.; Raftowicz, M. Community-Supported Agriculture as a Perspective Model for the Development of Small Agricultural Holding in the Region. *Sustainability* **2020**, *12*, 2656. [CrossRef]
63. Ellinger, M.; Braley, S. Urban Agriculture in Cuba. Race, Poverty & the Environment. *Urban Agric. Cuba Race Poverty Environ.* **2010**, *17*, 14–17.
64. Jung, Y.; Newman, A. An Edible Moral Economy in the Motor City: Food Politics and Urban Governance in Detroit. *Gastronomica* **2014**, *14*, 23–32. [CrossRef]
65. Cohen, N.; Reynolds, K.; Sanghwi, R. *Five Borough Farm: Seeding the Future of Urban Argiculture in New York City*; Design Trust for Public Spaces: New York, NY, USA, 2012.

Article

Optimizing the Environmental Profile of Fresh-Cut Produce: Life Cycle Assessment of Novel Decontamination and Sanitation Techniques

Miguel Vigil *, Maria Pedrosa Laza, Henar Moran-Palacios and JV Alvarez Cabal

Área de Proyectos de Ingenieria, Departamento de Explotación y Prospección de Minas, Universidad de Oviedo, Calle Independencia 13, 33004 Oviedo, Spain; maria.pedrosa@api.uniovi.es (M.P.L.); henar.moran@api.uniovi.es (H.M.-P.); valer@api.uniovi.es (J.A.C.)

* Correspondence: vigilmiguel@uniovi.es; Tel.: +34-985104272

Received: 13 March 2020; Accepted: 22 April 2020; Published: 2 May 2020

Abstract: Fresh-cut vegetables, namely those that undergo processes such as washing, sorting, or chopping while keeping their fresh state, constitute an important market element nowadays. Among those operations, the washing step becomes really important due both to the extensive use of water resources and to the utilization of controversial water sanitizing agents, such as chlorine. To ideally eliminate those chlorinated compounds while decreasing water consumption, four novel filtrating technologies (pulsed corona discharge combined with nanofiltration, NF-PCD; classical ultrafiltration, UF; nanofiltration membranes integrating silver nanoparticles, NF-AgNP; and microfiltration with cellulose acetate membranes containing chitin nanocrystals, ChCA) have been proposed to eliminate any contaminating agent in recirculated water. Here, we performed a life cycle assessment (LCA) to assess the environmental effects of introducing these new solutions and to compare those impacts with the burden derived from the current strategy. The novel technologies showed a decreased environmental burden, mainly due to the enhanced water recirculation and the subsequent decrease in energy consumption for pumping and cooling the water stream. The environmental gain would be maintained even if a certain amount of chlorine was still needed. This analysis could serve as an aid to decision-making while evaluating the introduction of new sanitizing techniques.

Keywords: fresh-cut vegetables; life cycle assessment; LCA; chlorine; filtering membranes; water recirculation

---

1. Introduction

In order to benefit from the well-known properties of fresh fruits and vegetables, many customers tend to favor the consumption of "fresh-cut" (FC) products, defined as "those fruits and vegetables that may have undergone procedures such as washing, sorting, trimming, peeling, slicing or chopping that do not affect their fresh life quality" [1]. These products play an important role in the present days, when the time allocated to cooking processes is in a clear decrease [2].

FC vegetables have shown an increased market size when compared with FC fruits, especially due to the sale of salad bags. The value of the European fresh-cut fruit and vegetable market is about 3.4 billion euros, of which salads, vegetables, and fruit account for 62%, 31%, and 7% of the market volume, respectively [3]. Regarding its production, harvested fresh vegetables typically undergo several unit operations to end up with the final FC products. These operations consist of trimming, slicing and shredding, washing, draining, weighing, and packing [4,5].

Washing unit operation is a key step in the production of FC vegetables [6]. Before packaging of shredded produce, it is necessary to remove dirt, pesticide residues, and microorganisms that may lead to quality loss along the shelf life of the final product [7]. This step is performed by immersing produce in tanks of washing water, which is partially recirculated to decrease the total cost of the operation [6].

Of crucial importance is the quality of water that is used for this operation, since water may paradoxically act as a contaminant agent when it is recirculated. The reused water is characterized by a high organic load, which provides nutrients supporting microbial growth [8]. There is, therefore, a need for using a sanitizing agent that can virtually eliminate any possible cross-contamination among water tanks [9]. The most widely used sanitizer is chlorine due to its low cost and effectiveness when eliminating contaminant bacteria through oxidation, although its utilization is controversial since its reaction byproducts have shown a carcinogenic potential [6,10,11]. Indeed, its use has been banned in several countries in the EU [2]. Besides, the organic compounds present in fresh vegetables generate a high chlorine demand, which leads to the rapid consumption of the free chlorine present in the recycled water [11]. Due to this fact, fresh solution of chlorine in water needs to be constantly added, and thus total recirculation of the water flux is not possible, which encourages extensive water usage in the process. This results both in the resource depletion and in the extensive use of energy for pumping and cooling, which is an important issue of the present society as it has been stated by the United Nations. Indeed, this organism claims in its 12th Sustainable Development Goal (SDG) the necessity for ensuring sustainable consumption and production patterns, promoting an efficient use of resources and energy [12].

In order to decrease the water consumption and production costs by increasing the recirculation rates, Fusi et al. proposed the introduction of a water filtering system in the production line of baby lamb leaves [13]. Membrane separation can be used to treat the process water before its recirculation, thus decreasing the organic particles and avoiding cross-contamination [14]. As stated by the authors of [13], the use of this technology would lead to the reduction of electricity needed for pumping water, as well as a general water saving and a further reduction in wastewater production, in line with the 12th SDG. Additionally, the introduction of filtering techniques would replace the controversial use of chlorine.

Membrane devices have become an alternative to traditional water purification processes. A membrane represents a thin physical interface that regulates the pass of certain species through it, depending on their physical and/or chemical properties [15]. Depending on the pore sizes, we can find several types of membranes that can be used for water treatment (Table 1).

Table 1. Classification of filtering membranes in terms of their pore size [15].

Membrane	Pore Size	Reject
Microfiltration (MF)	0.1–5	Large cells and bacteria, atmospheric dust
Ultrafiltration (UF)	0.01–0.1	Dissolved macromolecules and viruses
Nanofiltration (NF)	0.0001–0.01	Most organic molecules, viruses, divalent ions
Reverse osmosis (RO)	*	Low molecular weight species, aqueous inorganic solids, salts and ions

* RO membranes are so dense that they are considered as nonporous.

In the present work, we evaluated the environmental effects of introducing different membrane-based tools as water sanitizing agents in the FC industry. This work was carried out as part of the CEREAL project under the 7th Framework Program, which aimed to improve the resource efficiency throughout the postharvest chain of fresh-cut fruits and vegetables. In previous stages of the project, the consortium partners developed and/or evaluated from a technical point of view the suitability of several filtering devices for decontaminating FC produce washing waters. Their work resulted in valuable data, which were later used by the authors to evaluate the environmental impact of a hypothetical large-scale implementation of the developed techs through a life cycle assessment (LCA), with a special focus on the washing operation itself. LCA is a commonly used method to assess the environmental impact of a determined product through its whole life cycle, considering the extraction and processing of the raw materials, the manufacturing and distribution steps, the use and recycling by the consumer, and the final disposal [16]. This technique is broadly applicable to several fields, and food production is among them. It is possible to find publications performing LCA for meat [17], dairy [18,19], crops [20,21], and even edible insect [22] production industries. LCA has also been applied to the FC vegetables and fruit industry on several occasions [23–25], but none of them focused on the washing step of the process.

By means of the LCA, we were able to evaluate the potential decrease in the environmental impact of the washing step by the introduction of new sanitizing techniques. Ideally, these new tools would replace the need for chlorine as a sanitizing agent, thus avoiding both the potential issues of this compound related to human health and the costs derived from the infrastructure [26,27].

2. Materials and Methods

LCA was performed according to ISO 14040 and 14044 [28,29], taking into consideration the following stages: (1) goal and scope definition; (2) life cycle inventory analysis (LCI); (3) life cycle impact assessment (LCIA); and finally, life cycle interpretation.

2.1. Goal and Scope Definition

The global aim of the project was the assessment of the environmental profile regarding the washing of fresh-cut vegetables when introducing several membrane-based sanitizing techniques, and comparing them with the reference scenario.

Since cut and packaged lettuce dominate the market of FC vegetables, corresponding to the greatest part of the sales volume [3], the functional unit (FU) was set to 1 ton of cut lettuce to be washed. This FU provides the reference for the normalization of the LCA data and allows for the comparison between the different scenarios.

2.1.1. System Boundaries

Fresh-cut vegetable production is performed according to the steps in Figure 1. A cradle to grave approach would consider all the steps in it. However, in the LCA performed by [13], it was reported that the agricultural and processing phase contributes more than 80% to the total environmental loads in the 12 assessed impact categories, so downstream impacts could be negligible. What is more, the output of washed vegetables needs to meet the requirements for the maximal biological load independently of the sanitizing technique, so the way the washing is performed does not influence the downstream environmental loads. Similarly, we considered that the losses of produce will be equal under the use of the different technologies evaluated here, since they only affect the recirculated water and not the washed produce. Therefore, the impacts of the agricultural phase and the previous processing steps were considered equal among the five scenarios.

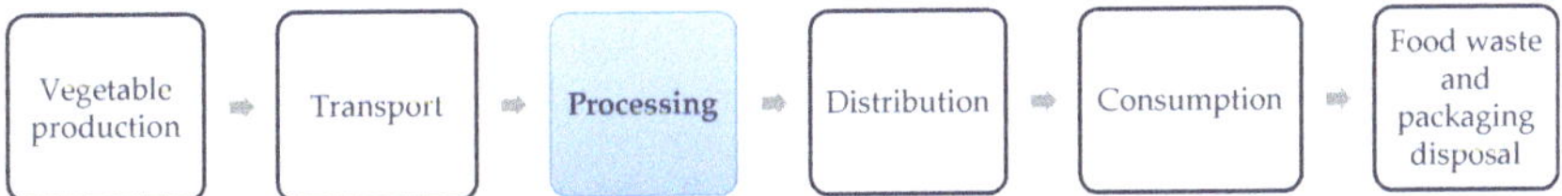

Figure 1. Cradle to grave production chain of fresh-cut (FC) vegetable production. In blue, the part of the process where this analysis focused.

Thus, we set the system boundaries in the washing phase itself, considering as the input the fresh vegetables and as the output the ready-to-pack vegetables. By limiting the model to one single production step, we avoided any potential errors due to the modeling of further operations, and thus diminished the uncertainty of our results.

2.1.2. Scenarios Definition

The current operational process for FC lettuce washing in the industry is summarized in Figure 2a [30]. This process is typically performed in three steps corresponding to three washing tanks. In this scenario, water is recirculated from the cleanest tank (the last one) to the previous one. Chlorine is added in the first and/or second tank as a sanitizing agent, while the third tank would perform a rinsing with potable water [30]. The water that is not recirculated to the previous tanks undergoes a wastewater treatment (WWT) before it is drained to the environment. The conceptualized model of this operation can be seen in Figure 2b. This was our reference scenario.

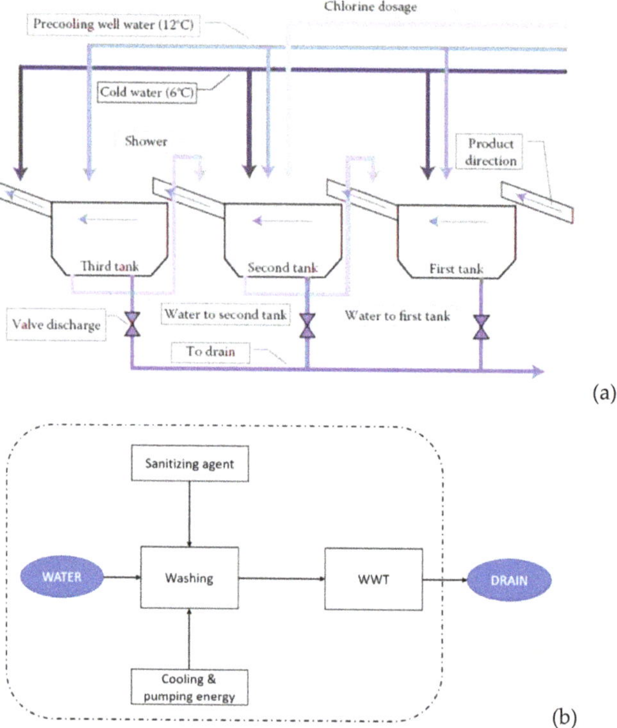

Figure 2. Description of how the washing step of the FC lettuce production is performed within the reference scenario. (**a**) Washing tanks structure, where FC produce enters by the right side of the picture and passes through the different washing tanks [31]. Water flow is schematically represented in (**b**), where the discontinuous line sets the boundaries of our scenario. The inputs of the system are freshwater, cooling and pumping energy, and the sanitizing agent, chlorine. The main output is wastewater, which is processed in the wastewater treatment (WWT) plant before it is returned to the technosphere. Another output of the process (not shown in the picture) is the unreacted chlorine.

In order to enhance water recirculation decreasing its consumption by 50% and to eliminate the chlorine as a sanitizing agent, the project consortium partners proposed and analyzed four membrane-based methods that can be used to remove part of the organic load within the reused water:

- A hybrid depuration system, based upon the utilization of ozone gas combined with inorganic filtering membranes. In this scenario, the oxidizing role of chlorine is replaced by ozone. The main drawback when using ozone as an oxidizing agent is its high cost, which can be diminished by ozonation using pulsed corona discharge (PCD), though it shows enhanced energy efficiency when compared with other methods [31,32]. As an active species, ozone oxidizes the organic compound present in water, but to a lesser extent than chlorine [33]. That is why ozonation has been combined with nanofiltration (NF) membranes in several studies [31,34,35]—to prevent membrane fouling by degrading the organic matter. The combination of these technologies as a means of water purification in the washing step of FC lettuce production has been proposed in [36] by a member of the CEREAL project consortium.

- Standard ultrafiltration (UF) membranes. When it comes to alternative water treatment processes, UF is one of the most widely used [37,38]. This technology has the ability to remove colloids, particles, bacteria, and viruses from water [39]. However, the major drawback of UF systems in a large-scale application is membrane fouling [40], which is treated through backwashes—pumping water backwards through the filters media [41]. The use of filtering membranes alone for the treatment of FC washing water has also been previously reported [42].
- Microfiltration (MF) membranes made of cellulose acetate (CA) and chitin nanocrystals. CA-based membranes are extensively used in industrial-scale applications since they are derived from an abundant natural polymer such as cellulose. However, they show poor mechanical strength and chemical and thermal stability [43]. Thus, this material needs to be reinforced in order to meet the requirements for its actual utilization. Chitin nanocrystals (ChNC) can be used for this aim. They are macromolecules that act as structural polymers in the exoskeleton of arthropods, in the cell walls of fungi and yeast, and in other microorganisms [44]. Besides their good mechanical properties, ChNC also possesses antifungal and antibacterial properties [45]. This behavior prevents the biofilm formation and the subsequent fouling of the membrane, providing a successful means of water filtering [45,46].
- Nanofiltration using ceramic membranes coated with biocide silver nanoparticles (AgNP). The use of fine-pore membranes is combined with silver, which has long been known to exhibit good antibacterial ability for a considerable range of microorganisms, and thus AgNP are commercialized as antimicrobial agents [47,48]. This combination is able to successfully treat water under an acceptable flux rate with excellent bacterial losses [49].

From this point to the end of the document, the four technologies will be denoted as NF-PCD (nanofiltration-pulsed corona display), UF (ultrafiltration), ChCA (chitin-cellulose acetate) and NF-AgNP (nanofiltration-silver nanoparticles), respectively.

By using these technologies, the conceptualized scenario changes, as described in Figure 3. The elimination of the sanitizing agent would also imply the displacement of the associated infrastructure. What is more, an enhanced water recirculation would decrease the energy consumption for cooling and pumping the freshwater input, while it would also decrease the stream to be treated by the WWT plant.

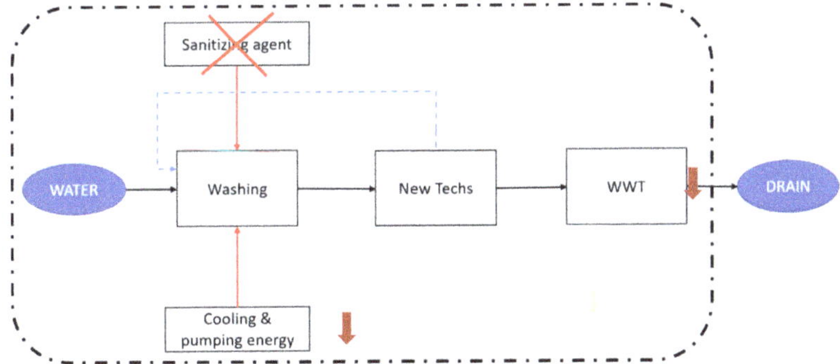

Figure 3. Schematic representation of the water flow in the new proposed scenarios. Water that exits the washing tanks is treated by means of the new technologies and recirculated to the tanks. Therefore, the flux to be processed by the WWT plant is decreased, as well as the cooling and pumping energy needs. Properly designed filtering techniques would dismiss the need for using a chlorine sanitizing agent. The discontinuous lines set the system boundaries, as in Figure 2b.

2.2. Life Cycle Inventory

Briefly, the data concerning the different washing elements across the five scenarios were obtained from three different sources:

- Primary data, which were kindly supplied by the different partners in the CEREAL project consortium. As it was stated in the introductory part, these data resulted from previous stages of this same project, where the consortium developed and/or evaluated at lab-scale the technologies here assessed, reaching conclusions such as the expected water saving and electricity usage.
- Secondary data retrieved from background databases. In this work, we used the Ecoinvent v3.2 database to gather the remaining missing data and to model the lacking processes [50]. This is a widely used database in the framework of LCA due to its three main strengths: the data's reliability, transparency, and the independence of the host institutions [51].
- Secondary data collected from a profound literature search. Fortunately, data concerning the manufacturing of the filtering devices had been previously reported and were here used for elaborating the inventory.

Ecoinvent unit processes were preferentially used for systems modeling. When any unit process was missing from that database, bespoke ones were compiled from scientific references. Processes were designed including the same factors and assumptions as of the equivalent Ecoinvent ones in order to ensure consistency across the whole LCI. Within the reference system, the inventory takes into account three fundamental items: (1) water intake from the general supply network; (2) chlorination infrastructure, considering both the sanitizing tanks and the purchase of the chemical compound; (3) energy supplies needed for water pumping and cooling, considering the Spanish electricity mixture. As it was previously stated, the output elements of the process are, on the one hand, the wastewaters that need to be treated and, on the other hand, the unreacted chlorine.

For the remaining evaluated scenarios, the application of the different filtering techniques eliminates the need for sanitizing chlorine. Thus, all the derived infrastructure is removed from the following inventories.

With regards to the NF-PCD membrane, the materials and energy intakes needed for the pulsed corona display (PCD) device construction were extracted from [36], and its disposal was based upon the recycling of the steel utilized for this purpose. This technology was combined with a standard nanofiltration membrane. Due to the lack of primary data and/or literature information, we modeled the production of the membrane using data from a reverse osmosis device recorded on Ecoinvent database (FILMTEC^{TM} SW30HR-380). The aforementioned process documented in Ecoinvent did not consider the device disposal, so we assumed that the ceramic NF membrane was disposed of in an inorganic residue landfill.

For the inventory regarding the NF-AgNP scenario, we lacked once again primary information in terms of membrane manufacturing. Thus, the filtering membrane serving as the basis of the device was the same Ecoinvent reverse osmosis standard as in the NF-PCD scenario, as well as its disposal. This membrane was coated with silver nanoparticles (AgNP), whose production was modeled based on the report in [52].

As for the ultrafiltration membrane system, the inventory relied on an already modeled system in Ecoinvent, which considered the production, utilization, and disposal of the membrane. Thus, the only output that needed to be taken into account was the wastewater that was subsequently treated.

Finally, the cellulose acetate (CA) membrane production was modeled basing on [53]. The process started with the pretreatment of the Kraft cellulose paste from corn starch, which subsequently underwent an acetylation or esterification step with acetic anhydride. This resulted in cellulose tri- or diacetate in the form of fine powder or flakes. The manufacturing of the final membrane from the cellulose acetate was also modeled thanks to the data in [53]. The membrane was finally coated with chitin bactericide nanocrystals, corresponding to 5% of the total weight [45]. The chitin was obtained from crab shell residues. The inventory regarding this step was modeled based on the data in [54]. As a byproduct, it generated a protein paste, which was employed as animal feeding, and therefore here allocated as avoided impact.

The summarized LCI can be found in Table 2. The full quantitative information of the inventories concerning each scenario can be viewed in Supplementary Materials.

Table 2. Elements acting as inputs and outputs of the different considered scenarios, namely nanofiltration-pulsed corona display scenario (NF-PCD), nanofiltration-silver nanoparticle filtration scenario (NF-AgNP), ultrafiltration scenario (UF), and chitin-cellulose acetate membrane scenario (ChCA). The items represent the components of the LCA inventories.

	Input		Output	
	Element	**Data Source**	**Element**	**Data Source**
Reference scenario	Sodium hypochlorite	[13]	Unreacted sodium hypochlorite	Author estimation
	Chlorination infrastructure	[55]	Wastewater treatment	Ecoinvent
	Washing water	Primary data		
	Energy consumption	[56]		
NF-PCD	PCD device	[36]	PCD device recycling	[37]
	NF membrane	Ecoinvent	Membrane disposal	Ecoinvent
	Washing water	Primary data	Wastewater treatment	Ecoinvent
	Energy consumption	[36,56]; Primary data		
NF-AgNP	NF membrane	[36]	Membrane disposal	Ecoinvent
	AgNP	[52]	Wastewater treatment	Ecoinvent
	Washing water	Primary data		
	Energy consumption	[56]; Primary data		
UF	UF membrane	[36]	Wastewater treatment	Ecoinvent
	Washing water	Primary data		
	Energy consumption	[56]; Primary data		
ChCA	CA membrane	[45,53]	Membrane disposal	Ecoinvent
	Chitin	[54]	Wastewater treatment	Ecoinvent
	Washing water	Primary data		
	Energy consumption	[56]; Primary data		

2.3. Life Cycle Impact Assessment

For modeling the life cycles within the different scenarios, we used SimaPro v8. In order to evaluate the environmental impacts, we used the ReCiPe method, whose primary objective is to transform the list of life cycle inventory results into a limited number of impact indicator scores, categorized in 18 midpoint indicators and 3 endpoint indicators [57]. Here, we assessed 12 ReCiPe midpoint indicators (same as [13]): climate change, ozone depletion, terrestrial acidification, freshwater eutrophication and ecotoxicity, marine eutrophication and ecotoxicity, human toxicity, photochemical oxidant formation, terrestrial ecotoxicity, and water and fossil depletion. We also considered the three endpoint indicators: damage to human health, damage to ecosystems, and damage to resource availability. The main strength of this methodology is that it ensures that the different impacts are not assessed more than once in different indicators, and thus ReCiPe scores are extensively used in the life cycle impact assessments [13,58–61].

2.4. Sensitivity Analysis

The introduction of these new techniques in industrial-scale applications is expected to create a water recirculation rate increased by 50% with respect to our reference scenario. However, this value might not be reached, since the conditions under which the washing operation is performed in the factory substantially differ from their application in laboratory conditions. In order to be able to properly assess if these membranes could decrease the environmental loads on a larger scale, we performed a sensitivity analysis. We evaluated whether the four new technologies would maintain a significant environmental gain in the case that the maximum water savings were 20%, 30%, or 40% of the consumption within the reference scenario.

On the other hand, we considered that in larger scales the use of a determined concentration of chlorine might still be needed in order to limit membrane fouling and maintain the standards of produce quality. To explore this possibility, another sensitivity analysis was performed concerning the chlorine addition in the proposed scenarios. We evaluated the environmental gain when the concentration of chlorine in the washing water was 100%, 50%, and 20% of the concentration used in our reference scenario. It should be noted that although in the first case the concentration of chlorine was the same as in our reference, the enhanced recirculation rate was maintained, so that a lesser freshwater input flux needed to be treated with the sanitizing agent. As a result, the net consumption of chlorine decreased.

Both analyses were performed focusing on the single score endpoint indicator, defined as the sum of the three ReCiPe endpoint impact scores. Besides, in the two analyses, we considered as significant a decrease in the single score indicator of 20%, compared to the reference scenario.

3. Results

3.1. Impacts Evaluation

The impact scores for each of the 12 ReCiPe midpoint categories calculated according to our reference scenario are presented in Figure 4. In almost all the considered categories, major impacts were due to the energy consumption, followed by the chlorination process and wastewater treatment as the principal environmentally damaging components.

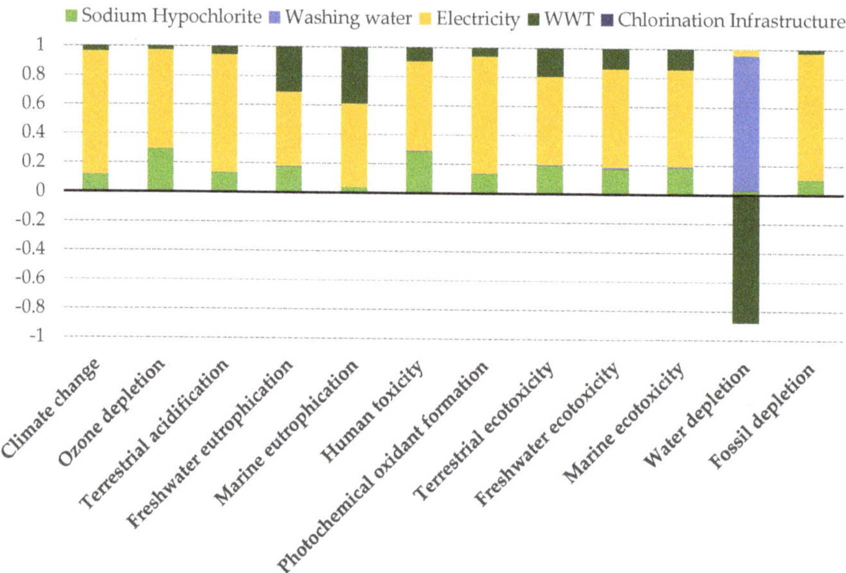

Figure 4. Relative contribution of the different components of the reference scenario to each midpoint impact indicator.

It is worth noting the negative contribution of WWT to the water depletion impact score. This is due to the fact that once the water is treated in the WWT plant it is returned to the technosphere, and thus the net water consumption is diminished.

The overall impact of the different elements in the reference scenario were evaluated according to the endpoint indicators, and the final result is presented in Figure 5. The electricity consumption is responsible for more than 80% of the total environmental burden.

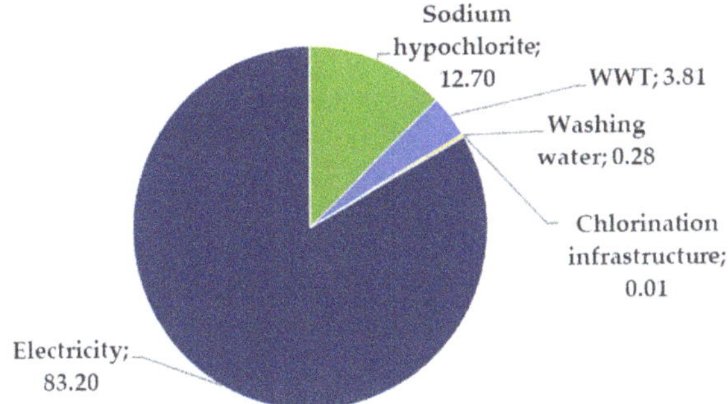

Figure 5. Contribution (in %) of each of the components of our reference scenario to the single score endpoint indicator. This single score is conceptualized as the sum of the three endpoint indicators stated in Section 2.3.

When introducing the different filtering techniques in the system, the impacts of 11 out of 12 midpoint categories decreased compared to our reference, independently of the considered scenario (Figure 6). Only NF scenarios showed an increased impact in the category of ozone depletion, due to the modeled membrane production. The data related to the membrane manufacturing were extracted from Ecoinvent, as stated in the inventory, and included the formation of chlorofluorocarbons, namely CFC-113, which is a major contributor to the ozone layer depletion.

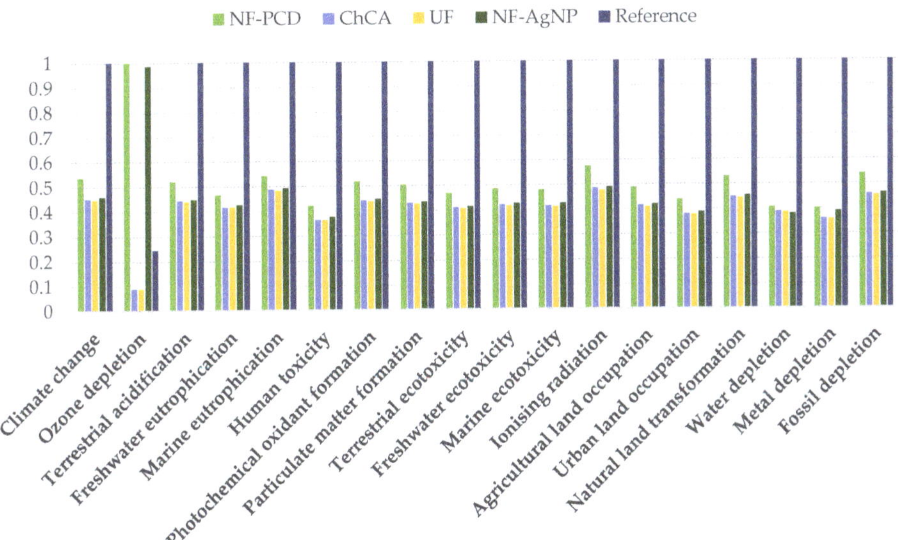

Figure 6. Midpoint impact indicators of the different scenarios within the 12 considered categories. The results are shown relative to the scenario where the indicator had its maximum value. NF-PCD: nanofiltration-pulsed corona display scenario; ChCA: chitin-cellulose acetate membrane scenario; UF: ultrafiltration scenario; NF-AgNP: nanofiltration-silver nanoparticles scenario.

In spite of it, the use of water purification membranes entailed an overall endpoint impact reduction of 55% in the case of NF-AgNP and UF scenario, whereas NF-PCD and ChCA filtration showed an impact reduction of 47% and 56%, respectively, when compared to the reference scenario.

When the endpoint impacts were allocated to the different elements of each of the scenarios, it was noticed that once again the major impacts were due to the energy consumption, followed by the wastewater treatment. The remaining elements entailed negligible impacts (Figure 7).

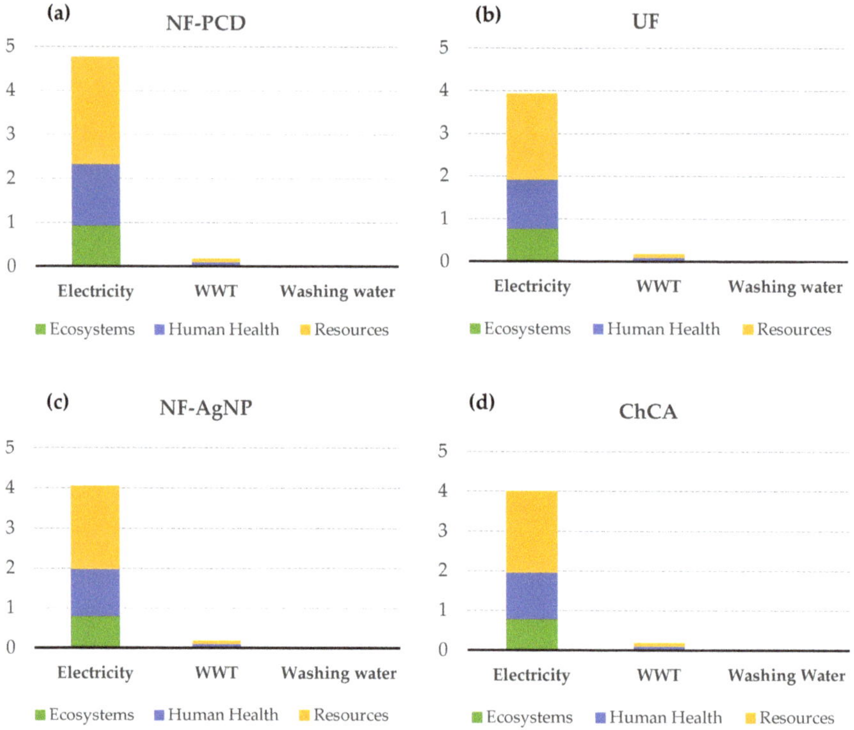

Figure 7. Absolute contribution to the different endpoint indicators of electricity consumption, WWT, and washing water usage in the case of pulsed corona display scenario (NF-PCD) (**a**), ultrafiltration scenario (UF) (**b**), silver nanoparticles scenario (NF-AgNP) (**c**) and chitin-cellulose acetate membrane scenario (ChCA) (**d**). The rest of the processes concerning membrane manufacturing, maintenance, and disposal showed negligible contributions and thus were not here represented.

3.2. *Sensitivity Analysis*

The sensitivity analysis on water consumption confirmed that the latter overall impact reduction, in terms of the ReCiPe single score endpoint indicator, was generally maintained when water recirculation rates were reduced (Figure 8). In all proposed scenarios but NF-PCD, the environmental gain was significant when compared to the reference for every considered recirculation rate, with reductions of the overall impact scores greater than 20%.

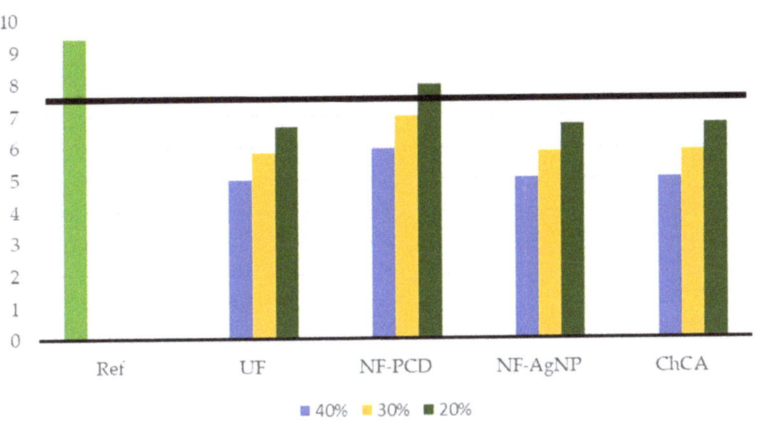

Figure 8. Sensitivity analysis on the water consumption decrease. We considered, on each of the proposed scenarios, a water consumption decrease of 40%, 30%, and 20% with respect to the reference scenario (where water saving is null). The black solid line denotes an environmental gain of 20% compared to the reference.

As for the sensitivity analysis concerning the chlorine addition, the results are shown in Figure 9. It can be noticed that the environmental gain is maintained even if the same concentration of chlorine as in the reference scenario is used to sanitize the recirculated washing water. This is due to the fact that the main contributor to the decrease in the environmental burden is once again the water saving. Even if we maintain the usage of chlorine, keeping a reduction of 50% in freshwater consumption is still nearly as beneficial as not using chlorine at all according to the ReCiPe indicators.

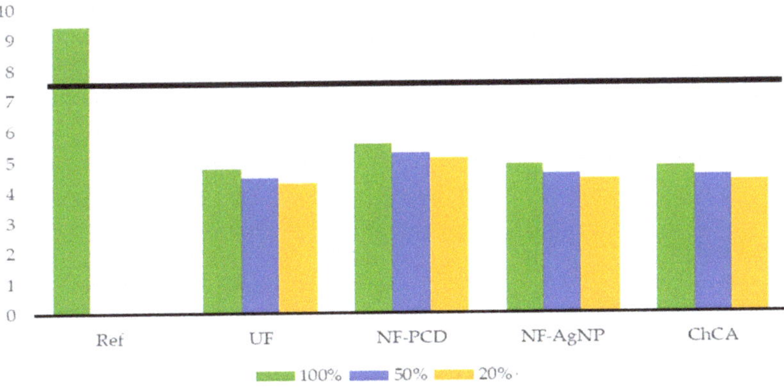

Figure 9. Result of the sensitivity analysis performed regarding the chlorine addition. It was considered here that an addition of chlorine concentrations of 100%, 50%, and 20%, referred to the added chlorine concentration on the reference scenario. The solid black line denotes our sensitivity limit of 20% reduction on the endpoint single score.

4. Discussion

The inclusion of filtering membranes as washing water sanitizing techniques appears to be clearly convenient from an environmental point of view. The major impact reduction was due to the energy savings derived from a decreased water stream to be cooled and pumped. This is consistent with the results reported by Fusi et al. [13], who reported a decrease in the environmental burden when introducing filtering techniques in the process. However, they did not evaluate the impact of the manufacturing of the filtering device. Considering the inventory here developed, we clarified that the impact due to this manufacturing and usage was negligible in the endpoint impact assessment.

This fact also validates the use of secondary data, both from articles and from the Ecoinvent database. The processes stored on databases and the data collected from articles entailed different aims, and consequently adapting those to our purposes may have led to imprecisions on the calculations of midpoint and endpoint impact indicators. However, the filtering infrastructure represented less than 0.0001% of the single-score endpoint indicator, so incorrectness in our procedures would not affect the main conclusions of the study.

The only issue related to modelization using Ecoinvent data was related to the ozone depletion indicator, which appeared to be a major issue in the nanofiltration scenarios. This could be due to the fact that Figure 6 plots the indicators relative to the scenario where their value is greatest, and the absolute value might not be enough high to entail actual harm. However, the modeled scenarios considered just an approximation of the real filtering device, since no NF system inventory was found in the secondary data retrieval. This component was replaced by a reverse osmosis device, present in the Ecoinvent database. The manufacturing of this membrane includes the use of polyester resin, a viscous liquid resin that is usually combined with fiberglass to serve as a supporting element [62], the use of which leads to CFC formation. Actual NF device manufacturing does not require the use of this compound, as it seems when analyzing the other membranes inventories, and thus the contribution to ozone depletion could be tackled. Thus, Ecoinvent databases are extremely useful but the LCA results reached with their data must be carefully analyzed.

The main reason for the reduction of the environmental burden in the four scenarios is the enhanced water recirculation, which results in decreases in the energy from water cooling and pumping. For the water being rinsed, it has been reported that 1–2 °C is the optimal water temperature for most FC products, in order to successfully remove traces of chlorine [4]. Here, we considered a cool water temperature of 6 °C, as suggested in [30]. Part of the savings on energy could be allocated to decrease even more the water temperature and enhance the final FC produce quality. It is worth pointing out that in our analysis the input water temperature was set to the average temperature in the Spanish general supply network, i.e., 20 °C. The input temperature will depend on the season of the year, on the country where the production takes place, or even on the temperature of alternative water sources, such as wells or rivers. As a result, even further energy saving could be achieved and this, together with a disminished water use, would contribute to the accomplishment of the 12th SDG proposed by the United Nations.

The possibility of saving water in the FC production chain had been previously explored in the literature. To date, water decontamination has been addressed using different physical methods (UV light [63], pulsed light [64], power ultrasound [65]), biological methods [66], or chemical methods (such as hydrogen peroxide [67] or citric acid [68]) As a drawback, most of these technologies have been reported to need long treatment times, which increases the turnover rate [69]. Furthermore, there are no studies so far considering these techniques from an environmental point of view, as we did here by means of the LCA. To decrease the residence times, most of these technologies have been combined together or with chlorine sanitizing agents, although this entails a high cost [70].

The membrane-based techniques here considered have the potential ability to totally replace the need for adding chlorine as a sanitizing agent, representing at the same time an environmentally friendly alternative to the current strategy. The wide use of chlorine in the process has been reported to entail some public health issues, including the formation of carcinogenic by-products such as chloroform or trihalomethanes, chloramines, and haloacetic acids [26,27]. Due to this fact, the use of these compounds in the production chain has already been prohibited in some European countries, namely Belgium, Denmark, Germany and The Netherlands [2]. The evaluated systems may offer an appropriate alternative for the FC vegetable washing at these locations.

However, laboratory conditions are indeed just an approximation of the real factory scenario. The efficiency of the system will clearly depend on the quality and dirtiness of the freshly harvested produce that enters the process. Ideally, the biocide or oxidant compounds that are combined with membranes within the techs will be sufficient to avoid membrane fouling, as it was reported in previous stages of the project. Lab-scale experiments showed that applying backwashes periodically would be sufficient to prevent the clogging of the system, but further experiments on a larger scale are needed to evaluate this issue.

Nevertheless, our sensitivity analysis showed that even if small amounts of chlorine were needed to keep the biological load standards for the washed produce, a significant environmental gain would still be achieved. Indeed, the differences between reducing up to 20% the chlorine concentration and keeping it as in our reference scenario are negligible for all the evaluated techniques when compared to the impacts derived from the current strategy. Furthermore, chlorine is a relatively cheap compound with an important oxidizing ability, so its combination with the new sanitizing techniques would probably increase the life span of their components and decrease the associated maintenance costs.

Indeed, this study has mainly focused on the environmental traits of the proposed filtering devices. The results are thus preliminary, and they just show the environmental viability of the devices. Further studies should be carried out in order to assess whether these technologies would be applicable on a larger scale. Special attention should be paid to the potential issues related to membrane fouling. Larger streams of water to be treated might carry some vegetable residues, which were not considered when testing the decontaminating ability of the membranes. The use of backwashes should be tested in larger scales and not only when washing leafy vegetables, but also some others whose residue may be smaller with higher clogging potential, such as broccoli.

Larger scale studies would also lead to new primary data that could be used to complement the LCA performed here. The use of an inventory mainly composed of direct primary data will lead to more robust conclusions, and will eliminate any possible imprecision due to estimations.

Moreover, this study has been limited to the environmental point of view. If these new techs are to be introduced in a real manufacturing process, also an economic analysis should be performed to confirm that the advantages of the membrane-based devices do not entail a disproportionate cost in the total operation.

5. Conclusions

In this work, we have presented four new membrane techniques that aim to address two major issues regarding FC vegetable washing operations: the current low water recirculation rate and the controversial use of chlorinated compounds as sanitizing agents. Introducing these devices on the FC production chain has been proven to be beneficial from an environmental point of view, since they are able to reduce the total impacts derived from the washing process. The environmental load reduction is mainly due to the decrease in water consumption, which subsequently implies decreased electricity consumption for water cooling and pumping. This environmental gain is maintained even if the increases in water recirculation rates are limited and/or the addition of small quantities of chlorine are still needed to ensure produce quality. On the whole, membrane-based sanitizing techniques appear to give a sustainable option for the sanitizing of FC washing water. Larger scale experiments should be performed to ensure technical viability. This environmental analysis would also complement an economic study of the decision-making process when assessing the alternatives to the current FC production chain strategy.

Supplementary Materials: The Supplementary Materials are available online at http://www.mdpi.com/2071-1050/12/9/3674/s1.

Author Contributions: Investigation, M.V., M.P.L., H.M.-P., and J.A.C. All authors have read and agreed to the published version of the manuscript.

Funding: This research was funded by the Plan of Science, Technology, and Innovation of the Principality of Asturias, grant number FC-GRUPIN-IDI/2018/000225.

Acknowledgments: The authors wish to thank the European Union's Seventh Programme for research and technological development FP7-ERANET-SUSFOOD "CEREAL Project" for the financial support given, along with the national Institute for Agricultural and Food Research and Technology Ref. ERA35-CEREAL-INIA.

Conflicts of Interest: The authors declare no conflict of interest.

References

1. Kader, A.A.; Gil, M.I. Fresh-cut fruit and vegetables. In *Improving the Health-Promoting Properties of Fruit and Vegetable Products*; Woodhead Publishing Series in Food Science, Technology and Nutrition; Woodhead Publishing: Cambridge, UK, 2008; pp. 475–504. ISBN 978-1-84569-184-4.
2. Baselice, A.; Colantuoni, F.; Lass, D.A.; Nardone, G.; Stasi, A. Trends in EU consumers' attitude towards fresh-cut fruit and vegetables. *Food Qual. Prefer.* **2017**, *59*, 87–96. [CrossRef]
3. Procentese, A.; Raganati, F.; Olivieri, G.; Elena Russo, M.; Marzocchella, A. Pre-treatment and enzymatic hydrolysis of lettuce residues as feedstock for bio-butanol production. *Biomass Bioenergy* **2017**, *96*, 172–179. [CrossRef]
4. Turatti, A. Process Design, Facility and Equipment Requirements. In *Advances in Fresh-Cut Fruits and Vegetables Processing*; Martin-Belloso, O., Soliva-Fortuny, R., Eds.; CRC Press: Boca Raton, FL, USA, 2011; pp. 339–360. ISBN 9781420071238.
5. Patrick, V.; Mazollier, J. Overview of the European Fresh-cut Produce Industry. In *Fresh-Cut Fruits and Vegetables: Science, Technology and Market*; Lamikanra, O., Ed.; CRC Press: Boca Raton, FL, USA, 2002; ISBN 9781498729949.
6. Gil, M.I.; Selma, M.V.; López-Gálvez, F.; Allende, A. Fresh-cut product sanitation and wash water disinfection: Problems and solutions. *Int. J. Food Microbiol.* **2009**, *134*, 37–45. [CrossRef] [PubMed]
7. Gil, M.I.; Allende, A.; Selma, M.V. Treatments to Ensure Safety of Fresh-cur Fruits and Vegetables. In *Advances in Fresh-Cut Fruits and Vegetables Processing*; Martin-Belloso, O., Soliva-Fortuny, R., Eds.; CRC Press: Boca Raton, FL, USA, 2011; pp. 221–230.
8. Teng, Z.; van Haute, S.; Zhou, B.; Hapeman, C.J.; Millner, P.D.; Wang, Q.; Luo, Y. Impacts and interactions of organic compounds with chlorine sanitizer in recirculated and reused produce processing water. *PLoS ONE* **2018**, *13*, e0208945. [CrossRef] [PubMed]
9. Van Haute, S.; Tryland, I.; Veys, A.; Sampers, I. Wash water disinfection of a full-scale leafy vegetables washing process with hydrogen peroxide and the use of a commercial metal ion mixture to improve disinfection efficiency. *Food Control* **2015**, *50*, 173–183. [CrossRef]
10. Weng, S.; Luo, Y.; Li, J.; Zhou, B.; Jacangelo, J.G.; Schwab, K.J. Assessment and speciation of chlorine demand in fresh-cut produce wash water. *Food Control* **2016**, *60*, 543–551. [CrossRef]
11. Luo, Y.; Zhou, B.; Van Haute, S.; Nou, X.; Zhang, B.; Teng, Z.; Turner, E.R.; Wang, Q.; Millner, P.D. Association between bacterial survival and free chlorine concentration during commercial fresh-cut produce wash operation. *Food Microbiol.* **2018**, *70*, 120–128. [CrossRef]
12. Goal 12 Sustainable Development Knowledge Platform. Available online: https://sustainabledevelopment.un.org/sdg12 (accessed on 6 April 2020).
13. Fusi, A.; Castellani, V.; Bacenetti, J.; Cocetta, G.; Fiala, M.; Guidetti, R. The environmental impact of the production of fresh cut salad: A case study in Italy. *Int. J. Life Cycle Assess.* **2016**, *21*, 162–175. [CrossRef]
14. Allende, A.; Selma, M.V.; López-Gálvez, F.; Villaescusa, R.; Gil, M.I. Impact of wash water quality on sensory and microbial quality, including Escherichia coli cross-contamination, of fresh-cut escarole. *J. Food Prot.* **2008**, *71*, 2514–2518. [CrossRef]
15. Lee, A.; Elam, J.W.; Darling, S.B. Membrane materials for water purification: Design, development, and application. *Environ. Sci. Water Res. Technol.* **2016**, *2*, 17–42. [CrossRef]
16. Ilgin, M.A.; Gupta, S.M. Environmentally conscious manufacturing and product recovery (ECMPRO): A review of the state of the art. *J. Environ. Manag.* **2010**, *91*, 563–591. [CrossRef] [PubMed]

17. Garcia-Gudiño, J.; Monteiro, A.N.T.R.; Espagnol, S.; Blanco-penedo, I.; Garcia-launay, F. Life Cycle Assessment of Iberian Traditional Pig Production System in Spain. *Sustainability* **2020**, *12*, 627. [CrossRef]
18. Pierucci, S.; Klemeš, J.J.; Piazza, L.; Bakalis, S.; Falcone, G.; De Luca, A.I.; Stillitano, T.; Iofrida, N.; Strano, A.; Piscopo, A.; et al. Shelf Life Extension to Reduce Food Losses: The Case of Mozzarella Cheese. In Proceedings of the Chemical Engeneering Transactions, Milano, Italy, 28–31 May 2017.
19. Partearroyo, T.; de Samaniego-Vaesken, M.L.; Ruiz, E.; Aranceta-Bartrina, J.; Gil, Á.; González-Gross, M.; Ortega, R.M.; Serra-Majem, L.; Varela-Moreiras, G. Current food consumption amongst the spanish anibes study population. *Nutrients* **2019**, *11*, 2663. [CrossRef] [PubMed]
20. Clement, O.; Olatayo Jekayinfa, S.; Jekayinfa, S.O.; Pecenka, R.; Jaiyeoba, F.; Ogunlade, C.A.; Oni, O. Life Cycle Assessment of Local Rice Production and Processsing in Nigeria. In Proceedings of the Internarional Commission of Agricultural and Biosystems Engineering, Ibadan, Nigeria, 22–25 October 2018.
21. Ghasemi-Mobtaker, H.; Kaab, A.; Rafiee, S. Application of life cycle analysis to assess environmental sustainability of wheat cultivation in the west of Iran. *Energy* **2020**, *193*, 116768. [CrossRef]
22. Halloran, A.; Roos, N.; Eilenberg, J.; Cerutti, A.; Bruun, S. Life cycle assessment of edible insects for food protein: A review. *Agron. Sustain. Dev.* **2016**, *36*. [CrossRef]
23. Moreno, J.; Pablos, C.; Marugán, J. Quantitative Methods for Life Cycle Assessment (LCA) Applied to the Vegetable Industry. In *Quantitative Methods for Food Safety and Quality in the Vegetable Industry*; Springer: New York, NY, USA, 2018; pp. 255–293.
24. Tasca, A.L.; Nessi, S.; Rigamonti, L. Environmental sustainability of agri-food supply chains: An LCA comparison between two alternative forms of production and distribution of endive in northern Italy. *J. Clean. Prod.* **2017**, *140*, 725–741. [CrossRef]
25. Ilari, A.; Duca, D. Energy and environmental sustainability of nursery step finalized to "fresh cut" salad production by means of LCA. *Int. J. Life Cycle Assess.* **2018**, *23*, 800–810. [CrossRef]
26. Bull, R.J.; Crook, J.; Whittaker, M.; Cotruvo, J.A. Therapeutic dose as the point of departure in assessing potential health hazards from drugs in drinking water and recycled municipal wastewater. *Regul. Toxicol. Pharmacol.* **2011**, *60*, 1–19. [CrossRef]
27. Legay, C.; Rodriguez, M.J.; Sérodes, J.B.; Levallois, P. Estimation of chlorination by-products presence in drinking water in epidemiological studies on adverse reproductive outcomes: A review. *Sci. Total Environ.* **2010**, *408*, 456–472. [CrossRef]
28. ISO/IEC ISO 14040:2006. *Environmental Management—Life Cycle Assessment—Principles and Framework*; International Organization for Standardization (ISO): Genève, Switzerland, 2006.
29. ISO/IEC ISO 14044:2006. *Environmental Management—Life Cycle Assessment—Requirements and Guidelines*; International Organization for Standardization (ISO): Genève, Switzerland, 2006.
30. Colelli, G.; Amodio, M.L. Factor Affecting Quality and Safety of Fresh-cut Fruits and Vegetables. In *Fresh-Cut Fruits and Vegetables: Technology, Physiology and Safety*; Pareek, S., Ed.; CRC Press: Boca Raton, FL, USA, 2016; pp. 194–222.
31. Ajo, P.; Preis, S.; Vornamo, T.; Mänttäri, M.; Kallioinen, M.; Louhi-Kultanen, M. Hospital wastewater treatment with pilot-scale pulsed corona discharge for removal of pharmaceutical residues. *J. Environ. Chem. Eng.* **2018**, *6*, 1569–1577. [CrossRef]
32. Ajo, P.; Kornev, I.; Preis, S. *Pulsed Corona Discharge in Water Treatment: The Effect of Hydrodynamic Conditions on Oxidation Energy Efficiency*; ACS Publications: Washington, DC, USA, 2015.
33. Williams, R.C.; Sumner, S.S.; Golden, D.A. Inactivation of Escherichia coli O157:H7 and Salmonella in Apple Cider and Orange Juice Treated with Combinations of Ozone, Dimethyl Dicarbonate, and Hydrogen Peroxide. *J. Food Sci.* **2006**, *70*, M197–M201. [CrossRef]
34. Ganiyu, S.O.; van Hullebusch, E.D.; Cretin, M.; Esposito, G.; Oturan, M.A. Coupling of membrane filtration and advanced oxidation processes for removal of pharmaceutical residues: A critical review. *Sep. Purif. Technol.* **2015**, *156*, 891–914. [CrossRef]
35. Arola, K.; Kallioinen, M.; Reinikainen, S.P.; Hatakka, H.; Mänttäri, M. Advanced treatment of membrane concentrate with pulsed corona discharge. *Sep. Purif. Technol.* **2018**, *198*, 121–127. [CrossRef]
36. Johansson, T.; Manttari, M.; Kallioinen, M. *Pulsed Corona Discharge and Membrane Filtration for Purification of Lettuce Washing Waters*; LUT University: Lappenranta, Finland, 2016.

37. Mierzwa, J.C.; da Silva, M.C.C.; Veras, L.R.V.; Subtil, E.L.; Rodrigues, R.; Li, T.; Landenberger, K.R. Enhancing spiral-wound ultrafiltration performance for direct drinking water treatment through operational procedures improvement: A feasible option for the Sao Paulo Metropolitan Region. *Desalination* **2012**, *307*, 68–75. [CrossRef]
38. Hernando, M.D.; Petrovic, M.; Radjenovic, J.; Fernández-Alba, A.R.; Barceló, D. Chapter 4.2 Removal of pharmaceuticals by advanced treatment technologies. In *Comprehensive Analytical Chemistry*; Elsevier: Amsterdam, The Netherlands, 2007; Volume 50, pp. 451–474. ISBN 9780444530523.
39. Tian, J.Y.; Ernst, M.; Cui, F.; Jekel, M. Correlations of relevant membrane foulants with UF membrane fouling in different waters. *Water Res.* **2013**, *47*, 1218–1228. [CrossRef]
40. Chew, C.M.; Aroua, M.K.; Hussain, M.A.; Ismail, W.M.Z.W. Evaluation of ultrafiltration and conventional water treatment systems for sustainable development: An industrial scale case study. *J. Clean. Prod.* **2016**, *112*, 3152–3163. [CrossRef]
41. Basu, O.D. Backwashing. In *Encyclopedia of Membranes*; Springer: Berlin/Heidelberg, Germany, 2014; pp. 1–3.
42. Nelson, H.; Singh, R.; Toledo, R.; Singh, N. The use of a submerged microfiltration system for regeneration and reuse of wastewater in a fresh-cut vegetable operation. *Sep. Sci. Technol.* **2007**, *42*, 2473–2481. [CrossRef]
43. Chung, H.Y.; Hall, J.R.B.; Gogins, M.A.; Crofoot, D.G.; Weik, T.M. Polymer, Polymer Microfiber, Polymer Nanofiber and Applications Including Filter Structures. U.S. Patent 7090715B2, 15 August 2006.
44. Rinaudo, M. Chitin and chitosan: Properties and applications. *Prog. Polym. Sci. Oxf.* **2006**, *31*, 603–632. [CrossRef]
45. Goetz, L.A.; Jalvo, B.; Rosal, R.; Mathew, A.P. Superhydrophilic anti-fouling electrospun cellulose acetate membranes coated with chitin nanocrystals for water filtration. *J. Membr. Sci.* **2016**, *510*, 238–248. [CrossRef]
46. Goetz, L.A.; Naseri, N.; Nair, S.S.; Karim, Z.; Mathew, A.P. All cellulose electrospun water purification membranes nanotextured using cellulose nanocrystals. *Cellulose* **2018**, *25*, 3011–3023. [CrossRef]
47. Morones, J.R.; Elechiguerra, J.L.; Camacho, A.; Holt, K.; Kouri, J.B.; Ramírez, J.T.; Yacaman, M.J. The bactericidal effect of silver nanoparticles. *Nanotechnology* **2005**, *16*, 2346–2353. [CrossRef] [PubMed]
48. Pinto, R.J.B.; Marques, P.A.A.P.; Neto, C.P.; Trindade, T.; Daina, S.; Sadocco, P. Antibacterial activity of nanocomposites of silver and bacterial or vegetable cellulosic fibers. *Acta Biomater.* **2009**, *5*, 2279–2289. [CrossRef] [PubMed]
49. Chen, L.; Peng, X. Silver nanoparticle decorated cellulose nanofibrous membrane with good antibacterial ability and high water permeability. *Appl. Mater. Today* **2017**, *9*, 130–135. [CrossRef]
50. Wernet, G.; Bauer, C.; Steubing, B.; Reinhard, J.; Moreno-Ruiz, E.; Weidema, B. The ecoinvent database version 3 (part I): Overview and methodology. *Int. J. Life Cycle Assess.* **2016**, *21*, 1218–1230. [CrossRef]
51. Ecoinvent—The World's Leading LCA Database Launches Version 3.0. Available online: https://www.psi.ch/en/media/our-research/ecoinvent-the-worlds-leading-lca-database-launches-version-30 (accessed on 10 February 2020).
52. Zhang, H.; Hortal, M.; Dobon, A.; Jorda-Beneyto, M.; Bermudez, J.M. Selection of Nanomaterial-Based Active Agents for Packaging Application: Using Life Cycle Assessment (LCA) as a Tool. *Packag. Technol. Sci.* **2017**, *30*, 575–586. [CrossRef]
53. Manda, B.M.K.; Worrell, E.; Patel, M.K. Innovative membrane filtration system for micropollutant removal from drinking water - Prospective environmental LCA and its integration in business decisions. *J. Clean. Prod.* **2014**, *72*, 153–166. [CrossRef]
54. Muñoz, I.; Rodríguez, C.; Gillet, D.M.; Moerschbacher, B. Life cycle assessment of chitosan production in India and Europe. *Int. J. Life Cycle Assess.* **2018**, *23*, 1151–1160. [CrossRef]
55. Dong, S.; Li, J.; Kim, M.H.; Park, S.J.; Eden, J.G.; Guest, J.S.; Nguyen, T.H. Human health trade-offs in the disinfection of wastewater for landscape irrigation: Microplasma ozonation: Vs. chlorination. *Environ. Sci. Water Res. Technol.* **2017**, *3*, 106–118. [CrossRef]
56. Rodrigo, A.; Wesche, M.; Llorca, I.; Scholl, S. Environmental assessment of new decontamination and sanitation techniques for fresh-cut products. In Proceedings of the Fouling and Cleaning in Food Processing, Cambridge, UK, 31 March–2 April 2014.
57. ReCiPe|PRé Sustainability. Available online: https://www.pre-sustainability.com/recipe (accessed on 28 January 2020).
58. Ernstoff, A.; Tu, Q.; Faist, M.; Del Duce, A.; Mandlebaum, S.; Dettling, J. Comparing the Environmental Impacts of Meatless and Meat-Containing Meals in the United States. *Sustainability* **2019**, *11*, 6235. [CrossRef]

59. Yelboga, M. LCA Analysis of Grafted Tomato Seedling Production in Turkey. *Sustainability* **2019**, *12*, 25. [CrossRef]
60. Vigil, M.; Marey-Pérez, M.F.; Martinez Huerta, G.; Álvarez Cabal, V. Is phytoremediation without biomass valorization sustainable?—Comparative LCA of landfilling vs. anaerobic co-digestion. *Sci. Total Environ.* **2015**, *505*, 844–850. [CrossRef] [PubMed]
61. García, S.G.; Montequín, V.R.; Fernández, R.L.; Fernández, F.O. Evaluation of the synergies in cogeneration with steel waste gases based on Life Cycle Assessment: A combined coke oven and steelmaking gas case study. *J. Clean. Prod.* **2019**, *217*, 576–583. [CrossRef]
62. Ouarhim, W.; Zari, N.; Bouhfid, R.; Qaiss, A. El kacem Mechanical performance of natural fibers–based thermosetting composites. In *Mechanical and Physical Testing of Biocomposites, Fibre-Reinforced Composites and Hybrid Composites*; Woodhead Publishing: Cambridge, UK, 2019; pp. 43–60.
63. Selma, M.V.; Allende, A.; López-Gálvez, F.; Conesa, M.A.; Gil, M.I. Disinfection potential of ozone, ultraviolet-C and their combination in wash water for the fresh-cut vegetable industry. *Food Microbiol.* **2008**, *25*, 809–814. [CrossRef] [PubMed]
64. Manzocco, L.; Ignat, A.; Bartolomeoli, I.; Maifreni, M.; Nicoli, M.C. Water saving in fresh-cut salad washing by pulsed light. *Innov. Food Sci. Emerg. Technol.* **2015**, *28*, 47–51. [CrossRef]
65. Elizaquível, P.; Sánchez, G.; Selma, M.V.; Aznar, R. Application of propidium monoazide-qPCR to evaluate the ultrasonic inactivation of Escherichia coli O157:H7 in fresh-cut vegetable wash water. *Food Microbiol.* **2012**, *30*, 316–320. [CrossRef] [PubMed]
66. Spricigo, D.A.; Bardina, C.; Cortés, P.; Llagostera, M. Use of a bacteriophage cocktail to control Salmonella in food and the food industry. *Int. J. Food Microbiol.* **2013**, *165*, 169–174. [CrossRef]
67. Huang, Y.; Ye, M.; Chen, H. Efficacy of washing with hydrogen peroxide followed by aerosolized antimicrobials as a novel sanitizing process to inactivate Escherichia coli O157:H7 on baby spinach. *Int. J. Food Microbiol.* **2012**, *153*, 306–313. [CrossRef] [PubMed]
68. Park, S.-H.; Choi, M.-R.; Park, J.-W.; Park, K.-H.; Chung, M.-S.; Ryu, S.; Kang, D.-H. Use of Organic Acids to Inactivate Escherichia coli O157:H7, Salmonella Typhimurium, and Listeria monocytogenes on Organic Fresh Apples and Lettuce. *J. Food Sci.* **2011**, *76*, M293–M298. [CrossRef]
69. Manzocco, L.; Ignat, A.; Bot, F.; Calligaris, S.; Valoppi, F. Efficient management of the water resource in the fresh-cut industry: Current status and perspectives. *Trends Food Sci. Technol.* **2015**, *46*, 286–294. [CrossRef]
70. Meireles, A.; Giaouris, E.; Simões, M. Alternative disinfection methods to chlorine for use in the fresh-cut industry. *Food Res. Int.* **2016**, *82*, 71–85. [CrossRef]

Review

"What a Waste"—Can We Improve Sustainability of Food Animal Production Systems by Recycling Food Waste Streams into Animal Feed in an Era of Health, Climate, and Economic Crises?

Gerald C. Shurson

Department of Animal Science, University of Minnesota, St. Paul, MN 55108, USA; shurs001@umn.edu

Received: 11 July 2020; Accepted: 28 August 2020; Published: 30 August 2020

Abstract: Food waste has been a major barrier to achieving global food security and environmental sustainability for many decades. Unfortunately, food waste has become an even bigger problem in many countries because of supply chain disruptions during the COVID-19 pandemic and African Swine Fever epidemic. Although Japan and South Korea have been leaders in recycling food waste into animal feed, countries that produce much greater amounts of food waste, such as the United States and the European Union, have lagged far behind. Concerns about the risk of transmission of bacteria, prions, parasites, and viruses have been the main obstacles limiting the recycling of food waste streams containing animal-derived tissues into animal feed and have led to government regulations restricting this practice in the U.S. and EU. However, adequate thermal processing is effective for inactivating all biological agents of concern, perhaps except for prions from infected ruminant tissues. The tremendous opportunity for nitrogen and phosphorus resource recovery along with several other environmental benefits from recycling food waste streams and rendered animal by-products into animal feed have not been fully appreciated for their substantial contribution toward solving our climate crisis. It is time to revisit our global approach to improving economic and environmental sustainability by more efficiently utilizing the abundant supply of food waste and animal tissues to a greater extent in animal feed while protecting human and animal health in food animal production systems.

Keywords: biosecurity; carcass disposal; food waste; greenhouse gas emissions; life cycle assessment; nitrogen; pathogens; phosphorus; rendered animal by-products

---

1. Introduction

Crises often lead to change [1]. For far too long, food waste has been the greatest contributor to inefficiency of resource use and our inability to achieve greater global food security and sustainability. More than 1.3 billion tonnes of edible food material are wasted annually around the world, which represents about one third of the total food produced and is enough to feed more than one billion people [2]. The amount and types of food waste vary between countries where 44% of global food waste occurs in less-developed countries during the post-harvest and processing stages of the food supply chain, while the remaining 56% of these losses, of which 40% occur at the pre- and post-consumer stages, are attributed to developed countries in Europe, North America, Oceania, Japan, South Korea, and China [2,3]. As a result, the United Nations (UN) has deemed food waste reduction as a global priority and included it in the list of sustainability goals [4]. Specifically, food waste reduction has significant implications for several of the UN Sustainable Development Goals including: 2. Zero hunger; 12. Responsible consumption and production; 13. Climate action; 14. Life below water; and 15. Life on land [4].

Crises often accelerate existing trends [1] and the COVID-19 (novel coronavirus SARS-CoV-2) pandemic is redefining the concept of sustainability [5]. The COVID-19 pandemic has caused major disruptions in food supply chains and caused huge shifts in food access, food security, and food losses due to changes in food flow and distribution patterns [6]. Food supply chains are complex and most operate in a "just-in-time" mode where minor disruptions can have dramatic consequences [7]. When workers were required to stay at home, and all businesses except those deemed essential were closed, consumer demand for food shifted from food services (e.g., restaurants, hotels, schools, and institutions) to retail grocery stores [7]. Although ample supplies of food were available, existing food distribution networks were unable to quickly respond to these changes, which resulted in increased food waste [6]. For example, short-term disruptions in eating habits during the early stages of the COVID-19 outbreak in Spain resulted in a 12% increase in food loss and waste [8]. Furthermore, increased shortages of agricultural and food processing workers caused by illness or fear of becoming ill led to fruit and vegetable crops being destroyed [7], and closures or reduced processing capacity of animal slaughter plants [9–11]. This severely restricted access for market-ready livestock and poultry and resulted in the unfortunate need to humanely euthanize and dispose of millions of animals originally destined to enter the food chain [12,13]. Economic losses due to COVID-19 disruptions have been estimated to be at least USD 13.6 billion for U.S. cattle producers and USD 5 billion for U.S. pork producers, with 30% less meat available to consumers at a projected 20% increase in price [12]. In addition to these economic losses, lack of sufficient rendering capacity for disposal of market-ready animals has required using other less desirable methods of disposal, which are detrimental to the environment and cause inefficiencies in resource use (i.e., land, water, nitrogen, phosphorus, labor) while increasing biosecurity risks [13]. As a result of the COVID-19 pandemic, researchers have proposed rethinking and redefining sustainability as the intersection of the economy, environment, society, and human health [5]. Furthermore, a more holistic approach that includes climate, economics, and nutrition is needed to improve food supply chain efficiency by reducing food loss and improving waste management of food supply chains adversely affected by changes in consumption patterns caused by pandemics [8]. In fact, the European Union has already indicated plans to use knowledge gained from COVID-19 impacts on food supply chains to revise the Farm to Fork subsection of the Green Deal reforms [14]. Now, more than ever before, it is time for researchers and food sector experts to accelerate efforts for developing more sustainable and modern food systems by reducing the cost of food waste recovery and reutilization in the food chain [15]. However, a very important component of food loss that has not been considered in all of these proposals, which also has dramatic effects on food security and sustainability, are mortalities caused by animal disease epidemics.

The African Swine Fever epidemic in China caused estimated losses of 220 to 300 million pigs that were originally destined for the food chain in 2019 [16,17]. This enormous number of pigs represents 25–35% of the total world pig population that died or were depopulated from infected farms [16,17]. Because of the lack of infrastructure to manage the disposition of millions of pigs, the capabilities to recover nutrients from carcasses through rendering was not possible, and carcass burial and disposal in landfills were used at great environmental costs and biosecurity risks [18]. In addition, highly pathogenic avian influenza outbreaks in many countries around the world have resulted in losses of millions of chickens due to mortality and depopulation [19]. Unfortunately, the likelihood of future disruptions in global food animal production caused by animal disease epidemics is increasing because of increased global trade and travel, urbanization, exploitation of natural resources, and changes in land use [20–23].

These unprecedented food losses due to disruptions in global food supply chains have created an urgent need to reevaluate the intertwining of resource recovery, environmental impacts, and biosafety of various food waste streams and animal carcasses to achieve the greatest value. This is essential because animal-derived foods provide about one third of total human protein consumption [24], but their production requires about 75% of arable land [25] and 35% of grain resources, while contributing about 14.5% to total greenhouse gas emissions [26]. Reimagining recovery of nutrients from food waste and animal carcasses, and subsequent recycling of these valuable nutrients into animal feed, can provide tremendous opportunities to use less arable land and rely less on global grain supplies to produce meat, milk, and eggs, while reducing animal agriculture's contribution to greenhouse gas emissions. Therefore, the purpose of this review is to summarize the current knowledge of the benefits and limitations of recycling various pre-harvest to post-harvest food animal-derived waste sources, as well as retail to post-consumer food waste sources, into animal feeds to achieve greater food security and sustainability during an era of escalating food losses throughout the entire food supply chain.

2. Maximizing Resource Recovery and Value of Waste Streams

2.1. Food Waste

Food waste disposal options have been characterized in a hierarchical order of priority based on achieving the greatest value from resource recovery while minimizing negative impacts on the environment [27]. The best solution and highest priority are to minimize or eliminate food waste, followed by redistributing food to hungry people. The next greatest priority is to convert food waste into animal feed, which is preferable to composting, anaerobic digestion to produce biogas, and disposal in landfills [27]. Food waste has been fed to pigs in every country for centuries, but since 2001 it has been banned in the European Union due to illegal feeding of uncooked food waste, which was associated with the foot-and-mouth disease outbreak in the United Kingdom [28]. Concerns about pathogen transmission as well as an abundant supply of relatively low-cost corn and soybean meal in the United States has also limited feeding food waste to pigs, which has been banned in 18 states [16]. In contrast, Japan (2001), South Korea (1997), and Taiwan (2003) have developed tightly regulated policies and invested in substantial infrastructure using adequate thermal processing to promote the conversion of 35–43% of food waste into animal feed [28].

The wide disparity in government policies among countries regarding recycling of food waste into animal feed has severely limited the ability to reuse the valuable nutrients, reduce negative environmental impacts, and improve sustainability of pig production in the United States and the European Union, which produce much greater quantities of food waste than Japan and South Korea. Furthermore, these Asian countries have demonstrated during the past 20 years that biosafety risks can be adequately managed. Now that social and consumer pressure is increasing to produce food with a lower carbon footprint and conserve resources [29–31], recycling food waste into animal feed needs to be revisited as a viable option in all countries around the world if adequate biosafety processes can be implemented and regulated.

2.2. Carcass Rendering

Options for managing mass carcass disposal vary by country or region, but in the United States an environmental impact statement and guidelines for mass carcass management options during a national health emergency have been developed [32]. Approved options include unlined burial, open-air burial, composting, offsite rendering, landfill, and fixed-facility incineration. Other disposal options that may be considered include alkaline hydrolysis, anaerobic digestion, microwave sterilization, and gasification [32], but these options do not currently provide sufficient capacity to dispose of large numbers of carcasses and are not available in many locations. In the European Union, Animal By-Product Regulations (1069/2009) classify sources and characteristics of animal by-products into one of three categories [33]. Feeding animal by-products to terrestrial animals other than fur animals of the same species, and feeding catering waste to farmed animals is prohibited [33] because of perceived risks of incomplete destruction of pathogens and prions (proteins causing transmissible spongiform encephalopathy) entering the animal feed supply chain [34]. However, approved disposal methods, depending on classification category, include incineration, burial in authorized landfills following processing, composting, and biogas production [33]. Differences in interpretation of relative environmental and biosecurity risks of various carcass disposal options have led to different legal requirements and regulations in various countries [34]. More information is needed for all carcass disposal methods so that more comprehensive environmental life cycle analyses and biosecurity risk assessments can be conducted to determine the methods that are least detrimental to the environment and biosafety while providing the greatest resource recovery value [34].

3. Comparison of Environmental Impacts of Alternative Disposal Methods

3.1. Food Waste

Although recycling food waste into animal feed is a higher value alternative with fewer negative environmental impacts than composting, anaerobic digestion, and landfill disposal, it is surprising that more comprehensive and comparative studies of disposal methods have not been conducted. A summary of nine published studies that compared the environmental impacts of using food waste as wet or dry animal feed with the alternatives of anaerobic digestion for biogas production, composting, incineration, and landfill is shown in Table 1. In general, results from these studies show greater environmental benefits from using food waste as animal feed compared to the other disposal alternatives, but have mainly focused on estimating impacts on global warming using greenhouse gas (GHG) emissions as indicators, with limited evaluation of impacts on the use of such resources as energy, land, and water. Furthermore, several of these researchers indicated that results obtained under the scope, scenarios, and assumptions used in each study may not apply to broader applications and suggested that more studies are needed using harmonized assessment approaches. Another key finding of these studies was that the nutritional composition of food waste sources affects the extent of GHG reduction. Nutritional composition also determines whether recycling a specific type of food waste into animal feed was the most beneficial option. For example, Eriksson et al. [35] showed that bread waste had the greatest potential for reducing GHG emission, followed by chicken, beef, and bananas, with lettuce having the lowest potential. These results suggest that food waste sources that contain high energy and dry matter content are more suitable for use as animal feed than less nutritionally dense sources.

Table 1. Summary of Life Cycle Assessment studies comparing environmental impacts of food waste disposal options with converting food waste to animal feed.

Food Waste Source/Reference	Disposal Methods Compared	Environmental Indicators	Key Results
Household and catering food waste [36]	Composting Incineration Landfill [1] Dry animal feed	Global warming Human toxicity Acidification Eutrophication Ecotoxicity	Feed manufacturing had: • lowest global warming potential • lowest human toxicity except for incineration • less acidification potential than composting and incineration • highest eutrophication potential • less ecotoxicity than composting
Kitchen and food factory waste [37]	Wet animal feed (sterilization) Dry animal feed Incineration	Global warming Energy consumption Water use	• GHG emissions (g CO_2 equivalent/kg DM): • wet feed (268) < incineration (1066) and dry feed (1073) • Energy consumption (MJ/kg DM): • wet feed (3.9) < incineration (14.5) < dry feed (16.7) • Water use (L/kg DM): • wet feed (2.9) < dry feed (5.1) < incineration (1035)
Household and catering food waste [38]	Composting Dry animal feed Wet animal feed Landfill	Global warming	1 tonne of food waste contributed: • 61 kg CO_2 equivalent from wet feeding • 123 kg CO_2 equivalent from composting • 200 kg CO_2 equivalent from dry feeding • 1010 kg CO_2 equivalent from landfilling
Household and catering food waste [39]	Anaerobic digestion Composting Dry animal feed Wet animal feed Incineration Landfill	Global warming Energy/resource recovery	Environmental benefit/cost ratio was: • 0.42 for wet feeding • 0.26 for dry feeding • 0.26 for dryer incineration • 0.22 for composting • 0.11 for anaerobic digestion • 0.04 for landfilling

Table 1. *Cont.*

Food Waste Source/Reference	Disposal Methods Compared	Environmental Indicators	Key Results
Catering, retail, and manufacturing food waste [40]	Anaerobic digestion Machine composting Windrow composting Dry animal feed Wet animal feed	Global warming Economics	1 tonne of food waste contributed: • −126 kg CO_2 equivalent for dry feeding • −48 kg CO_2 equivalent for anaerobic digestion • −48 kg CO_2 equivalent for wet feeding • +5 kg CO_2 equivalent for windrow composting • +45 kg CO_2 equivalent for machine-integrated composting
Evaluated by-products used in animal feeds including distiller's waste, rapeseed cake, whey permeate, fodder milk, and bakery residues [41]	Anaerobic digestion Anaerobic digestion with portion diverted to animal feed	Global warming	All industrial organic by-products evaluated are suitable for biogas production, provide substantial reduction in GHG compared with fossil fuels, but contribute to eutrophication and acidification potential. If these by-products are used as animal feed, the reduction would be significantly less.
Retail food waste [42]	Anaerobic digestion Dry animal feed (bread) fraction combined with anaerobic digestion of remaining fraction	18 environmental and health impact categories	More environmental benefits were obtained by converting the bread waste portion into animal feed than by using only anaerobic digestion to produce heat and electricity.
Compared banana, chicken, lettuce, beef, bread [35]	Anaerobic digestion Dry animal feed Composting Donations Incineration Landfill	Global warming	• Anaerobic digestion and food donations had the greatest effect on reducing GHG emissions, followed by incineration and animal feed • Composting and landfills increased kg CO_2 equivalent/kg • Bread had the greatest benefit for reducing GHG emissions when converted to animal feed, followed by chicken and beef
Household and catering food waste [43]	Anaerobic digestion Composting Dry animal feed Wet animal feed	14 environmental and health impact categories	• Wet feed had the highest score for 13/14 impacts • Dry feed had second best score for 12/14 impacts • Overall average ranking of disposal technologies was: • 1.1 for wet feed • 2.2 for dry feed • 3.3 for anaerobic digestion • 3.4 for composting

[1] Landfill with and without electricity generation.

3.2. Animal Mortalities and Carcass Residuals

Environmentally sustainable and biosecure disposal of animal carcasses resulting from on-farm mortality or from inedible components of carcasses after slaughter is an important function of food animal production supply chains. Globally, the most common methods for disposal of animal mortalities include burial, burning, incineration, rendering, composting, anaerobic digestion, and alkaline hydrolysis [34]. Some of these disposal methods are prohibited in certain countries because of real and perceived biosafety and environmental risks associated with them [34]. Gwyther et al. [34] conducted a comprehensive review of the socioeconomic, human health, biosecurity, and environmental impacts of various carcass disposal methods and a summary of relative environmental impacts is shown in Table 2. It is unfortunate and somewhat surprising to discover the lack of information for one or more environmental impact indicators for each of these disposal methods. Despite this incomplete environmental impact information for various carcass disposal methods, rendering is considered to have moderate effects on odor and water pollution, good impact on reducing GHG emissions, and very good ranking for impacts on soil and vegetation. Only anaerobic digestion was considered a better alternative than rendering for minimizing odor and GHG emissions. Although various other disposal methods ranked higher than rendering in specific categories, it is important to remember that rendering represents much greater value for resource (nutrient) recovery than all of the other disposal methods, which is important from an environmental sustainability perspective. Gooding and Meeker [44] compared the GHG emissions, resource recovery, and biosecurity differences between using anaerobic digestion, composting, and rendering of animal carcasses. Results from their analysis showed that rendering resulted in at least three times greater economic value than products resulting from anaerobic digestion and at least five times greater value than composting, and concluded that rendering is the most sustainable method for handling large quantities of animal carcasses. Currently, about 85% of rendered animal by-products in the U.S. are used in animal feeds, with the remaining 15% used in biofuels production and other industrial products [44].

4. Potential Amounts of Food Waste Streams That Can Be Used as Animal Feed Ingredients

Globally, about 6 billion tonnes of feed (dry matter basis) is consumed by food-producing animals annually, of which 72% is comprised of roughages consumed by ruminants (i.e., cattle, goats, and sheep) [45]. Of the 1.57 billion tonnes of grain, grain by-products, and oilseed meals consumed, 65% (about 1 billion tonnes) are fed to swine and poultry [45]. To put this in perspective, more than 1.3 billion tonnes of edible food material is wasted annually around the world [2], which is 3 million tonnes more than the global consumption of all cereal grains, by-products, and oilseed meals by swine and poultry combined. In addition, about 60 million tonnes of rendered animal by-products are produced annually from the global meat-processing and animal production industry [46]. Therefore, there is tremendous opportunity to recycle energy and nutrients from various food waste sources into animal feed, especially for swine and poultry because they are unable to efficiently utilize fiber in roughages and require diets that are more energy- and nutrient-dense than those for ruminants. By repurposing a greater proportion of food waste into animal feed, there would be much less pressure on land and water use for agricultural purposes, as well as less dependence on global crop production for animal feed. In fact, zu Ermgassen et al. [28] estimated that if the European Union were to adopt regulated and centralized systems for safely recycling food waste into animal feed, similar to those being used successfully in Japan and South Korea, it would result in a 21.5% reduction in land use (1.8 million hectares) for EU pork production. Furthermore, if 39% of the total amount of food waste in the EU was used in pig feeds, it could replace 8.8 million tonnes of edible grains currently fed to pigs, which is equivalent to 70.3 million tonnes of annual cereal consumption by EU citizens [47]. These conservative estimates do not include the additional benefits from processing and using more rendered animal by-products in animal feed, but they clearly show the enormous potential to improve recovery of energy, nitrogen, and phosphorus by diverting these valuable resources toward feed use in food animal production systems.

Table 2. Summary of relative environmental impacts [1] of various livestock mortality disposal methods used routinely around the world (adapted from [34]).

Disposal Method	Pollution and Contamination					
	Odor	**GHG Emission**	**Air**	**Soil and Vegetation**	**Water**	**Land Application of Waste**
Burial	-	-	Very low	High	Moderate	NA
Burning	Very high	??	??	??	??	??
Incineration, on-farm [a]	Low	High	Low [b]	Low [b]	Low [b]	??
Incineration, large central facility	Very low	High	Moderate [b]	Moderate [b]	Moderate [b]	??
Rendering	Moderate	Low	??	Very low	Moderate	??
Composting [c]	Low	Low	??	Moderate	??	Low
Anaerobic digestion	Low	Very low	Very low	??	??	Low
Alkaline hydrolysis	Moderate	??	??	Low	Moderate	Moderate

[1] Relative impacts are ranked based on Very high, High, Moderate, Low, Very low environmental impacts in each category; ?? = more research is needed due to limited information; NA = not applicable. [a] Assumes use of afterburners. [b] Omits pre-incineration handling and storage of carcasses that may result in biosecurity risks. [c] Assumes unlined static pile with no forced aeration.

5. Urgent Need to Achieve Greater Global Nitrogen, Phosphorus, Carbon Resource Recovery

Food production requires the use of vast amounts of resources, including land, water, energy, fertilizer, and other inputs that contribute to climate change, soil and water degradation, and loss of biodiversity and habitats [48]. Global nitrogen (N), phosphorus (P), carbon (C), and water cycles are the major biogeochemical components required for sustaining life, which were stable, self-sustained, and regularly recycled on various temporal scales on the planet at one time in history, but now these cycles are disrupted [49]. In fact, the planetary boundaries of N and P biogeochemical flows have been exceeded [50] and natural P deposits are becoming depleted.

Food animal production plays a critical role in these global N and P cycles [51] because animal consumption represents more than 80% of the total N and P harvested, but unfortunately only 20% of N and P is converted into edible products for human consumption [52]. Therefore, improving the efficiency of N and P use is essential for achieving global food security and sustainability [52–54]. Recovering and recycling greater amounts of energy and nutrients from food waste and animal carcasses into animal feed will not only improve the efficiency of energy, N, and P use in animal protein production, but also has the potential to simultaneously reduce GHG emissions from food waste disposal in landfills. Using livestock and poultry to recycle these nutrients into animal-derived food products is a prudent and practical strategy for recovering these wasted resources.

5.1. Nitrogen

The consumption of animal-derived foods contributes 18% of calories and 25% of protein to the world human population [55]. Demand for pork and poultry meat is expected to increase with future population growth and rising incomes [55], which will increase demand for feed protein. Currently, there is an insufficient quantity of high protein feed ingredients to support current levels of global animal production, and this deficit will become worse as animal production increases in the future to meet increased consumer demand for meat, milk, and eggs [56].

5.2. Phosphorus

Phosphorus is a critical resource that is essential for achieving global food security in the future [57]. Most of the global phosphate rock is mined for use in food production [58], but it is a finite resource with no substitutes and cannot be produced in greater amounts than what already exist in the Earth's deposits. Estimates of the rate and timeline of when high quality global phosphate reserves will be depleted are highly variable [59–65], but only 20% of the amount of mined P is ultimately consumed as food [59]. As a result, the convergence of the increasing depletion of global P reserves for use in food production and the excessive P losses to aquatic ecosystems, which cause eutrophication in lakes and coastal ecosystems [66,67], requires implementation of more comprehensive practices that prevent P losses and improve the efficiency of P use in food production [68].

6. Most Food Waste and Rendered Animal By-Products Are Concentrated Sources of Energy, Nitrogen, and Phosphorus

6.1. Energy, Protein, and Phosphorus in Animal Nutrition

Swine and poultry require nutritionally dense diets. Energy, protein (amino acids), and phosphorus are the three most expensive components of animal diets [69]. Gross energy (GE) content represents total calories in a feedstuff, and the metabolizable energy (ME) content is an estimate of the proportion of GE retained by the animal after accounting for energy losses from digestion (feces) and metabolism (urine) [69]. The crude protein content of a food or feed ingredient is estimated by determining the nitrogen content and multiplying by a factor of 6.25 [69]. Although the nitrogen content varies among foods, the factor of 6.25 is derived from the assumption that an average of 16 g of nitrogen is present in 100 g of protein [69]. Most plant-based feed ingredients (i.e., grains, grain by-products, oilseed by-products) contain relatively low concentrations of phosphorus, and much of it is present in the chemical form of phytate, which is indigestible for pigs and poultry [69]. In contrast, animal-derived feed ingredients contain relatively high concentrations of phosphorus that is also highly digestible.

6.2. Nutritional Composition of Food Waste Sources

Several studies have been conducted to determine the energy and nutrient composition of various food waste sources [70], especially for restaurant and cafeteria waste [71,72]. Dou et al. [48] summarized nutrient composition data from several food waste sources and reported that the average crude protein content among sources was moderately high (19.2%), lipid content was very high (21.5%), and crude fiber content was low (6.2%), indicating that many food waste sources are rich sources of energy and protein and suitable for use in swine and poultry diets. Similarly, Truong et al. [73] summarized studies that evaluated the nutritional composition and feeding value of various food waste sources for poultry and concluded that all were suitable for use in broiler and laying hen diets at appropriate diet inclusion rates.

As expected, there is considerable variability in nutrient content among samples within and between food waste sources [48]. Managing variability in nutrient content and digestibility of feed ingredients is one of the greatest challenges in optimizing diet formulations in precision nutrition animal feeding programs [74]. Fung et al. [75,76] reported that using nutrient composition data of some food waste sources in selected energy prediction equations can provide accurate estimates of digestible and metabolizable energy content for swine. Therefore, the development and use of accurate prediction equations to estimate ME and digestible nutrient content are a potential practical solution for managing variability in nutrient content among food waste sources and formulating precision nutrition diets.

Knowing the general proximate analysis of food waste sources is useful, but it is more important to obtain accurate estimates of ME, digestible amino acids, and digestible phosphorus content for accurate diet formulation. Accurate and precise diet formulation is essential for optimizing caloric and nutritional efficiency of animal growth as well as minimizing nutrient excretion in manure. Unfortunately, only a few studies have been conducted to determine the ME, digestible amino acids, and digestible phosphorus content of food waste sources for poultry and swine [73,75,76]. Therefore, additional studies are needed to develop more robust and comprehensive ME and digestible nutrient composition databases of various food waste sources for swine and poultry to encourage animal nutritionists to fully capture the nutritional and economic value of food waste sources when formulating nutritionally adequate and cost-effective complete animal feeds.

Although a few studies have shown nutritional benefits from feeding food waste to ruminants [77–80] and fish [81], most studies have been conducted with poultry and swine (Table 3). All of these studies not only report the nutrient composition of the various food waste sources being evaluated, but also provide information on appropriate diet inclusion rates to achieve optimal growth performance and, in some cases, meat quality, when fed to poultry and swine.

Table 3. Published studies evaluating the nutritional composition and feeding value of various types of food waste in swine and poultry diets.

Food Waste Source	Poultry Feeding Value References	Swine Feeding Value References
Bakery by-product/breakfast cereal	[82–86]	[87–89]
Fish waste	[90]	[76]
Fruit and vegetable waste	[91–94]	[76]
Household waste, dried	[95–97]	[75]
Meat meal	[98]	-
Municipal waste	[99]	[75,100]
Restaurant and cafeteria waste, dried	[101]	[75,102–106]
Supermarket waste	[107]	[75,76]

6.3. Nutritional Efficiency of Food Waste Sources Can be Equivalent or Greater Than Corn and Soybean Meal

Nutritional efficiency can be defined as the proportion of GE and nutrients in a feed ingredient that is digested, absorbed, and used by an animal for productive purposes [69] (i.e., meat, milk, and egg production). The most nutritionally and economically valuable feed ingredients are those that contain high concentrations of ME, digestible amino acids (nitrogen), and digestible phosphorus [69]. Globally, corn is generally considered to be the reference standard for comparing grain-based energy sources, while soybean meal is used as the standard for comparing protein sources because they are the most widely used and economical energy and protein sources in animal feeds [108]. A comparison of the energy, N, and P efficiency of several food waste sources fed to swine, with corn and soybean meal used as energy and protein standards, is shown in Table 4. Except for food waste from the transfer station and fruit and vegetable waste, all other food waste sources contained more ME than corn and soybean meal and had comparable ME:GE content. The amount (g/kg DM) of digestible N in fish waste exceeded that in corn and soybean meal, whereas supermarket waste contained about four times the amount in corn, but less than soybean meal. Lastly, the digestible P contents (g/kg dry matter) in fish (17.4), supermarket (3.0), and fruit and vegetable (2.0) wastes were greater than in corn (0.99), and fish waste exceeded the digestible N content of soybean meal. Furthermore, fish, supermarket, and fruit and vegetable wastes far exceeded the digestible P content in corn, and contained 460%, 80%, and 53%, respectively, of the digestible P content of soybean meal. These results clearly show that the nutritional value of using common food waste sources in swine diets can have a significant impact on recycling and conserving energy, nitrogen, and phosphorus resources while minimizing the dependence on corn and soybean meal as feed ingredients in swine and poultry diets.

Table 4. Gross energy (GE), metabolizable energy (ME), and crude protein (CP) and phosphorus (P) content and digestibility of food waste sources compared with corn and dehulled soybean meal for swine (dry matter basis).

Ingredient	GE, kcal/kg	ME [1], kcal/kg	ME:GE	CP [2], %	Digestible N [3], g/kg	P [4], %	Digestible P [5], g/kg
Corn [a]	4454	3844	0.86	9.33 (80)	11.9	0.29 (34)	0.99
Dehulled soybean meal [b]	4730	3660	0.77	53.05 (87)	73.8	0.79 (48)	3.79
Food waste source							
Supermarket [c]	5909	4832	0.82	25.51	-	0.64	-
University dining hall [c]	5419	4188	0.77	18.90	-	0.30	-
Transfer station [c]	4829	3198	0.66	17.71	-	0.46	-
Household source separated organics [c]	4455	4114	0.92	13.53	-	0.31	-
Fish waste [d]	6376	4820	0.76	62.49 (95)	95.0	2.95 (59)	17.4
Supermarket [d]	6316	4922	0.78	29.42 (89)	41.9	0.37 (82)	3.03
Fruit and vegetable [d]	4123	2460	0.60	10.13 (11)	1.78	0.27 (74)	2.00

[1] Metabolizable energy was calculated based on prediction equation from [109]. [2] Values in parentheses are standardized ileal digestibility (%) [76]. [3] Digestible nitrogen (N) was calculated by assuming crude protein = nitrogen content × 6.25, multiplying by the digestibility coefficient for each ingredient and converting to a g/kg of dry matter basis. [4] Values in parentheses are standardized total tract digestibility (%) [76]. [5] Digestible phosphorus (P) was calculated by multiplying total P content by the respective digestibility coefficient for each ingredient and converting to a g/kg of dry matter basis. [a] Corn was used as a standard of reference because it is the predominant grain and energy source used in animal feeds globally due to its high ME:GE content, compared with other cereal grains. Values were obtained from [69]. [b] Soybean meal was used as a standard of reference because it is the predominant protein source used in animal feeds globally due to its high crude protein content, digestibility, and desirable amino acid profile, compared with other high protein ingredients. [c] Values obtained from [75]. [d] Values obtained from [76].

6.4. Nutrition and Technical Challenges Limiting Use of Food Waste in Animal Feed

Although it is clear that many food waste streams are rich sources of energy as well as of digestible nitrogen and phosphorus, and can serve a valuable function in conserving resources and reducing environmental impacts, there are several nutritional and technical limitations that need to be addressed to optimize their use in animal feeds. Nutritional challenges include (1) managing variability in energy and nutrient content and digestibility within and among sources, (2) accurate ME, digestible amino acid, and digestible phosphorus content of food waste sources being fed so that the amount of supplementation of other ingredients and additives to formulate nutritionally balanced diets can be determined, and (3) potential feed and food biosafety risks, including bacterial and viral pathogens, parasites, and prions associated with transmissible spongiform encephalopathies (TSEs).

Along with these nutritional challenges, there are also technical challenges for utilizing food waste as animal feed. Depending on the source, seasonality and other reasons for inconsistent supplies may occur that may require storage capabilities. Handling characteristics (bulkiness, moisture content, powdery texture), and the need for further processing such as grinding, drying, and thermal treatment for pathogen inactivation, require use of specialized equipment. Finally, except for countries like Japan and South Korea, most countries lack infrastructure, except for rendering, which has limited the development of recycling food waste into animal feed.

6.5. Nutritional Efficiency of Rendered Animal By-Products Can be Equivalent or Greater Than Corn and Soybean Meal

Similar to some of the food waste sources described in Table 4, some rendered animal protein by-products such as blood meal, chicken by-product meal, and feather meal contain greater ME content for swine than corn and soybean meal (Table 5). Furthermore, the ME:GE of blood meal, chicken meal, and chicken by-product meal is comparable to soybean meal, indicating that similar caloric efficiency can be achieved by using rendered animal by-products as partial replacements for corn and soybean meal in swine diets. In addition, all animal protein by-products listed in Table 5, except for poultry meal, contain substantially more digestible N content than soybean meal, with blood meal containing more than twice the amount of digestible N as soybean meal. Digestible P is also three to five times greater in many rendered animal protein by-products (except for blood meal and feather meal), compared to soybean meal, and they contain even greater amounts than the digestible P content in corn. Although the chemical composition and digestibility varies within and among sources, digestible and metabolizable energy can be accurately predicted for pigs using specific nutritional components, which is essential when accurately formulating diets in precision swine feeding programs [110]. Therefore, using any of the food waste and rendered animal protein by-products as complete or partial replacements for corn and soybean meal in swine diets could dramatically reduce dependence on corn and soybean meal use in swine diets and other animal feeds, and reduce land and water use while conserving N and P resources.

Table 5. Gross energy (GE), metabolizable energy (ME), and crude protein (CP) and phosphorus (P) content and digestibility of rendered animal protein by-products compared with corn and dehulled soybean meal for swine (dry matter basis).

Ingredient	GE, kcal/kg	ME, kcal/kg	ME:GE	CP, %	Digestible N [3], g/kg	P, %	Digestible P [4], g/kg
Corn [1]	4454	3844	0.86	9.33 (65)	9.70	0.29 (26)	0.75
Dehulled soybean meal [1]	4730	3660	0.77	53.05 (82)	69.6	0.79 (39)	3.08
Animal protein by-product [2]							
Blood meal	5789	4618	0.80	97.09 (93)	144	0.20 (99)	1.98
Chicken meal	5015	3719	0.74	69.52 (91)	101	3.26 (42)	13.7
Chicken by-product meal	5521	4204	0.76	69.20 (87)	96.3	1.84 (63)	11.6
Feather meal	5809	4031	0.69	88.86 (80)	114	0.32 (59)	1.89
Meat meal	4732	3034	0.64	57.97 (81)	75.1	3.49 (38)	13.6
Meat and bone meal	4469	2620	0.59	56.14 (82)	73.7	4.46 (33)	14.7
Poultry meal	4183	2508	0.60	49.26 (80)	63.1	4.51 (37)	16.7
Poultry by-product meal	4,381	3,038	0.69	58.04 (87)	80.7	4.67 (34)	15.9

[1] Values were obtained from [69]. Values in parentheses for CP and P are apparent digestibility values (%). [2] Values were obtained from [110]. Values in parentheses for CP and P are apparent digestibility values (%). [3] Digestible nitrogen (N) was calculated by assuming crude protein = nitrogen content × 6.25, multiplying by the digestibility coefficient for each ingredient and converting to a g/kg of dry matter basis. [4] Digestible phosphorus (P) was calculated by multiplying total P content by the respective digestibility coefficient for each ingredient and converting to a g/kg of dry matter basis.

7. Using Food Waste and Rendered Animal By-Products as Animal Feed Ingredients Can Substantially Reduce Several Environmental Impacts of Food Animal Production

Global food animal production contributes about 14.5% (7.1 gigatonnes of CO_2 equivalent) of the total human-induced GHG emissions per year, with the greatest proportion from beef (35.3%) and dairy cattle (30.1%), followed by swine (9.5%) and poultry (8.7%) [26]. Of the total GHG emissions attributed to global food animal production, about 46.7% is associated with the production, processing, and transport of feed, followed by about 39.1% attributed to enteric methane emissions from ruminants, with about 9.5% attributed to methane and N_2O emissions from manure storage [26]. Because feed production, processing, and transport represent the greatest proportion of GHG emissions in all food animal sectors, the greatest opportunity to significantly reduce GHG emissions is represented by using feed ingredients that have less environmental impact.

Historically, decisions on feed ingredient selection and use were determined almost exclusively on price and the economics of animal production (i.e., feed cost/kg of body weight gain, margin over feed cost) [111,112]. Although the economic value of feed ingredients will always be the primary consideration when selecting feed ingredients, consideration for their contribution toward reducing the environmental impacts of animal production is a rapidly emerging trend in the global feed and food animal production industries [113]. This has led to the development of life cycle assessment (LCA) databases of feed ingredients to use in developing "eco-nutrition" feeding programs.

Life cycle assessment of environmental impacts of food production systems has become a widely accepted reference method for guiding decisions and transitioning toward more globally sustainable production and consumption patterns [114]. However, although methodology for LCA has been standardized and guidelines have been published for evaluating environmental performance of animal feed supply chains (101), LCAs have some limitations [115,116].

A limited number of databases of feed ingredients with LCA indicator estimates have been developed and generally only include types of ingredients that are approved for use in the European Union. Because of regulatory restrictions on the use of food waste and rendered animal by-products in animal feeds in the EU, LCA estimates are either absent or limited to only a few by-products. The largest LCA database (962 feed ingredients) with the most LCA indicators (n = 18) and the greatest global application (EU, U.S., Canada) was developed by the Global Feed LCA Institute (GFLI) (https://tools.blonkconsultants.nl/tool/16/). However, it has estimates only for spray dried blood meal, animal protein meals, and animal fats derived from beef, pigs, and poultry with no other food waste sources. Although the animal by-products described in this database are somewhat generic, they do provide a general basis for comparing rendered animal protein by-products with various other feed ingredients using the 18 environmental impact indicators. For example, in comparison with Brazilian soybean meal, using these animal protein meals in animal feed would have less impact on global warming, including land-use change, land use, human carcinogenic and non-carcinogenic toxicity, terrestrial ecotoxicity, human health and terrestrial ecosystem effects from ozone formation, mineral resource scarcity, freshwater eutrophication, and marine eutrophication.

In contrast, the Feed Print database [117] includes 274 feed ingredients used under conditions in the Netherlands, and include LCA estimates of only seven environmental indicators for three sources of food waste (ground biscuits, bakery meal, and potato crisps; Figure 1) and four sources of rendered by-products (meat meal, meat and bone meal, hydrolyzed feather meal, and spray dried blood meal; Figure 2). Users of this database specify the level of calculations (supplier of milk products and compound feeds from a feed mill, supplier of by-products and roughage without a feed mill, or farm production including downstream emissions on farm). In addition, the method of land-use change allocation (area-specific or crop-specific) and the allocation method (economic, mass balance, or gross energy) used for the production of feed is specified. The LCA values presented in Figures 1 and 2 are based on calculations for a supplier of by-products without a feed mill, crop-specific land-use change allocation, and the economic allocation method for the production of feed. With the exception of meat meal, all other food waste and rendered animal by-products have less environmental impact on climate change with land-use change; marine, freshwater, and terrestrial eutrophication; acidification; mineral, fossil fuel, and renewable resource depletion; and land surface use than corn and soybean meal. Therefore, despite the limited LCA data available, using most sources of food wastes and common types of rendered animal by-products as partial or complete replacements for corn and soybean meal can dramatically reduce multiple environmental impacts and provide significant advantages as the global feed and food animal production industries begin implementing eco-nutrition feeding programs. However, much research is needed to expand the list of food wastes and rendered animal by-products in global LCA feed ingredient databases.

8. Real and Perceived Biosafety Risks of Rendered Animal By-Products and Food Waste

Biosafety of feed ingredients and biosecurity of feed supply chains have been an important part of feed ingredient sourcing and feed manufacturing for many decades. The feed industry plays an important role in minimizing the risk of transmission of prions, parasites, bacteria, and viruses from feed to animals and, in some cases, from animals to human food. The risk of transmission of hazardous biological agents is one of the main reasons that food waste and rendered animal by-products have not been used to their fullest potential in animal feeds around the world. Despite global access to all published research information on feed safety risks, some countries like Japan and South Korea have regulations and processes in place to minimize the risk of disease transmission and promote the use of food waste in animal feeds, while other regions like the European Union have regulations that are very restrictive and reflect a different perception of feed safety risks of using food waste and rendered animal by-products in animal feed. It seems that now is an appropriate time to reevaluate the potential biosafety risks and determine if the utilization of these abundant and underutilized food waste and rendered animal by-product nutritional resources can be increased in animal feeds by improving biosecurity process controls, managing critical biosafety risk factors, and overcoming unwarranted feed safety concerns. This process begins with a review of biosafety risks of various food waste and carcass disposal methods.

8.1. Comparison of Biosafety Risks of Carcass Disposal Methods

Prevention of transmission of disease-causing biological agents to animals and humans is an essential consideration when determining the most appropriate disposal method for food waste and animal carcasses. Unfortunately, no reviews have been published to compare the biosafety risks of various food waste disposal methods, but Gwyther et al. [34] conducted a comprehensive review of the socioeconomic, human health, biosecurity, and environmental impacts of various carcass disposal methods, which have relevance to animal-derived food waste sources. As shown in Table 6, there is limited research information available to completely assess the benefits and limitations of various carcass disposal methods for biosecurity, but based on this summary, the most effective carcass disposal methods for preventing transmission of pathogens and prions (biological agents in ruminant animal tissues associated with TSE) are the use of on-farm incineration and alkaline hydrolysis.

Unfortunately, the use of incineration and alkaline hydrolysis methods for disposal of large amounts of animal carcasses is infeasible, particularly during disease epidemics when millions of animal mortalities require immediate disposal. Rendering, composting, anaerobic digestion, and burial methods are more appropriate for handling large volumes of animal mortalities than incineration and alkaline hydrolysis. Optimal carcass disposal should be based on multiple criteria using a holistic assessment of economics, value and extent of resource recovery, biosecurity and risk of disease transmission, and environmental impacts. The Food and Agriculture Organization of the United Nations has suggested that restricting the use of rendered animal by-products in animal feed may result in severe economic and environmental problems, and facilitate disease spread to animals and humans and loss of valuable nutrients [118]. Furthermore, according to Hamilton [46], the United Kingdom Department of Health reported a relative comparison of the potential human health risks from carcass disposal using rendering, incineration, landfill, pyre, and burial, and noted that rendering had minimal impact on every health hazard except prions associated with TSE (Table 7).

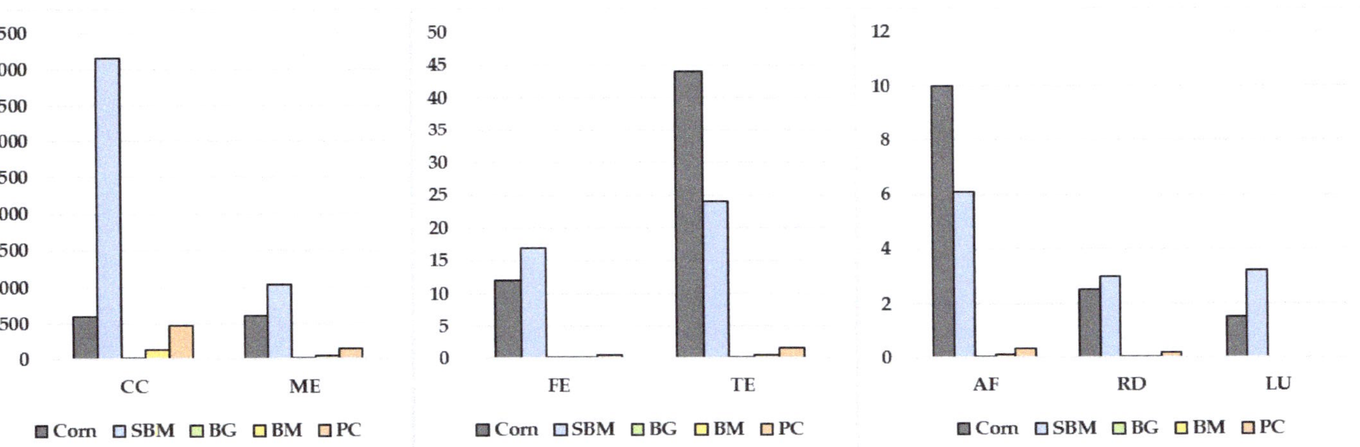

Figure 1. Comparison of environmental impacts of climate change with land-use change potential (CC, g CO_2 equivalent/kg); marine eutrophication potential (ME, mg N equivalent/kg); freshwater eutrophication potential (FE, mg P equivalent/kg); terrestrial eutrophication (TE, mmol N equivalent/kg); acidification (AF, mmol H^+ equivalent/kg); mineral, fossil fuel, and renewable resource depletion potential (RD, mg SB equivalent/kg); and land surface use (LU, m^2/kg) for corn, soybean meal (SBM), ground biscuits (BG), bread meal (BM), and potato crisps (PC) from the FeedPrint LCA ingredient database [118]. The LCA values are based on calculations for a supplier of by-products without a feed mill, crop-specific land-use change allocation, and the economic allocation method for the production of feed.

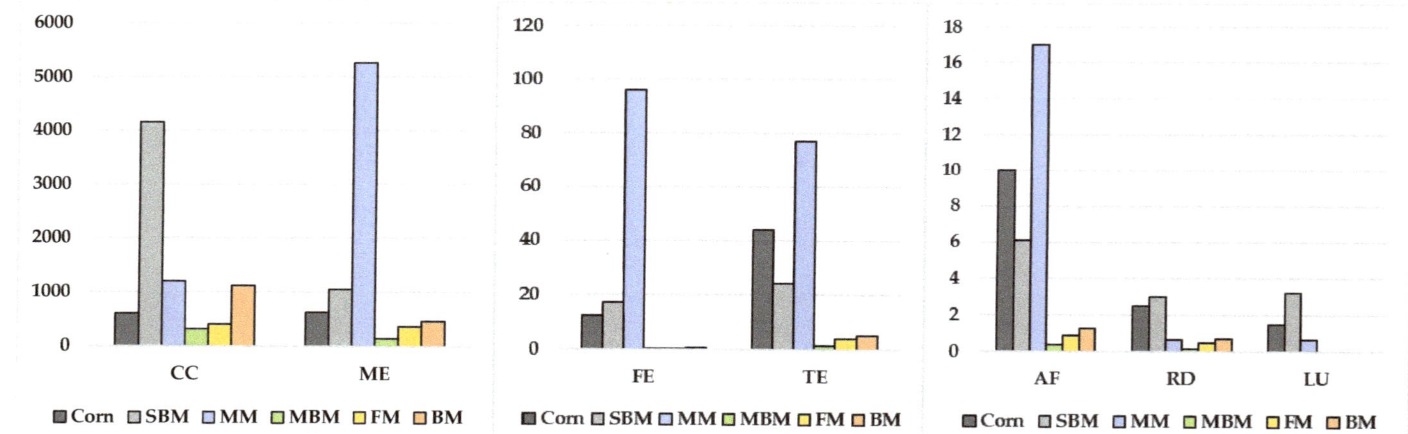

Figure 2. Comparison of environmental impacts of climate change with land-use change potential (CC, g CO_2 equivalent/kg); marine eutrophication potential (ME, mg N equivalent/kg); freshwater eutrophication potential (FE, mg P equivalent/kg); terrestrial eutrophication (TE, mmol N equivalent/kg); acidification (AF, mmol H^+ equivalent/kg); mineral, fossil fuel, and renewable resource depletion potential (RD, mg SB equivalent/kg); and land surface use (LU, m^2/kg) for corn, soybean meal (SBM), meat meal (MM), meat and bone meal (MBM), feather meal (FM), and blood meal (BM) from the FeedPrint LCA ingredient database [118].]. The LCA values are based on calculations for a supplier of by-products without a feed mill, crop-specific land-use change allocation, and the economic allocation method for the production of feed.

Table 6. Summary of relative biosecurity risks [1] of various livestock mortality disposal methods routinely used around the world (adapted from [34]).

Disposal Method	Human Health	Pathogen Contamination					
	Dioxins/Furans	Air	Soil and Vegetation	Water	Land Application of Waste	Transport of Animals Off-Farm	Prion Destruction
Burial	Very low	Low	Moderate	??	NA	Very low	Very high
Burning	High	??	??	??	??	Very low	Moderate
Incineration, on-farm [a]	Low	Very low [b]	Very low [b]	Very low [b]	??	Very low	Very low
Incineration, large central facility	Moderate	Very low [b]	Very low [b]	Very low [b]	??	Very high	Very low
Rendering	??	Very low	NA	??	NA	Very high	Low
Composting [c]	??	Moderate	Moderate	??	??	Very low	Moderate
Anaerobic digestion	??	Low	Moderate	Moderate	??	Very low	High
Alkaline hydrolysis	??	Very low	Very low	Very low	Very low	Very low	Very low

[1] Relative impacts are ranked based on Very high, High, Moderate, Low, Very low environmental impacts in each category; ?? = more research is needed due to limited information; NA = not applicable. [a] Assumes use of afterburners. [b] Omits pre-incineration handling and storage of carcasses that may result in biosecurity risks. [c] Assumes unlined static pile with no forced aeration.

Table 7. Summary of potential health risks of various animal carcass and tissue disposal methods (adapted from [46]).

Hazardous Agent	Rendering	Incineration	Landfill	Pyre	Burial
Campylobacter, Escherichia coli, Listeria, Salmonella, Bacillus anthracis, Clostridium botulinum, Leptospira, Mycobaterium tuberculosis var *bovis, Yersinia*	Low	Low	Some	Low	High
Cryptosporidium, Giardia	Low	Low	Some	Low	High
Clostridium tetani	Low	Low	Some	Low	High
Prions for transmissible spongiform encephalopathies	Some	Low	Some	Some	High
Methane, Carbon dioxide	Low	Low	Some	Low	High
Fuel-specific chemicals, Metal salts	Low	Low	Low	High	Low
Particulates, sulfur dioxide, nitrous oxide, nitrous particles	Low	Some	Low	High	Low
Polycyclic aromatic hydrocarbons, Dioxins	Low	Some	Low	High	Low
Disinfectants, Detergents	Low	Low	Some	Some	High
Hydrogen sulfide	Low	Low	Some	Low	High
Radiation	Low	Some	Low	Some	Some

Gooding and Meeker [44] compared differences in biosecurity, greenhouse gas emissions, resource recovery, and environmental regulations from using anaerobic digestion, composting, and rendering processes to dispose of large quantities of animal carcasses. Results of their analysis showed that the economic value of rendered by-products was at least three times greater than if carcasses were processed using anaerobic digestion, and at least five times greater than if carcasses were composted. Using estimates of the carbon footprint of the rendering process [119], rendering contributed the least amount to total greenhouse gas emissions (CO_2 equivalent) from processing 1000 kg of carcass by-products, where 2500–4000 kg CO_2 equivalent was produced from composting, 60–500 kg CO_2 equivalent was generated from anaerobic digestion, and only 200 kg CO_2 equivalents were produced from rendering [44]. Furthermore, rendering has less biosecurity risk compared with composting, where rendering has been shown to reduce prion infectivity by two logs [120], while composting does not destroy prions, spore-forming bacteria, and other pathogens [121,122]. More research is needed to determine if the use of anaerobic digesters is adequate for destroying prions and pathogens in carcass materials [123]. Despite limited information on the biosecurity of various carcass disposal methods, rendering has many advantages over all other methods, especially for biosecurity, impacts on climate change, economics, and the potential for effectively managing large volumes of animal carcasses resulting from the COVID-19 pandemic and African Swine Fever epidemic.

8.2. Potential Biosafety Risks of Feeding Rendered Animal By-Products to Food-Producing Animals

Historically, the two main biosafety concerns involving feeding rendered animal by-products to food-producing animals are the risks of transmission of *Salmonella* and TSEs, especially bovine spongiform encephalopathy (BSE) [46]. However, in recent years widespread animal disease epidemics such as Porcine Epidemic Diarrhea Virus (PEDV) and African Swine Fever Virus (ASFV) have caused increased scrutiny of biosecurity risks of animal disease transmission through animal-derived feed ingredients and led to their restricted use or elimination in some swine diets [124]. The ultimate determination of the biosafety of feeding rendered animal by-products to animals is primarily dependent on the capability of the thermal treatment conditions used during the rendering process to inactivate pathogenic bacteria, viruses, parasites, and prions.

8.2.1. Adequate Thermal Processing Minimizes Feed Safety Risks of Rendered Animal By-Products

In the United States, dry rendering is the most common process used in either batch or continuous systems, where heat (120 °C to 135 °C) produced by steam condensation is applied and uniformly distributed to ground carcass material for 45 min to 1.5 h under pressure (2.8–4.2 bar), while wet rendering uses high pressure and 140 °C [125]. Gwyther et al. [34] indicated that the European Union requires that for carcasses at high risk for BSE, rendering must be done under processing conditions of 133 °C for 20 min at 300 kPa or equivalent. When comparing these thermal processing conditions used in the rendering process with the temperature and time required to inactivate various prions, parasites, and pathogenic bacteria and viruses in animal tissue matrices (Table 8), it is clear that these hazardous biological agents are destroyed and are no longer infective, except for prions associated with TSE in mammals.

Table 8. Time and temperature to inactivate disease-causing biological agents (prions, parasites, bacteria, and viruses) in animal tissue matrices.

Biological Agent	Temperature and Time for Inactivation	Reference
Prions		
Bovine spongiform encephalitis	136–138 °C for 18 min at 2 bar (29.4 psi)	[125]
Parasites		
Trichinella spiralis	55 °C for 6 min 60 °C for 2 min	[126]
Toxoplasma gondii	60 °C for 1 min	[127]
Bacteria		
Salmonella	80 °C for 30 min	[128]
Escherichia coli	65 °C for 20 min	[128]
Viruses		
African swine fever virus	56 °C for 70 min or 60 °C for 20 min	[129]
Classical swine fever virus	65.5 °C for 30 min or 71 °C for 1 min	[129]
Highly pathogenic avian influenza virus H5 and H7	74 °C for 3.5 s	[129]
Newcastle disease virus	56 °C for 3 h or 60 °C for 30 min	[129]
Foot-and-mouth disease virus	70 °C for 30 min	[129]

Although increasing the time, temperature, and pressure necessary to completely inactivate prions and their infectivity seems possible, it is unclear why this approach has not been adopted in the rendering industry, other than perhaps because of an increase in cost, reduced throughput, and potential loss of some nutritional value of by-products. In fact, Taylor [130] reported that prions are completely inactivated when rendered materials are subjected to 132 °C for 4.5 h at 3 bar (45 psi).

8.2.2. Salmonella

For many decades, *Salmonella* has been one of the most important pathogens to manage in feed ingredients [131,132] and has served as an example of the interrelationships between animal feed safety, food animal production, food processing, public health, and global trade. The first documented evidence of bacterial contamination in the U.S. was in 1948 when *Salmonella* was detected in poultry feed [133]. Since then, numerous studies have documented the presence of *Salmonella* in contaminated feeds around the world [134]. Because feed ingredients can be a potential source of *Salmonella* infection in food-producing animals, regulations to control contamination in feed have existed for many decades in some countries [135]. However, despite many years of research involving factors causing contamination and mitigation strategies to inactivate and control *Salmonella* in feed supply and manufacturing chains, it continues to be a challenge to achieving a *Salmonella* negative standard for animal feeds [135]. Although microbial contamination has been shown in many types of feed ingredients and animal feeds, animal by-product meals have generally been considered to be at highest risk [136].

Concerns about the presence of *Salmonella* in rendered animal by-products (fats and proteins) have played a historic role in government regulations and use of animal by-products in animal feeds in many countries. Beginning as early as 1958, multiple studies have shown that many different serotypes of *Salmonella* were identified in feeds containing animal by-products [137–141]. This subsequently led to applying preventative controls based on Hazard Analysis and Critical Control Point principles to the manufacturing process [142]. However, published surveys from several different countries and time periods show that *Salmonella* contamination can occur not only in animal protein by-products, but also in grains, fish meal, and plant protein sources (Table 9).

Table 9. Summary of surveys reporting percentage of samples contaminated with Salmonella in various types of feed ingredients.

Country	Animal Proteins	Plant Proteins	Grains	Fishmeal	Reference
Canada	20	18	5	22	[143]
Germany	6	26	3	-	[144]
Netherlands	6	3	-	-	[145]
United Kingdom	3	7	1	22	[146]
United States	33	10	0	10	[147]

Furthermore, although *Salmonella*-contaminated feed is one of many risk factors for introduction on swine farms, non-feed sources represent much greater risk [135]. In fact, results from field trials have shown that raw materials contaminated with foodborne pathogens, including *Salmonella* spp., *Listeria monocytogenes*, *Campylobacter jejuni*, and *Clostridium perfringens*, were inactivated using the time and temperature conditions employed during the rendering process (Table 10) [148]. Davies and Funk [149] reported that there is a general perception that animal-derived by-products pose the greatest risk of *Salmonella* contamination, but plant-based ingredients can also be contaminated with *Salmonella*. Fortunately, Salmonella can be easily inactivated by processing ingredients at 55 °C for 1 h or at 60 °C for 15–20 min [143].

Table 10. Effectiveness of the rendering process to inactivate pathogenic bacteria (adapted from [148]).

	Percentage of Positive Samples [a], %	
Pathogen	**Unprocessed Raw Material**	**Rendered Final Product**
Campylobacter jejuni	20.0	0
Campylobacter spp.	29.8	0
Clostridium perfringens	71.4	0
Listeria monocytogenes	8.3	0
Listeria spp.	76.2	0
Salmonella spp.	84.5	0

[a] Samples were collected from 17 different rendering facilities in the United States during summer and winter seasons.

8.2.3. Bovine Spongiform Encephalitis

The risk of bovine spongiform encephalitis (BSE) transmission through ruminant-derived feed ingredients has had dramatic effects on limiting the use of rendered animal by-products and food waste sources containing animal tissues in the global feed industry. Bovine spongiform encephalopathy is part of a group of TSEs that are fatal degenerative diseases of the brain in adult cattle caused by consuming prions (an abnormal form of prion protein attached to the surface of nerve cells) from ruminant by-products such as meat and bone meal containing nervous tissue derived from infected animals [150–152].

The first case of BSE occurred in the United Kingdom in 1985, and it was widely accepted to have resulted from feeding meat and bone meal of ruminant origin to cattle [153]. Shortly thereafter, the European Union implemented regulations that prohibited the feeding of animal protein by-products to all food-producing animals, despite no evidence indicating that pigs can be infected with TSE through the consumption of infected ruminant-derived meat and bone meal [154], and no evidence that BSE can be transmitted between pigs if they were fed brain tissue from cattle [155]. Furthermore, Cutlip et al. [156] conducted a study to determine if feeding rendered meat and bone meal and tallow from scrapie-infected sheep would cause BSE in cattle during a one- to eight-year feeding period. No clinical signs, lesions, or presence of prion protein were detected in the spinal cords and brains of calves fed rendered meat and bone meal and tallow at maximum recommended diet inclusion rates in this study [156]. However, despite these favorable results, the perception of risk of BSE transmission through animal by-products has led to government regulations and restrictions regarding the use of animal by-products in various countries today [157].

The presence of BSE is geographically limited to the European Union, United Kingdom, and North America [158–160]. Although a case of BSE in cattle was previously reported in Japan, it was assumed to have been caused by feeding contaminated meat and bone meal imported from the European Union [161]. The United States prohibits the feeding of ruminant-derived rendered animal protein by-products to ruminants, while allowing the feeding of these ruminant-derived by-products to swine and poultry. Australia and New Zealand do not have TSEs but allow using ruminant-derived protein meals in some monogastric animal feeds by controlling ingredient imports, enforcing strict feeding regulations, and using proactive surveillance methods [162].

8.2.4. Swine Corona Viruses

More recently, additional skepticism about the biosafety of animal-derived by-products, especially those of porcine origin, relative to the risk of transboundary transmission of foreign animal disease viruses, has limited the use of animal protein by-products in swine diets in North America. Swine diseases of major concern for transboundary transmission include ASFV, PEDV, Classical Swine Fever, foot-and-mouth disease (FMD), and porcine reproductive and respiratory syndrome (PRRS) viruses because they have spread across country borders [163]. The PEDV outbreak that occurred in North America in 2013 and 2014 created initial concerns regarding transmission of viral pathogens through feed, especially animal protein by-products such as spray dried porcine plasma (SDPP). Although it was initially suspected that PEDV was introduced to a swine farm from a common feed source containing contaminated SDPP [164–168], the definitive source and route of PEDV introduction into Canada or the United States has not been determined [169]. However, because of the high swine mortality and economic losses [170] caused by the PEDV epidemic, the use of SDPP and other porcine-derived by-products in weaned pig diets was significantly reduced because they were perceived as high risk for PEDV transmission. As a result of the PEDV epidemic, several studies were immediately conducted to evaluate survival of PEDV and two other corona viruses, Porcine Delta Coronavirus (PDCoV) and Transmissible Gastroenteritis virus (TGEV), in various feed ingredients if they are contaminated [171]. No differences were observed in the number of days to reduce PEDV concentration by one log (delta value) among animal protein by-products (spray dried porcine plasma, blood meal, meat meal, and meat and bone meal), nor were there differences in delta values between these animal protein by-products and plant-based ingredients (i.e., corn, soybean meal, and corn-dried distillers grains with solubles; Table 11). Furthermore, the number of days to achieve a one log reduction of PDCoV and TGEV were much greater for soybean meal than for all animal protein by-products. Results from this study clearly showed that survival of PEDV is not different among animal protein by-products and common plant-based ingredients, and more strikingly, corn and soybean meal are greater risks to PDCoV and TGEV survival than animal-based feeding ingredients if they are contaminated.

8.2.5. African Swine Fever Virus

The continual spreading of the ASFV outbreak in Eastern Europe, China, and other major swine-producing countries in Southeast Asia has become a major threat to global pork production and food security [172]. Although there are no published data showing natural contamination of ASFV in complete feed and feed ingredients, if certain ingredients such as soybean meal or choline chloride were to be contaminated with ASFV, the virus has been shown to survive and be infective under the time and environmental conditions of trans-Pacific and trans-Atlantic shipping models [173]. The USDA-APHIS conducted a qualitative assessment of the likelihood of the ASFV virus entering the United States from legal and illegal transboundary movements of potentially infected animals and contaminated products from ASFV-affected countries or regions [174]. This assessment suggested that feed ingredients of either animal or plant origin were associated with moderate likelihood for ASFV entry, but with high uncertainty due to lack of data on virus survival throughout the supply chain necessary to cause infection if contaminated. These results suggest that due to lack of data, risk of ASFV transmission was considered similar between rendered animal by-products and plant-based ingredients [174].

Table 11. Comparison of days necessary to reduce Porcine Epidemic Diarrhea Virus (PEDV), Porcine Delta Coronavirus (PDCoV), and Transmissible Gastroenteritis Virus (TGEV) concentration by one log (delta value) in complete feed and various feed ingredients (adapted from [171]).

Feed Ingredient	PEDV	PDCoV	TGEV
Spray dried porcine plasma	1.14	3.25 ^{a}	19.18 ^{a}
Blood meal	2.84	1.23 ^{a}	2.15 ^{a}
Meat meal	3.87	2.82 ^{a}	1.04 ^{a}
Meat and bone meal	4.90	6.22 ^{a}	0.99 ^{a}
Corn	2.25	25.60 ^{b}	11.78 ^{a}
Soybean meal	7.50	42.04 ^{c}	41.94 ^{b}
Low oil DDGS [1]	0.70	6.23 ^{a}	1.04 ^{a}
Medium oil DDGS	7.32	3.76 ^{a}	1.66 ^{a}
High oil DDGS	0.56	8.80 ^{a}	0.78 ^{a}
Complete feed	1.12	2.29 ^{a}	3.20 ^{a}

[1] DDGS = dried distillers grains with solubles. ^{a,b,c} Means with uncommon superscripts differ ($p < 0.05$).

8.3. Different Perspectives of Potential Biosafety Risks of Recycling Food Waste into Animal Feed

Historically, various forms of uncooked ("garbage") or cooked ("swill") food waste have been fed to livestock, especially swine, in many countries for many centuries [175]. However, during the past few decades, specific disease outbreaks have occurred in a few countries that resulted in different biosecurity perspectives and regulations among countries for recycling food waste into animal feed. Countries such as Japan, South Korea, Taiwan, and Thailand have proactively embraced the nutritional and environmental benefits of recycling a high proportion of food waste into animal feed by developing appropriate regulations and infrastructure to accomplish this while minimizing biosafety risks [176]. In Japan, initial government regulations were implemented in 2001, and were later revised in 2007, that prioritize recycling of food waste into animal feed, compared with other disposal options [40]. In fact, about 40% of food waste in Japan from pre-consumer sources (food manufacturing facilities, wholesale, and retail grocery stores) along with lesser amounts from post-consumer sources (restaurants, households) is thermally processed, recycled into animal feed, and trademarked as "EcoFeed" [177]. Similarly, South Korea has implemented regulations, infrastructure, and processes to convert a high proportion of food waste into safe animal feed. Disposal of food waste in landfills was banned in South Korea in 2005 [48], and about 45% is recycled into animal feed, 45% is composted, and the remaining 10% is disposed through anaerobic digestion, vermicomposting, and co-digestion with sewage sludge [39]. Both Japan and South Korea have demonstrated that recycling a large proportion of domestic food waste into animal feed can be successfully accomplished by developing and implementing appropriate regulations and oversight to ensure adequate thermal treatment, storage, and transport of processed food waste to minimize its biosafety risks as a feed ingredient for animals [28,178].

In contrast, feeding food waste to food-producing animals in the European Union was banned in 2002 [179] because the outbreak of foot-and-mouth disease (FMD) that occurred in the UK in 2001 was associated with feeding uncooked food waste to pigs. However, it is interesting to note that the FMD outbreaks that occurred in Japan (2010) and South Korea (2010–2011) had no connections to feeding food waste [180,181]. At about the same time, the BSE crisis occurred in the European Union and resulted in banning the use of all processed animal protein by-products in animal feeds [182]. However, no evidence has ever been observed or reported showing that pigs, poultry, or fish are capable of naturally developing or transmitting prions attributed to TSEs such as BSE [183]. Therefore, from a feed safety perspective, feeding food waste containing animal-derived food products from any species to pigs, poultry, and fish should not be a concern for controlling TSE transmission. As a result of these regulations, only about 3 million tonnes of the estimated total of 102 million tonnes of food waste generated in the European Union is fed to animals [48]. However, if a food waste source can be verified to have no risk of contamination with meat, fish, or other animal by-products, it can be approved for feeding animals. In addition to the influence of FMD and BSE on European Union regulations, concerns about the risk of spreading of avian influenza and Newcastle disease to poultry have also been attributed to feeding food waste in the United Kingdom [184]. Some researchers have questioned whether the European Union goal of eliminating disposal of biodegradable waste in landfills by 2025, as specified in the EU Waste Framework Directive, can be achieved without diverting more food waste into safe animal feed [28]. Dou et al. [48] proposed that by using adequate thermal processing technologies and revising current regulations, the conversion of food waste into animal feed is the only viable option among all disposal alternatives to reduce environmental impact, conserve resources, and improve food security.

In the United States, feeding uncooked or improperly heated processed food waste contaminated with *Salmonella*, *Campylobacter*, *Mycobacterium*, *Trichinella*, *Toxoplasma* [103], and *Clostridium* [185] to swine was viewed as an initial public health concern several decades ago. In fact, *Trichinellosis* was one of the most devastating parasites in pigs and humans in the early 1930s to 1950s and was associated with the feeding of food waste containing meat scraps [186]. However, the United States pork industry is now free of *Trichinella* and the detection of positive cases has been maintained at 0% [187]. Because of the risks of parasite and pathogen transmission from feeding uncooked food waste to pigs, the Swine Health Protection Act was implemented in 1980 that requires food waste containing animal tissues to be thermally processed at 100 °C for 30 min at licensed facilities to destroy these biological hazards before feeding to swine [188]. In addition, feeding food waste containing mammalian tissues to ruminants is prohibited due to the risk of BSE transmission [189]. Current laws and regulations for feeding food waste to swine vary among states [190] and are primarily based on feed safety concerns involving animal-derived food products. Although several states have allowed licensed operations to feed heat-treated food waste to swine in the past, new national efforts are underway to increase processing of food waste into safe animal feed [191].

However, heightened concerns about the potential risk of introduction of African Swine Fever and other foreign animal disease viruses into the United States from food waste containing animal-derived food products obtained from international airlines and cruise ships have become a barrier to achieving this goal [163]. The USDA-APHIS [174] conducted a qualitative assessment of the likelihood of ASFV entering the United States from legal regulated garbage and concluded that there was a low likelihood and moderate uncertainty of ASFV entry from outside the U.S. Biosecurity concerns about feeding food waste to swine have been based on historical references to the risks of transmission of ASFV [192], Classical Swine Fever [193], and swine vesicular disease [194] from feeding food waste collected from international airports that contained infected pork products to swine. Feeding uncooked food waste has also been associated with ASFV transmission in traditional backyard and free-range pig production systems globally [195–197]. The ASFV has been shown to survive in pork meat, fat, and skin for many months [198]. However, by diverting these potentially high-risk food waste sources to non-feed resource recovery and disposal, or by implementing and enforcing strict regulations to ensure adequate thermal processing for complete virus inactivation, these potential biosecurity risks can be avoided.

Adequate Thermal Processing Minimizes Feed Safety Risks of Food Waste

Food waste that has not been adequately heat-processed may potentially contain pathogenic bacteria, prions, and viruses. As a result, countries that allow the recycling and processing of food waste into animal feeds have implemented regulations that require thermal treatment to destroy various biological hazards. Government regulations in the United States require heating food waste for 30 min at 100 °C [188]. In Japan, Ecofeed manufacturers are required to heat food waste containing meat for at least 30 min at 70 °C or for 3 min at 80 °C [178]. In South Korea, all types of food waste must be heat-treated for 30 min at a temperature of at least 80 °C. If the food waste is used in wet feed production, it is first heated to at least 80 °C and then mixed with corn or rice husks to standardize moisture content to about 70–80% [28]. For dry feed production, food waste is dehydrated using hot air (390 °C) for sterilization, which increases shelf life and minimizes nutritional losses from spoilage [28]. In contrast, current EU regulations allow food waste to be used as animal feed only if it can be guaranteed that there is no risk of contamination with meat, fish, or other animal products [28].

With the exception of prions, the time (30 min) and temperature (100 °C) required for thermally processing food waste in the United States are adequate for inactivating the major parasites and pathogenic bacteria and viruses of greatest concern (Table 8). García et al. [70] confirmed that heating food waste at 65 °C for 20 min has been shown to be adequate for reducing *Salmonella*, *Escherichia coli*, and *Staphylococcus aureus* below levels deemed safe for animal feed. Therefore, use of appropriate thermal treatment protocols can adequately inactivate all of the parasites, bacteria, and viruses that may be present in various sources of food waste to minimize feed and food safety risks, even if these sources contain animal products that are of greatest concern.

8.4. Solutions to Overcome Biosafety Concerns of Using Rendered Animal By-Products and Thermally Treated Food Waste in Animal Feed

Managing risk of microbial contamination in all feed ingredient supply chains continues to increase in importance for preventing the spread of animal diseases domestically and internationally, but it requires further research and process development [199]. Minimizing pathogen contamination in feed ingredients is a dynamic process that has changed over time and can involve various frameworks for decision-making [200]. In the United States, the implementation of the Food Safety Modernization Act has shifted the focus of regulating human and animal food safety from responding to contamination toward preventing it [201]. This requires the use of a Hazard Analysis and Risk-Based Preventative Controls (HARPC) system to identify preventative controls to avoid reasonable or foreseeable food safety hazards introduced or present in the human or animal food supply chain from either a domestic or an international source. The HARPC system is broader in scope than the traditional Hazard Analysis and Critical Control Points (HACCP) system, and not only focuses on process preventative controls, but also includes other preventative controls such as supply chain and sanitation preventative controls. Although HARPC plans for animal feed are focused on preventing physical, chemical, and biological food safety hazards, they do not specifically require consideration of viral contamination in feed. However, most principles and preventative controls including process, supply chain, sanitation, and sanitary transport preventative controls can be effectively applied to prevent and control viral contamination in feed ingredients.

As described in this review, the risk of transmission of TSEs from ruminant-derived carcass tissues in rendered by-product meals and food waste is perhaps the greatest concern. It is now possible to accurately identify raw materials that contain TSEs to substantially minimize this risk. Several research groups around the world have developed analytical methodologies to identify species-specific DNA using polymerase chain reaction (PCR) in samples where the DNA may be partially degraded [202,203]. In addition, enzyme-linked immunosorbent assays (ELISAs) that are capable of differentiating skeletal muscle in protein meals from other tissues have been developed [204]. Furthermore, a Surround Optical Fiber Immunoassay (SOFIA) and specific monoclonal antibodies for brain prion protein from hamsters, sheep, and deer appear to be capable of detecting extremely low concentrations of brain prion protein molecules [205]. Therefore, there are several emerging opportunities to provide surveillance of the presence of prions in rendered animal by-products and animal-derived food waste sources to determine if cross contamination occurred in feed supply chains.

New technologies have also been developed and are available to thermally process high moisture food waste sources into dry, pathogen-free feed ingredients. For example, patented technologies, including enzymatic digestion at 55–57 °C, pasteurization at 75–77 °C, filtering, and pH stabilization (2.8 to 3.0), are being used to process small and large particles of supermarket food waste to preserve nutritional quality and digestibility while eliminating pathogens [206].

9. Next Steps

A renaissance is needed to reimagine and redesign all stages of our global food supply chains to better cope with the increased losses of food resources caused by global human, animal, and climate health crises. As described in this review, there is enormous potential to significantly contribute to achieving the UN Sustainability Development Goals of responsible consumption and production, reducing climate change impacts, improving life below water, and improving life on land by repurposing food waste streams from pre-harvest to post-consumer stages of supply chains. Although there is ample justification and incentive to do this, government policies and regulations must be reformed using a more holistic approach that will mandate recovery and recycling of greater amounts of valuable nutrients from various food waste streams into animal feed. Governments could provide economic incentives or initial subsidies to encourage entrepreneurs to develop the necessary modern infrastructure to facilitate collection, provide adequate capacity and modern thermal processing equipment to ensure biosafety of dehydrated waste streams and create market channels that connect these supplies with commercial animal feed manufacturers. High-risk food waste sources that may potentially be contaminated with disease-causing biological agents can be identified and diverted toward other useful recycling processes such as biofuels and biogas production or composting to avoid possible disease transmission.

As the global animal feed industry continues to evolve toward sourcing and using feed ingredients with high nutritional value and low environmental impact, additional Life Cycle Analysis determinations are needed for various sources of dehydrated food waste and rendered animal by-products that animal nutritionists can use when formulating eco-nutrition feeding programs for food-producing animals. However, additional animal nutrition studies are urgently needed to develop more robust and comprehensive ME and digestible nutrient composition databases, and to develop accurate prediction equations of various food waste sources for swine and poultry, to encourage animal nutritionists to fully capture the nutritional and economic value of food waste sources when formulating nutritionally adequate and cost-effective complete animal feeds. Furthermore, new risk assessments should be conducted, and extensive biosecurity protocols should be developed based on best biosafety practices, especially for pathogenic viruses, to minimize risk of pathogen and prion transmission through processed food waste sources used as animal feed. Finally, governments, citizens, entrepreneurs, and all sectors of food supply chains need the courage to build food waste collection and processing infrastructure that is economically and environmentally sustainable, using life cycle assessments as well as regulated and certifiable biosafety conditions to create a new model of food sustainability.

Funding: This research received no external funding.

Conflicts of Interest: The author declares no conflict of interest.

References

1. Kuzmanova, M.; Ivanov, I. Relation between change management and crisis management: Survey evidence. *Int. Conf. Knowl. Based Organ.* **2015**, *25*, 255–260. [CrossRef]
2. Gustavsson, J.; Cederberg, C.; Sonesson, U.; Global Food Losses and Food Waste. Rome (Italy): Food and Agriculture Organization of the United Nations. 2011. Available online: http://www.fao.org/docrep/014/mb060e/mb060e00.pdf (accessed on 6 July 2020).
3. Lipinski, B.; Hanson, C.; Lomax, J.; Kitinoja, L.; Waite, R.; Searchinger, T. Reducing Food Loss and Waste. Available online: http://pdf.wri.org/reducing_food_loss_and_waste.pdf (accessed on 15 August 2020).
4. United Nations. Transforming our world: The 2030 agenda for sustainable development. 2015. Available online: http://www.un.org/ga/search/view_doc.asp?symbol=A/RES/70/1&Lang=E (accessed on 6 July 2020).
5. Hakovirta, M.; Denuwara, N. How COVID-19 redefines the concept of sustainability. *Sustainability* **2020**, *12*, 3727. [CrossRef]

6. ReFED 2020 COVID-19, U.S. Food System Review. Available online: https://www.refed.com/content-hub/refeds-covid-19-u-s-food-system-review/ (accessed on 16 August 2020).
7. Jámbor, A.; Czine, P.; Balogh, P. The impact of the Coronavirus on agriculture: First evidence based on global newspapers. *Sustainability* **2020**, *12*, 4535. [CrossRef]
8. Aldaco, R.; Hoehn, D.; Laso, J.; Margallo, M.; Ruiz-Salmón, J.; Cristobal, J.; Kahhat, R.; Villanueva-Rey, P.; Bala, A.; Batlle-Bayer, L.; et al. Food waste management during the COVID-19 outbreak: A holistic climate, economic and nutritional approach. *Sci. Total Environ.* **2020**, *742*, 140524. [CrossRef] [PubMed]
9. COVID-19 Response Team. *Update: COVID-19 Among Workers in Meat and Poultry Processing Facilities—United States, April–May 2020*; Morbidity and Mortality Weekly Report; US Department of Health and Human Services, Centers for Disease Control and Prevention: Atlanta, GA, USA, 2020; Volume 69, pp. 887–892.
10. Hadrich, J.; Roberts, M.; Tuck, B. The role of hog farmers in Minnesota's rural economy. In *Hog Economic Report 2020*; University of Minnesota Extension: St. Paul, MN, USA, 2020; pp. 1–6.
11. McEwan, K.; Marchand, L.; Shang, M.; Bucknell, D. Potential implications of COVID-19 on the Canadian pork industry. *Can. J. Agric. Econ.* **2020**, *68*, 201–206. [CrossRef]
12. Roembke, J. 7 Ways COVID-19 has Impacted the US Feed Industry. Feed Strategy. June 2020. pp. 5–9. Available online: www.FeedStrategy.com (accessed on 6 July 2020).
13. Rogers, E.; Rozeboom, D.; Zangaro, C. In Times of Supply Chain Disruption, how do I Appropriately Dispose of My Livestock Mortalities? Michigan State University Extension Newsletter. 2020. Available online: https://www.canr.msu.edu/news/in-times-of-supply-chain-disruption-how-do-i-appropriately-dispose-of-my-livestock-mortalities (accessed on 15 August 2020).
14. Penrod, E. How COVID-19 will Reshape Feed Industry Supply Chains. Feed Strategy. June 2020. pp. 14–17. Available online: www.FeedStrategy.com (accessed on 6 July 2020).
15. Galanakis, C.M. The food systems in the era of the coronavirus (COVID-19) pandemic crisis. *Foods* **2020**, *9*, 523. [CrossRef]
16. Gu, H.; Daly, T. China has Shown 'Shortcomings' in Bid to Contain African Swine Fever. Reuters. 3 July 2019. Available online: https://uk.reuters.com/article/us-china-swinefever-policy/china-has-shown-shortcomings-in-bid-to-contain-african-swine-fever-cabinet-idUKKCN1TY15E (accessed on 15 August 2020).
17. Pan, C. African Swine Fever Affects China's Pork Consumption. Rabobank. June 2019. Available online: https://research.rabobank.com/far/en/sectors/animal-protein/african-swine-fever-affects-china-s-pork-consumption.html (accessed on 15 August 2020).
18. Daly, J.; Birtles, B. China Struggles to Contain African Swine Fever, Resorts to Mass Live-Pig Burials, Millions of Culls. ABC Rural News, 29 May 2019. Available online: https://www.abc.net.au/news/rural/rural-news/2019-05-30/mass-live-pig-burials-millions-culled-china-african-swine-fever/11146642 (accessed on 6 July 2020).
19. Capua, I.; Marangon, S. Control of avian influenza in poultry. *Emerg. Infect. Dis.* **2006**, *12*, 1319–1324. [CrossRef]
20. Morse, S.S. Factors in the emergence of infectious diseases. *Emerg. Infect. Dis.* **1995**, *1*, 7–15. [CrossRef]
21. Jones, K.E.; Patel, N.G.; Levy, M.A.; Storeygard, A.; Balk, D.; Gittleman, J.L.; Daszak, P. Global trends in emerging infectious diseases. *Nature* **2008**, *451*, 990–993. [CrossRef]
22. Madhav, N.; Oppenheim, B.; Gallivan, M.; Mulembakani, P.; Rubin, E.; Wolfe, N. Pandemics: Risks, impacts, and mitigation. In *Disease Control Priorities*, 3rd ed.; Improving Health and Reducing Poverty; International Bank for Reconstruction and Development; The World Bank: Washington, DC, USA, 2017; Volume 9, pp. 315–345.
23. Pudenz, C.C.; Schulz, L.L.; Tonsor, G.T. Adoption of secure pork supply plan biosecurity by U.S. swine producers. *Front. Vet. Sci.* **2019**, *6*, 146. [CrossRef]
24. Herrero, M.; Thornton, P.K.; Gerber, P.; Reid, R.S. Livestock, livelihoods and the environment: Understanding the trade-offs. *Curr. Opin. Environ. Sustain.* **2009**, *1*, 111–120. [CrossRef]
25. Foley, J.A.; Ramankutty, N.; Brauman, K.A.; Cassidy, E.S.; Gerber, J.S.; Johnston, M.; Mueller, N.D.; O'Connell, C.; Ray, D.K.; West, P.C.; et al. Solutions for a cultivated planet. *Nature* **2011**, *478*, 337–342. [CrossRef] [PubMed]
26. Gerber, P.J.; Steinfeld, H.; Henderson, B.; Mottet, A.; Opio, C.; Dijkman, J.; Falcucci, A.; Tempio, G. *Tackling Climate Change Through Livestock—A Global Assessment of Emissions and Mitigation Opportunities*; Food and Agriculture Organization of the United Nations (FAO): Rome, Italy, 2013.
27. Papargyropoulou, E.; Lozano, R.; Steinberger, J.K.; Wright, J.; Ujang, N.; Bin, Z. The food waste hierarchy as a framework for the management of food surplus and food waste. *J. Clean. Prod.* **2014**, *76*, 106–115. [CrossRef]
28. Zu Ermgassen, E.K.H.J.; Phalen, B.; Green, R.E.; Balmford, A. Reducing the land use of EU pork production: Where there's swill, there's a way. *Food Policy* **2016**, *58*, 35–48. [CrossRef] [PubMed]

29. Gardner, G.; Assadourian, E.; Sarin, R. The State of Consumption Today. State of the World. 2004. Available online: http://erikassadourian.com/wp-content/uploads/2013/07/SOW-04-Chap-1.pdf (accessed on 6 July 2020).
30. Gandenberger, C.; Garrelts, H.; Wehlau, D. Assessing the effects of certification networks on sustainable production and consumption: The cases of FLO and FSC. *J. Consum. Policy* **2011**, *34*, 107–126. [CrossRef]
31. Grunert, K.G.; Hieke, S.; Wills, J. Sustainability labels on food products: Consumer motivation, understanding and use. *Food Policy* **2014**, *44*, 177–189. [CrossRef]
32. USDA-APHIS. *Carcass Management during a Mass Animal Health Emergency*; Final Programmatic Environmental Impact Statement, December 2015; National Center for Animal Health Emergency Management, Veterinary Services, Animal and Plant Health Inspection Service; U.S. Department of Agriculture: Riverdale, MD, USA, 2015; p. 252.
33. European Union Regulation 1069/2009. Available online: https://eur-lex.europa.eu/legal-content/EN/TXT/PDF/?uri=CELEX:02009R1069-20191214&from=EN (accessed on 6 July 2020).
34. Gwyther, C.L.; Williams, A.P.; Golyshin, P.N.; Edwards-Jones, G.; Jones, D.L. The environmental and biosecurity characteristics of livestock carcass disposal methods: A review. *Waste Manag.* **2011**, *31*, 767–778. [CrossRef]
35. Eriksson, M.; Strid, I.; Hansson, P. Carbon footprint of food waste management options in the waste hierarchy—A swedish case study. *J. Clean. Prod.* **2015**, *93*, 115–125. [CrossRef]
36. Lee, S.-H.; Choi, K.-I.; Osako, M.; Dong, J.-I. Evaluation of environmental burdens caused by changes of food waste management systems in Seoul, Korea. *Sci. Total Environ.* **2007**, *387*, 42–53. [CrossRef]
37. Ogino, A.; Hirooka, H.; Ikeguchi, A.; Tanaka, Y.; Waki, M.; Yokoyama, H.; Kawashima, T. Environmental impact evaluation of feeds prepared from food residues using life cycle assessment. *J. Environ. Qual.* **2007**, *36*, 1061–1068. [CrossRef]
38. Kim, M.-H.; Kim, J.-W. Comparison through a LCA evaluation analysis of food waste disposal options from the perspective of global warming and resource recovery. *Sci. Total Environ.* **2010**, *408*, 3998–4006. [CrossRef] [PubMed]
39. Kim, M.-H.; Song, Y.-E.; Song, H.-B.; Kim, J.-W.; Hwang, S.-J. Evaluation of food waste disposal options by LCC analysis from the perspective of global warming: Jungnang case, South Korea. *Waste Manag.* **2011**, *31*, 2112–2120. [CrossRef] [PubMed]
40. Takata, M.; Fukushima, K.; Kino-Kimata, N.; Nagao, N.; Niwa, C.; Toda, T. The effects of recycling loops in food waste management in Japan: Based on the environmental and economic evaluation of food recycling. *Sci. Total Environ.* **2012**, *432*, 309–317. [CrossRef] [PubMed]
41. Tufvesson, L.M.; Lantz, M.; Borjesson, P. Environmental performance of biogas produced from industrial residues including competition with animal feed—life cycle calculations according to different methodologies and standards. *J. Clean Prod.* **2013**, *53*, 214–223. [CrossRef]
42. Vandermeersch, T.; Alvarenga, R.A.F.; Ragaert, P.; Dewulf, J. Environmental sustainability assessment of food waste valorization options. *Resour. Conserv. Recycl.* **2014**, *87*, 57–64. [CrossRef]
43. Salemdeeb, R.; zu Ermgassen, E.K.H.J.; Kim, M.H.; Balmford, A.; Al-Tabbaa, A. Environmental and health impacts of using food waste as animal feed: A comparative analysis of food waste management options. *J. Clean. Prod.* **2017**, *140*, 871–880. [CrossRef]
44. Gooding, C.H.; Meeker, D.L. Review: Comparison of 3 alternatives for large-scale processing of animal carcasses and meat by-products. *Prof. Anim. Sci.* **2016**, *32*, 259–270. [CrossRef]
45. Mottet, A.; de Haan, C.; Falcucci, A.; Tempio, G.; Opio, C.; Gerber, P. Livestock: On our plates or eating at the table? A new analysis of the feed/food debate. *Global Food Secur.* **2017**, *14*, 1–8. [CrossRef]
46. Hamilton, C.R. Real and perceived issues involving animal protein. In *Protein Sources for the Animal Feed Industry, Proc. Expert Consultation and Workshop, Bangkok, 28 April–3 May, 2002*; Food and Agriculture Organization of the United Nations: Rome, Italy, 2004.
47. FAO. FAOSTAT: Statistical Databases. 2020. Available online: http://www.fao.org/faostat/en/#data/ (accessed on 6 July 2020).
48. Dou, Z.; Toth, J.D.; Westendorf, M.L. Food waste for livestock feeding: Feasibility, safety, and sustainability. *Global Food Secur.* **2018**, *17*, 154–161. [CrossRef]
49. Food and Agriculture Organization. *Nutrient Flows and Associated Environmental Impacts in Livestock Supply Chains: Guidelines for Assessment (Version 1)*; Livestock Environmental Assessment and performance (LEAP) Partnership; FAO: Rome, Italy, 2018; p. 196.

50. Steffen, W.; Richardson, K.; Rockström, J.; Cornell, S.E.; Fetzer, I.; Bennett, E.M.; Biggs, R.; Carpenter, S.R.; de Vries, W.; de Wit, C.A. Planetary boundaries: Guiding human development on a changing planet. *Science* **2015**, *347*, 1259855. [CrossRef]
51. Bouwman, L.; Goldewijk, K.K.; Van Der Hoek, K.W.; Beusen, A.H.W.; Van Vuuren, D.P.; Willems, J.; Rufino, M.C.; Stehfest, E. Exploring global changes in nitrogen and phosphorus cycles in agriculture induced by livestock production over the 1900–2050 period. *Proc. Natl. Acad. Sci. USA* **2013**, *110*, 20882–20887. [CrossRef] [PubMed]
52. Sutton, M.A.; Bleeker, A.; Howard, C.; Bekunda, M.; Grizzetti, B.; de Vries, W.; van Grinsven, H.; Abrol, Y.; Adhya, T.; Billen, G.; et al. Our nutrient world: The challenge to produce more food and energy with less pollution. In *Global Overview of Nutrient Management*; Sutton, M.A., Bleeker, A., Howard, C.M., Bekunda, M., Grizzetti, B., de Vries, W., van Grinsven, H.J.M., Abrol, Y.P., Adhya, T.K., Billen, G., et al., Eds.; CEH/UNEP: Edinburgh, UK, 2013.
53. Meena, V.S.; Verma, J.P.; Meena, S.K. Towards the current scenario of nutrient use efficiency in crop species. *J. Clean. Prod.* **2015**, *102*, 556–557. [CrossRef]
54. Oenema, O.; Ju, X.; de Klein, C.; Alfaro, M.; del Prado, A.; Lesschen, J.P.; Zheng, X.; Velthof, G.; Ma, L.; Gao, B.; et al. Reducing nitrous oxide emissions from the global food system. *SI Syst. Dyn. Sustain.* **2014**, *9–10*, 55–64. [CrossRef]
55. Alexandratos, N.; Bruinsma, J. *World Agriculture towards 2030/2050: The 2012 Revision*; ESA Workshop Paper 12-03; Agricultural Development Economics Division; FAO: Rome, Italy, 2012.
56. Kim, S.W.; Less, J.F.; Wang, L.; Yan, Y.; Kiron, V.; Kaushik, S.J.; Lei, Z.G. Meeting global feed protein demand: Challenge, opportunity, and strategy. *Annu. Rev. Anim. Biosci.* **2019**, *7*, 221–243. [CrossRef] [PubMed]
57. Cordell, D.; White, S. Life's bottleneck: Sustaining the world's phosphorus for a food secure future. *Ann. Rev. Environ. Resour.* **2014**, *39*, 161–188. [CrossRef]
58. Smil, V. Phosphorus in the environment: Natural flows and human interferences. *Annu. Rev. Energy Environ.* **2000**, *25*, 53–88. [CrossRef]
59. Cordell, D.; Drangert, J.-O.; White, S. The story of phosphorus: Global food security and food for thought. *Global Environ. Chang.* **2009**, *19*, 292–305. [CrossRef]
60. Fixen, P. Phosphorus: Worldwide supplies and efficiency. In Proceedings of the Presented at the Manitoba Agronomists Conference, Winnipeg, MB, Canada, 16 December 2009.
61. van Vuuren, D.P.; Bouwman, A.F.; Beusen, A.H.W. Phosphorus demand for the 1970–2100 period: A scenario analysis of resource depletion. *Global Environ. Chang.* **2010**, *20*, 428–439. [CrossRef]
62. van Kauwenbergh, S.J. *World Phosphorus Rock Reserves and Resources*; Tech. Bull. T-75; International Fertilizer Development Center (IFDC): Muscle Shoals, AL, USA, 2010.
63. Cordell, D.; White, S.; Lindström, T. Peak Phosphorus: The crunch time for humanity? *Sustain. Rev.* **2011**, *2*, 1. Available online: https://www.thesustainabilityreview.org/articles/peak-phosphorus-the-crunch-time-for-humanity (accessed on 29 August 2020).
64. Mohr, S.; Evans, G. Projections of Future Phosphorus Production. PHILICA.COM, Article 380, 9 July 2013. Available online: http://www.resilience.org/wp-content/uploads/articles/General/2013/09_Sep/peak-phosphorus/Phosphorus%20Projections.pdf (accessed on 6 July 2020).
65. Walan, P. Modeling of Peak Phosphorus a Study of Bottlenecks and Implications. Master's Thesis, Department of Earth Sciences, Uppsala University, Uppsala, Sweden, 2013.
66. Carpenter, S.; Caraco, N.F.; Correll, D.L.; Howarth, R.W.; Sharpley, A.N.; Smith, V.H. Nonpoint pollution of surface waters with phosphorus and nitrogen. *Ecol. Appl.* **1998**, *8*, 559–568. [CrossRef]
67. Smith, V.; Schindler, D. Eutrophication science: Where do we go from here? *Trends Ecol. Evol.* **2009**, *24*, 201–207. [CrossRef] [PubMed]
68. Leinweber, P.; Bathmann, U.; Buczko, U.; Douhaire, C.; Eichler-Löbermann, B.; Frossard, E.; Ekardt, F.; Jarvie, H.; Krämer, I.; Kabbe, C.; et al. Handling the phosphorus paradox in agriculture and natural ecosystems: Scarcity, necessity, and burden of P. *Ambio* **2018**, *47* (Suppl. 1), S3–S19. [CrossRef]
69. NRC. *Nutrient Requirements of Swine*, 11th ed.; Natl. Acad. Press: Washington, DC, USA, 2012.
70. García, A.J.; Esteban, M.B.; Márquez, M.C.; Ramos, P. Biodegradable municipal solid waste: Characterization and potential use as animal feedstuffs. *Waste Manag.* **2005**, *25*, 780–787. [CrossRef]
71. Jin, Y.; Chen, T.; Li, H. Hydrothermal treatment for inactivating some hygienic microbial indicators from food waste-amended animal feed. *J. Air Waste Manag. Assoc.* **2012**, *62*, 810–816. [CrossRef]

72. Chen, T.; Jin, Y.; Shen, D. A safety analysis of food waste-derived animal feeds from three typical conversion techniques in China. *Waste Manag.* **2015**, *45*, 42–50. [CrossRef] [PubMed]
73. Truong, L.; Morash, D.; Liu, Y.; King, A. Food waste in animal feed with a focus on use for broilers. *Int. J. Recycl. Organic Waste Agric.* **2019**, *8*, 417–429. [CrossRef]
74. Pomar, C.; Remus, A. Precision pig feeding: A breakthrough toward sustainability. *Anim. Front.* **2019**, *9*, 52–59. [CrossRef]
75. Fung, L.; Urriola, P.E.; Baker, L.; Shurson, G.C. Estimated energy and nutrient composition of difference sources of food waste and their potential use in sustainable swine feeding programs. *Transl. Anim. Sci.* **2019**, *3*, 143–152. [CrossRef]
76. Fung, L.; Urriola, P.E.; Shurson, G.C. Energy, amino acid, and phosphorus digestibility and energy prediction of thermally processed food waste sources for swine. *Transl. Anim. Sci.* **2019**, *3*, 676–691. [CrossRef]
77. Angulo, J.; Mahecha, L.; Yepes, S.A.; Yepes, A.M.; Bustamente, G.; Jaramillo, H.; Valencia, E.; Villamil, T.; Gallo, J. Nutritional evaluation of fruit and vegetable waste as a feedstuff for diets of lactating Holstein cows. *J. Environ. Manage. Environ. Risks Probl. Strateg. Reduce Biotechnol. Eng.* **2012**, *95*, S210–S214. [CrossRef]
78. Ishida, K.; Yani, S.; Kitagawa, M.; Oishi, K.; Hirooka, H.; Kumagai, H. Effects of adding food by-products mainly including noodle waste to total mixed ration silage on fermentation quality, feed intake, digestibility, nitrogen utilization and ruminal fermentation in wethers. *Anim. Sci. J.* **2012**, *83*, 735–742. [CrossRef] [PubMed]
79. Paek, B.H.; Kang, S.W.; Chi, Y.M.; Cho, W.M.; Yang, C.J.; Yun, S.G. Effects of substituting concentrate with dried leftover food on growth and carcass characteristics of Hanwoo steers. *Asian Australas. J. Anim. Sci.* **2005**, *18*, 209–213. [CrossRef]
80. Summers, J.D.; Macleod, G.K.; Warner, W.C. Chemical composition of culinary wastes and their potential as a feed for ruminants. *Anim. Feed Sci. Technol.* **1980**, *5*, 205–214. [CrossRef]
81. Cheng, Z.; Mo, W.Y.; Man, Y.B.; Lam, C.L.; Choi, W.M.; Nie, X.P.; Liu, Y.H.; Wong, M.H. Environmental mercury concentrations in cultured low-trophic-level fish using food waste-based diets. *Environ. Sci. Pollut. Res.* **2015**, *22*, 495–507. [CrossRef] [PubMed]
82. Damron, B.; Waldroup, P.; Harms, R. Evaluation of dried bakery products for use in broiler diets. *Poult. Sci.* **1965**, *44*, 1122–1126. [CrossRef]
83. Farhat, A.; Normand, L.; Chavez, E.R.; Touchburn, S.P. Comparison of growth performance, carcass yield and composition, and fatty acid profiles of Pekin and Muscovy ducklings fed diets based on food wastes. *Can. J. Anim. Sci.* **2001**, *81*, 107–114. [CrossRef]
84. Al-Tulaihan, A.; Najib, H.; Al-Eid, S. The nutritional evaluation of locally produced dried bakery waste (DBW) in the broiler diets. *Pak. J. Nutr.* **2004**, *3*, 294–299.
85. Ayanwale, B.; Aya, V. Nutritional evaluation of cornflakes waste in diets of broilers. *Pak. J. Nutr.* **2006**, *5*, 485–489.
86. Stefanello, C.; Vieira, S.; Xue, P.; Ajuwon, K.; Adeola, O. Age-related energy values of bakery meal for broiler chickens determined using the regression method. *Poult. Sci.* **2016**, *95*, 1582–1590. [CrossRef]
87. Kwak, W.S.; Kang, J. Effect of feeding food waste-broiler litter and bakery by-product mixture to pigs. *Bioresour. Technol.* **2006**, *97*, 243–249. [CrossRef]
88. Almeida, F.N.; Petersen, G.I.; Stein, H.H. Digestibility of amino acids in corn, corn coproducts, and bakery meal fed to growing pigs. *J. Anim. Sci.* **2011**, *89*, 4109–4115. [CrossRef] [PubMed]
89. Rojas, O.J.; Liu, Y.; Stein, H.H. Phosphorus digestibility and concentration of digestible and metabolizable energy in corn, corn coproducts, and bakery meal fed to growing pigs. *J. Anim. Sci.* **2013**, *91*, 5326–5335. [CrossRef] [PubMed]
90. Hammoumi, A.; Faid, M.; El Yachioui, M.; Amarouch, H. Characterization of fermented fish waste used in feeding trials with broilers. *Process. Biochem.* **1997**, *33*, 423–427. [CrossRef]
91. Joshi, V.; Gupta, K.; Devrajan, A.; Lal, B.; Arya, S. Production and evaluation of fermented apple pomace in the feed for broilers. *J. Food Sci. Technol.* **2000**, *37*, 609–612.
92. Wadhwa, W.; Bakshi, M.; Makkar, H. *Utilization of Fruit and Vegetable Wastes as Livestock Feed and as Substrates for Generations of other Value-Added Products*; FAO: Rome, Italy, 2013.
93. Fard, S.; Toghyani, M.; Tabeidian, S. Effect of oyster mushroom wastes on performance, immune responses and intestinal morphology of broiler chickens. *Int. J. Recycl. Org. Waste Agric.* **2014**, *3*, 141–146. [CrossRef]
94. Bakshi, M.; Wadhwa, M.; Makkar, P. Waste to worth: Vegetable wastes as animal feed. *CAB Rev. Perspect. Agric. Vet. Sci. Nutr. Nat. Resour.* **2016**, *11*, 1–26. [CrossRef]

95. Cho, Y.; Lee, G.; Jang, J.; Shin, I.; Myung, K.; Choi, K.; Bae, I.; Yang, C. Effects of feeding dried leftover food on growth and body composition of broiler chicks. *Asian-Aust. J. Anim. Sci.* **2004**, *17*, 386–393. [CrossRef]
96. Kojima, S. Dehydrated kitchen waste as a feedstuff for laying hens. *Int. J. Poult. Sci.* **2005**, *4*, 689–694.
97. Asar, E.; Raheem, H.; Daoud, J. Using dried leftover food as nontraditional feed for Muscovy duck diet. *Assiut. Vet. Med. J.* **2018**, *64*, 107–114.
98. Navidshad, B.; Adibmoradi, M.; Seifdavati, J. Effect of dietary levels of a modified meat meal on performance and small intestine morphology of broiler chicken. *Afr. J. Biotechnol.* **2009**, *8*, 5620–5626.
99. Chen, H.-L.; Chang, H.-J.; Yang, C.-K.; You, S.-H.; Jeng, H.-D.; Yu, B. Effect of dietary inclusion of dehydrated food waste products on Taiwan native chicken (Taishi No.13). *Asian Australas. J. Anim. Sci.* **2007**, *20*, 754–760. [CrossRef]
100. Kornegay, E.T.; Vander Noot, G.W.; Barth, K.M.; McGrath, W.S.; Welch, J.G.; Purkhiser, E.D. Nutritive value of garbage as a feed for swine. I. Chemical composition, digestibility and nitrogen utilization of various types of garbage. *J. Anim. Sci.* **1965**, *24*, 319–324. [CrossRef] [PubMed]
101. Hossein, S. Growth Performances, Carcass Yield and Meat Quality of Free-Range Village Chickens Fed on Diet Containing Dehydrated Processed Food Waste. Master's Thesis, Universiti Putra Malaysia, Seri Kembangan, Malaysia, 2015.
102. Westendorf, M.L.; Schuler, T.; Zirkle, E.W. Nutritional quality of recycled food plate waste in diets fed to swine. *Prof. Anim. Sci.* **1999**, *15*, 106–111. [CrossRef]
103. Westendorf, M.L.; Zirkle, E.W.; Gordon, R. Feeding food or table waste to livestock. *Prof. Anim. Sci.* **1996**, *12*, 129–137. [CrossRef]
104. Myer, R.O.; Brendemuhl, J.H.; Johnson, D.D. Evaluation of dehydrate restaurant food waste products as feedstuffs for finishing pigs. *J. Anim. Sci.* **1999**, *77*, 685–692. [CrossRef]
105. Chae, B.J.; Choi, S.C.; Kim, Y.G.; Kim, C.H.; Sohn, K.S. Effects of feeding dried food waste on growth and nutrient digestibility of growing-finishing pigs. *Asian Australas. J. Anim. Sci.* **2000**, *13*, 1304–1308. [CrossRef]
106. Choe, J.; Moyo, K.M.; Park, K.; Jeong, J.; Kim, H.; Ryu, Y.; Kim, J.; Kim, J.-M.; Lee, S.; Go, G.-W. Meat quality traits of pigs finished on food waste. *Korean J. Food Sci. Anim. Resour.* **2017**, *37*, 690–697. [CrossRef]
107. Ruttanavut, J.; Yamauchi, K.; Thongwittaya, N. Utilization of eco-feed containing mugwort microorganism compounds as a feed ingredient source for layer hens. *Am. J. Anim. Vet. Sci.* **2011**, *6*, 35–39. [CrossRef]
108. Food and Agriculture Organization. 1. Basic Foodstuffs. Available online: http://www.fao.org/3/Y4343E/y4343e02.htm (accessed on 6 July 2020).
109. Noblet, J.; Perez, J.M. Prediction of digestibility of nutrients and energy values of pig diets from chemical analysis. *J. Anim. Sci.* **1993**, *71*, 3389–3398. [CrossRef]
110. Kerr, B.J.; Jha, R.; Urriola, P.E.; Shurson, G.C. Nutrient composition, digestible and metabolizable energy content, and prediction of energy for animal protein byproducts in finishing pig diets. *J. Anim. Sci.* **2017**, *95*, 2614–2626. [CrossRef]
111. Ferguson, N. Commercial application of integrated models to improve performance and profitability in pigs and poultry. In *Nutritional Modelling in Pigs and Poultry*; Sakmoura, N., Gous, R., Kyriazakis, I., Hauschild, L., Eds.; CABI: Wallingford, Oxfordshire, UK, 2014; pp. 141–156.
112. Saddoris-Clemons, K.; Schneider, J.; Feoli, C.; Cook, D.; Newton, B. Cost-effective feeding strategies for grow-finish pigs. *Adv. Pork Prod.* **2011**, *22*, 187–194.
113. FAO. *Livestock's Long Shadow—Environmental Issues and Options*; Food and Agriculture Organization of the United Nations: Rome, Italy, 2006.
114. LEAP. *Environmental Performance of Animal Feed Supply Chains: Guidelines for Assessment*; Livest. Environ. Assess. Perform. Partnership; Food and Agriculture Organization: Rome, Italy, 2015.
115. Notarnicola, B.; Sala, S.; Anton, A.; McLaren, S.J.; Saouter, E.; Sonesson, U. The role of life cycle assessment in supporting sustainable agri-food systems: A review of the challenges. *J. Clean. Prod.* **2017**, *140*, 399–409. [CrossRef]
116. Sala, S.; Anton, A.; McLaren, S.J.; Notarnicola, B.; Saouter, E.; Sonesson, U. In quest of reducing the environmental impacts of food production and consumption. *J. Clean. Prod.* **2017**, *140*, 387–398. [CrossRef]
117. Feed Print Database. Available online: http://webapplicaties.wur.nl/software/feedprintNL/index.asp (accessed on 6 July 2020).
118. FAO. *Executive Summary: Expert Consultation and Workshop on Protein Sources for the Animal Feed Industry*; Food and Agriculture Organization in association with the International Feed Industry Federation: Bangkok, Thailand, 2002.

119. Gooding, C.H. Data for the carbon footprinting of rendering operations. *J. Indust. Ecol.* **2012**, *16*, 223–230. [CrossRef]
120. Taylor, D.; Woodgate, S.; Atkinson, M. Inactivation of the bovine spongiform encephalopathy agent by rendering procedures. *Vet. Rec.* **1995**, *137*, 605–610. [PubMed]
121. Wichuk, K.; McCartney, D. A review of the effectiveness of current time-temperature regulations on pathogen inactivation during composting. *J. Environ. Eng. Sci.* **2007**, *6*, 573–586. [CrossRef]
122. Franke-Whittle, I.; Insam, H. Treatment alternatives of slaughterhouse wastes, and their effect on the inactivation of different pathogens: A review. *Crit. Rev. Microbiol.* **2013**, *39*, 139–151. [CrossRef]
123. Masse, D.; Talbot, G.; Gilbert, Y. On farm biogas production: A method to reduce GHG emissions and develop more sustainable livestock operation. *Anim. Feed Sci. Technol.* **2011**, *166–167*, 436–445. [CrossRef]
124. Dee, S.A.; Niederwerder, M.C.; Patterson, G.; Cochrane, R.; Jones, C.; Diel, D.; Brockhoff, E.; Nelson, E.; Spronk, G.; Sundberg, P. The risk of viral transmission in feed: What do we know, what do we do? *Transbound. Emerg. Dis.* **2020**, 1–7. [CrossRef]
125. Auvermann, B.; Kalbasi, A.; Ahmed, A. Rendering, Chapter 4. In *Carcass Disposal: A Comprehensive Review*; National Agricultural Biosecurity Center Consortium; USDA APHIS Cooperative Agreement Project; Carcass Disposal Working Group; Kansas State University: Manhattan, KS, USA, 2004; p. 76.
126. Kotula, A.W.; Murrell, K.D.; Acosta-Stein, L.; Lamb, L.; Douglass, L. Trichinella spiralis: Effect of high temperature on infectivity in pork. *Exp. Parasitol.* **1983**, *56*, 15–19. [CrossRef]
127. Alizadeh, A.M.; Jazaeri, S.; Shemshadi, B.; Hashempour-Baltork, F.; Sarlak, Z.; Pilevar, Z.; Hosseini, H. A review on inactivation methods of Toxoplasma gondii in foods. *Pathogens Glob. Health* **2018**, *112*, 306–319. [CrossRef] [PubMed]
128. Duong, T.M. Qualitative Assessment of Pig Health Risks Related to the Use of Food Waste for Pig Production in Sub-Urban Area of Hanoi Capital. Master's Thesis, Kasetsart University, Bangkok, Thailand, 2016.
129. Knight, A.I.; Haines, J.; Zuber, S. Thermal inactivation of animal virus pathogens. *Curr. Top. Virol.* **2013**, *11*, 103–119.
130. Taylor, D.M. Inactivation of transmissible degenerative encephalopathy agents: A review. *Vet. J.* **2000**, *159*, 10–17. [CrossRef] [PubMed]
131. Burr, W.E.; Helmboldt, C.F. Salmonella species contaminants in three animal by-products. *Avian Dis.* **1962**, *6*, 441–443. [CrossRef]
132. Veldman, A.; Vahl, H.A.; Borggreve, G.J.; Fuller, D.C. A survey of the incidence of Salmonella species and Enterobacteriaceae in poultry feeds and feed components. *Vet. Rec.* **1995**, *136*, 169–172. [CrossRef]
133. Edwards, P.R.; Bruner, D.W.; Moran, A.B. The genus *Salmonella*: Its occurrence and distribution in the United States. Kentucky Agric. *Exp. Sta. Bull.* **1948**, *525*, 1–60.
134. Crump, J.A.; Griffin, P.M.; Angulo, F.J. Bacterial contamination of animal feed and its relationship to human foodborne illness. *Clin. Infect. Dis.* **2002**, *35*, 859–865. [CrossRef]
135. Davies, P.R. *The Role of Contaminated Feed in the Epidemiology of Salmonella in Modern Swine Production*; Proc. CDC Animal Feeds Workshop/Symposium: Atlanta, GA, USA, 2004.
136. Sapkota, A.R.; Lefferts, L.Y.; McKenzie, S.; Walker, P. What do we feed to food-production animals? A review of animal feed ingredients and their potential impacts on human health. *Environ. Health Perspect.* **2007**, *115*, 663–670. [CrossRef]
137. Boyer, C.I.; Bruner, D.W.; Brown, J.A. Salmonella organisms isolated from poultry feed. *J. Avian Dis.* **1958**, *2*, 396. [CrossRef]
138. Watkins, J.R.; Flowers, A.I.; Grumbles, L.C. Salmonella organisms in animal products used in poultry feeds. *Avian Dis.* **1959**, *3*, 290. [CrossRef]
139. Pomeroy, B.S.; Grady, M.K. *Salmonella Organisms Isolated from Feed Ingredients*; Proc. 65th Annual Meeting of the U.S. Livestock Sanitation Association; MacCrellish & Quigley Co.: Trenton, NJ, USA, 1962; Volume 65, pp. 449–452.
140. Harris, I.T.; Fedorka-Cray, P.J.; Gary, J.T.; Thomas, L.A. Prevalence of *Salmonella* organisms in swine feed. *J. Am. Vet. Med. Assoc.* **1997**, *210*, 382–385. [PubMed]
141. Franco, D.A. A survey of *Salmonella* serovars and most probable numbers in rendered animal protein meals: Inferences for animal and human health. *J. Environ. Health* **2005**, *67*, 18–22.
142. Franco, D.A. The genus salmonella. In Proceedings of the Animal Protein Producers Industry; Inst. Contin. Education: Washington, DC, USA, 1999; pp. 1–22.

143. Canadian Food Inspection Agency. 2005. Available online: https://www.inspection.gc.ca/food-safety-for-industry/archived-food-guidance/fish-and-seafood/manuals/standards-and-methods/eng/1348608971859/1348609209602?chap=7#s19c7 (accessed on 6 July 2020).
144. Sreenivas, P.T. Salmonella-control strategies for the feed industry. *Feed Mix.* **1988**, *6*, 8. Available online: https://agris.fao.org/agris-search/search.do?recordID=US201302898369 (accessed on 6 July 2020).
145. Beumer, H.; van der Poel, A.F.B. Effects on hygienic quality of feeds examined. *Feedstuffs* **1997**, *13*, 13–15.
146. Brooks, P. *Technical Service Publication, 1989*; National Renderers Association, Inc.; Canadian Food Inspection Agency: Ottawa, ON, Canada, 1989.
147. Isa, J.M.; Boycott, B.R.; Broughton, E. A survey of Salmonella contamination in animal feeds and feed constituents. *Can. Vet. J.* **1963**, *4*, 41–43. [PubMed]
148. Troutt, H.F.; Schaeffer, D.; Kakoma, I.; Pearl, G.G. *Prevalence of Selected Foodborne Pathogens in Final Rendered Products*; Fats and Proteins Research Foundation, Inc.: Alexandria, VA, USA, 2001.
149. Davies, P.R.; Funk, J.A. *Proc. 3rd International Symposium on the Epidemiology and Control of Salmonella in Pork*; Safe Pork Conference, Iowa State University: Ames, IA, USA, 1999; pp. 1–11.
150. Wilesmith, J.W. 1998. Manual on Bovine Spongiform Encephalopathy. FAO, Rome. Available online: http://www.fao.org/3/W8656E/W8656E00.htm (accessed on 6 July 2020).
151. Prince, M.J.; Bailey, J.A.; Barrowman, P.R.; Bishop, K.J.; Campbell, G.R.; Wood, J.M. Bovine spongiform encephalopathy. *Rev. Sci. Tech. Off. Int. Épizooties* **2003**, *22*, 37–60. [CrossRef]
152. Hörnlimann, B.; Bachmann, J.; Bradley, R. Portrait of bovine spongiform encephalopathy in cattle and other ungulates. In *Prions in Humans and Animals*; Hörnlimann, B., Riesner, D., Kretschmar, H., Eds.; De Gruyter: Berlin, Germany, 2007; pp. 233–249.
153. Kimberlin, R.H. Scrapie and possible relationships with viroids. *Se. Virol.* **1990**, *1*, 153–162.
154. Jahns, H.; Callanan, J.J.; Sammin, D.J.; McElroy, M.C.; Bassett, H.F. Survey for transmissible spongiform encephalopathies in Irish pigs fed meat and bone meal. *Vet. Rec.* **2006**, *159*, 137–142. [CrossRef]
155. Wells, G.A.; Hawkins, S.A.; Austin, A.R.; Ryder, S.J.; Done, S.H.; Green, R.B.; Dexter, I.; Dawson, M.; Kimberlin, R.H. Studies of the transmissibility of the agent of bovine spongiform encephalopathy to pigs. *J. Gen. Virol.* **2003**, *84*, 1021–1031. [CrossRef]
156. Cutlip, R.C.; Miller, J.M.; Hamir, A.N.; Peters, J.; Robinson, M.M.; Jenny, A.L.; Lehmkuhl, H.D.; Taylor, W.D.; Bisplinghoff, F.D. Resistance of cattle to scrapie by the oral route. *J. Infect. Dis.* **2001**, *169*, 814–820. [CrossRef]
157. Franco, D.A. *An Introduction to the Prion Diseases of Animals: Assessing the History, Risk Inferences, and Public Health Implications in the United States*; National Renderers Association: Alexandria, VA, USA, 2005; pp. 1–32.
158. Hill, A.F.; Desbruslais, M.; Joiner, S.; Sidle, K.C.; Gowland, I.; Collinge, J.; Doey, L.J.; Lantos, P. The same prion strain causes vCJD and BSE. *Nature* **1997**, *389*, 448–450. [CrossRef] [PubMed]
159. Chesboro, B. A fresh look at BSE. *Science* **2004**, *305*, 1918–1921. [CrossRef] [PubMed]
160. Richt, J.A.; Hall, S.M. BSE case associated with prion protein gene mutation. *PLoS Pathog.* **2008**, *4*, e1000156. [CrossRef]
161. Kamisato, T. BSE crisis in Japan: A chronological overview. *Environ. Health Prevent. Med.* **2005**, *10*, 295–302. [CrossRef] [PubMed]
162. Food Standards Australia and New Zealand. 2020. Available online: http://www.foodstandards.gov.au/industry/bse/beefexport/pages/default.aspx (accessed on 6 July 2020).
163. Beltran-Alcrudo, D.; Falco, J.R.; Raizman, E.; Dietze, K. Transboundary spread of pig diseases: The role of international trade and travel. *BMC Vet. Res.* **2019**, *15*, 64. [CrossRef]
164. Dee, S.; Clement, T.; Schelkopf, A.; Nerem, J.; Knudsen, D.; Christopher-Hennings, J.; Nelson, E. An evaluation of contaminated complete feed as a vehicle for porcine epidemic diarrhea virus infection of naïve pigs following consumption via natural feeding behavior: Proof of concept. *BMC Vet. Res.* **2014**, *10*, 176. [CrossRef]
165. Pasick, J.; Berhane, Y.; Ojkic, D.; Maxie, G.; Embury-Hyatt, C.; Swekla, K.; Handel, K.; Fairles, J.; Alexandersen, S. Investigation into the role of potentially contaminated feed as a source of the first-detected outbreaks of Porcine Epidemic Diarrhea in Canada. *Transbound. Emerg. Dis.* **2014**, *61*, 397–410. [CrossRef]
166. Pasma, T.; Furness, M.C.; Alves, D.; Aubry, P. Outbreak investigation of porcine epidemic diarrhea in swine in Ontario. *Can. Vet. J.* **2016**, *57*, 84–88.
167. Aubry, P.; Thompson, J.L.; Pasma, T.; Furness, M.C.; Tataryn, J. Weight of the evidence linking feed to an outbreak of porcine epidemic diarrhea in Canadian swine herds. *J. Swine Health Prod.* **2017**, *25*, 69–72.

168. Perri, A.M.; Poljak, Z.; Dewey, C.; Harding, J.C.S.; O'Sullivan, T.L. An epidemiological investigation of the early phase of the porcine epidemic diarrhea (PED) outbreak in Canadian swine herds in 2014: A case-control study. *Prev. Vet. Med.* **2018**, *150*, 101–109. [CrossRef]
169. Scott, A.; McCluskey, B.; Brown-Reid, M.; Grear, D.; Pitcher, P.; Ramos, G.; Spencer, D.; Singrey, A. Porcine epidemic diarrhea virus introduction into the United States: Root cause investigation. *Preventive Vet. Med.* **2016**, *123*, 192–201. [CrossRef] [PubMed]
170. Schulz, L.L.; Tonsor, G.T. Assessment of the economic impacts of porcine epidemic diarrhea virus in the United States. *J. Anim. Sci.* **2015**, *93*, 5111–5118. [CrossRef] [PubMed]
171. Trudeau, M.P.; Verma, H.; Sampedro, F.; Urriola, P.E.; Shurson, G.C.; Goyal, S.M. Environmental persistence of porcine coronaviruses in feed and feed ingredients. *PLoS ONE* **2017**, *12*, e0178094. [CrossRef] [PubMed]
172. Dixon, L.K.; Stahl, K.; Jori, F.; Vial, L.; Pfeiffer, D.U. African swine fever epidemiology and control. *Annu. Rev. Anim. Biosci.* **2020**, *8*, 221–246. [PubMed]
173. Dee, S.A.; Bauermann, F.V.; Niederwerder, M.C.; Singrey, A.; Clement, T.; de Lima, M.; Long, C.; Patterson, G.; Sheahan, M.A.; Stolan, A.M.M.; et al. Survival of viral pathogens in animal feed ingredients under transboundary shipping models. *PLoS ONE* **2018**, *13*, e0194509. [CrossRef] [PubMed]
174. USDA-APHIS-VS-Center for Epidemiology and Animal Health, Risk Assessment Team. Qualitative Assessment of the Likelihood of African Swine Fever Virus Entry to the United States: Entry Assessment. Fort Collins, CO, 8 pp. Available online: https://www.aphis.usda.gov/animal_health/downloads/animal_diseases/swine/asf-entry.pdf (accessed on 6 July 2020).
175. Westendorf, M.L. Food waste as animal feed: An introduction. In *Food Waste to Animal Feed*; Westendorf, M.L., Ed.; Iowa State University Press: Ames, IA, USA, 2000; pp. 91–111.
176. Menikpura, S.N.M.; Sang-Arun, J.; Bengtsson, M. Integrated solid waste management: An approach for enhancing climate co-benefits through resource recovery. *J. Clean. Prod.* **2013**, *58*, 34–42. [CrossRef]
177. Liu, C.; Hotta, Y.; Santo, A.; Hengesbaugh, M.; Watabe, A.; Totoki, Y.; Allen, D.; Bengstsson, M. Food waste in Japan: Trends, current practices and key challenges. *J. Clean. Prod.* **2016**, *133*, 557–564. [CrossRef]
178. Sugiura, K.; Yamatani, S.; Watahara, M.; Onodera, T. Ecofeed, animal feed produced from recycled food waste. *Vet. Ital.* **2009**, *45*, 397–404.
179. European Commission. Regulation (EC) No. 1774/2002 of the European Parliament and of the Council of 3 October 2002 Laying down Health Rules Concerning Animal By-Products not Intended for Human Consumption. Available online: https://op.europa.eu/en/publication-detail/-/publication/28ab554e-8e93-4976-89a9-8b6c9d17dfb4/language-en (accessed on 29 August 2020).
180. Muroga, N.; Hayama, Y.; Yamamoto, T.; Kurogi, A.; Tsuda, T.; Tstsui, T. The 2010 foot-and-mouth disease epidemic in Japan. *J. Vet. Med. Sci.* **2012**, *74*, 399–404. [CrossRef]
181. Park, J.-H.; Lee, K.-N.; Ko, Y.-J.; Kim, S.-M.; Lee, H.-S.; Shin, Y.-K.; Sohn, H.-J.; Park, J.-Y.; Yeh, J.-Y.; Lee, Y.-H.; et al. Control of foot-and-mouth disease during 2010-2011 Epidemic, South Korea. *Emerg. Infect. Dis.* **2013**, *19*, 655–659.
182. European Commission. Regulation (EC) No. 999/2001 of the European Parliament and of the Council of 22 May 2001 Laying down Rules for the Prevention, Control and Eradication of Certain Transmissible Spongiform Encephalopathies. Available online: https://www.legislation.gov.uk/eur/2001/999/contents# (accessed on 6 July 2020).
183. Andreoletti, O.; Budka, H.; Buncic, S.; Colin, P.; Collins, J.D.; De Koeijer, A.; Griffin, J.; Havelaar, A.; Hope, J.; Klein, G. Opinion of the scientific panel on biological hazards on a request from the European parliament on certain aspects related to the feeding of animal proteins to farm animals. *EFSA J.* **2007**, *576*, 1–41.
184. Whitehead, M.L.; Roberts, V. Backyard poultry: Legislation, zoonoses and disease prevention. *J. Small Anim. Pract.* **2014**, *55*, 487–496. [CrossRef] [PubMed]
185. Raymundo, D.; Carloto Gomes, D.; Boabaid, F.; Colodel, E.; Schmitz, M.; Correa, A.; Dutra, I.; Driemeier, D. Type C botulism in swine fed restaurant waste. *Pesqui. Vet. Brasil.* **2012**, *32*, 1145–1147. [CrossRef]
186. Zimmerman, J.J.; Karriker, L.A.; Ramirez, A.; Schwartz, K.J.; Stevenson, G.W. *Diseases of Swine*, 10th ed.; Wiley-Blackwell: Ames, IA, USA, 2012; p. 1008.
187. Gamble, H.R.; Murrell, K.D. Trichinellosis. In *Laboratory Diagnosis of Infectious Disease: Principles and Practice*; Balows, A., Hausler, W.J., Lennette, E.H., Eds.; Springer: New York, NY, USA, 1998; p. 1018.
188. U.S. Congressional Record. In Proceedings of the Swine Health Protection Act. H.R. 6593. Public Lay 96-468, 96th Congress, 17 October 1980. Available online: https://www.congress.gov/public-laws/96th-congress (accessed on 11 July 2020).

189. U.S. Food and Drug Administration. 2017. Available online: https://www.accessdata.fda.gov/scripts/cdrh/cfdocs/cfcfr/CFRSearch.cfm?CFRPart=589&showFR=1 (accessed on 6 July 2020).
190. Broad Leib, E.; Blakus, O.; Rice, C.; Maley, M.; Taneja, R.; Cheng, R.; Civita, N.; Alvoid, T. *Leftovers for Livestock: A Legal Guide for Using Excess Food as Animal Feed*; Harvard Food Law and Policy Clinic: Cambridge, MA, USA, 2016.
191. Rethink Food Waste through Economics and Data (ReFED). Roadmap to Reduce, U.S. Food Waste by 20 Percent. 2016. Available online: https://www.refed.com/downloads/Executive-Summary.pdf (accessed on 6 July 2020).
192. Sanchez-Vizcaino, J.M. African swine fever. In *Diseases of Swine*, 8th ed.; Mengeling, W.L., Ed.; Iowa State University Press: Ames, IA, USA, 1999; pp. 93–112.
193. House, J.A.; House, C.A. Vesicular diseases. In *Diseases of Swine*, 8th ed.; Mengeling, W.L., Ed.; Iowa State University Press: Ames, IA, USA, 1999; pp. 327–340.
194. van Oirschot, J.T. Classical swine fever (hog cholera). In *Diseases of Swine*, 8th ed.; Mengeling, W.L., Ed.; Iowa State University Press: Ames, IA, USA, 1999; pp. 159–172.
195. Costard, S.; Porphyre, V.; Messad, S.; Rakotondrahanta, S.; Vidon, H.; Roger, F.; Pfeiffer, D. Multivariate analysis of management and biosecurity practices in smallholder pig farms in Madagascar. *Prev. Vet. Med.* **2009**, *92*, 199–209. [CrossRef]
196. Kagira, J.M.; Kanyari, P.W.; Maingi, N.; Githigia, S.M.; Ng'ang'a, J.C.; Karuga, J.W. Characteristics of the smallholder free-range pig production system in western Kenya. *Trop. Anim. Health Prod.* **2010**, *42*, 865–873. [CrossRef]
197. Phengsavanh, P.; Ogle, B.; Stür, W.; Frankow-Lindberg, B.E.; Lindberg, J.E. Feeding and performance of pigs in smallholder production systems in Northern Lao PDR. *Trop. Anim. Health Prod.* **2010**, *42*, 1627–1633. [CrossRef]
198. Mazur-Panasiuk, N.; Żmudzki, J.; Woźniakowski, G. African swine fever virus -persistence in different environmental conditions and the possibility of its indirect transmission. *J. Vet. Res.* **2019**, *63*, 303–310. [CrossRef]
199. Karunasagar, I. International risk assessment leading to development of food safety standards. *Procedia Food Sci.* **2016**, *6*, 34–36. [CrossRef]
200. Dennis, S.B.; Miliotis, M.D.; Buchanan, R.L. *Hazard Characterization/Dose-Response Assessment. Microbiological Risk Assessment in Food Processing*; CRC Press: Boca Raton, FL, USA, 2002; Volume 301, pp. 77–99.
201. FDA. 2018. Available online: https://www.fda.gov/files/animal%20&%20veterinary/published/CVM-GFI--246-Hazard-Analysis-and-Risk-Based-Preventive-Controls-for-Food-for-Animals--Supply-Chain-Program.pdf (accessed on 6 July 2020).
202. Dalmasso, A.; Fontanella, E.; Piatti, P.; Civera, T.; Rosati, S.; Bottero, M.T. A multiplex PCR assay for the identification of animal species in feedstuffs. *Molec. Cell. Probes* **2004**, *18*, 81–87. [CrossRef]
203. Lahiff, S.; Glennon, M.O.B.L.; Lyng, J.; Smith, T.; Maher, M.; Shelton, N. Species-specific PCR for the identification of ovine, porcine and chicken species in meat and bone meal (MBM). *Molec. Cell. Probes* **2001**, *15*, 27–35. [CrossRef]
204. Schmidt, G.R.; Hossner, K.L.; Yemm, R.S.; Gould, D.H.; O'Callaghan, J.P. An enzyme-linked immunosorbent assay for glial fibrillary acidic protein as an indicator of the presence of brain or spinal cord in meat. *J. Food Protect.* **1999**, *62*, 394–397. [CrossRef] [PubMed]
205. Chang, B.; Gray, P.; Piltch, M.; Bulgin, M.S.; Sorensen-Melson, S.; Miller, M.W.; Davies, P.; Brown, D.R.; Coughlin, D.R.; Rubenstein, R. Surround optical fiber immunoassay (SOFIA): An ultra-sensitive assay for prion protein detection. *J. Virol. Methods* **2009**, *159*, 15–22. [CrossRef] [PubMed]
206. Jinno, C.; He, Y.; Morash, D.; McNamara, E.; Zicari, S.; King, A.; Stein, H.H.; Liu, Y. Enzymatic digestion turns food waste into feed for growing pigs. *Anim. Feed Sci. Technol.* **2018**, *242*, 48–58. [CrossRef]

MDPI
St. Alban-Anlage 66
4052 Basel
Switzerland
Tel. +41 61 683 77 34
Fax +41 61 302 89 18
www.mdpi.com

Sustainability Editorial Office
E-mail: sustainability@mdpi.com
www.mdpi.com/journal/sustainability

www.ingramcontent.com/pod-product-compliance
Lightning Source LLC
LaVergne TN
LVHW070141170726
843515LV00007B/1844

* 9 7 8 3 0 3 6 5 4 8 5 3 1 *